S0-AKY-004

Irradiation of Polymeric Materials

ACS SYMPOSIUM SERIES 527

Irradiation of Polymeric Materials

Processes, Mechanisms, and Applications

Elsa Reichmanis, EDITOR
AT&T Bell Laboratories

Curtis W. Frank, EDITOR
Stanford University

James H. O'Donnell, EDITOR
University of Queensland

Developed from a symposium sponsored
by the Polymer Division
of the Royal Australian Chemical Institute,
the National Science Foundation of the United States,
and the Department of Industry, Technology, and Commerce
of the Australian Government

American Chemical Society, Washington, DC 1993

Library of Congress Cataloging-in-Publication Data

Irradiation of polymeric materials: processes, mechanisms, and applications / Elsa Reichmanis, Curtis W. Frank, James H. O'Donnell, editors.

p. cm.—(ACS symposium series, ISSN 0097–6156; 527)

Includes bibliographical references and index.

ISBN 0–8412–2662–8

1. Polymers—Effect of radiation on—Congresses.

I. Reichmanis, Elsa, 1953– . II. Frank, C. W. III. O'Donnell, James H. IV. Series.

QD381.9.R3I78 1993
620.1'9204228—dc20 93–22431
CIP

The paper used in this publication meets the minimum requirements of American National Standard for Information Sciences—Permanence of Paper for Printed Library Materials, ANSI Z39.48–1984. ∞

PRINTED IN THE UNITED STATES OF AMERICA

1993 Advisory Board

ACS Symposium Series

M. Joan Comstock, *Series Editor*

Foreword

THE ACS SYMPOSIUM SERIES was first published in 1974 to provide a mechanism for publishing symposia quickly in book form. The purpose of this series is to publish comprehensive books developed from symposia, which are usually "snapshots in time" of the current research being done on a topic, plus some review material on the topic. For this reason, it is necessary that the papers be published as quickly as possible.

Before a symposium-based book is put under contract, the proposed table of contents is reviewed for appropriateness to the topic and for comprehensiveness of the collection. Some papers are excluded at this point, and others are added to round out the scope of the volume. In addition, a draft of each paper is peer-reviewed prior to final acceptance or rejection. This anonymous review process is supervised by the organizer(s) of the symposium, who become the editor(s) of the book. The authors then revise their papers according to the recommendations of both the reviewers and the editors, prepare camera-ready copy, and submit the final papers to the editors, who check that all necessary revisions have been made.

As a rule, only original research papers and original review papers are included in the volumes. Verbatim reproductions of previously published papers are not accepted.

M. Joan Comstock
Series Editor

Contents

Preface

THE EFFECTS OF RADIATION on polymeric materials is an area of rapidly increasing interest. Several high-technology industries require specialty polymers that exhibit a specific response upon exposure to radiation. For example, some applications require materials for which efficient radiation-induced scission or cross-linking effects a change in solubility. Highly radiation-stable materials are required for other applications. The design and development of appropriate chemistry for a given purpose requires full understanding of the effect of radiation-induced chemical processes occurring in polymers. With this background the technological advances required by today's industries can be realized.

Irradiation of polymers is widespread in many industries. Microlithography, an essential process in the fabrication of integrated circuits, involves modification by radiation of the solubility or volatility of thin polymer resist films. The role of radiation sterilization of medical and pharmaceutical items, many of which are manufactured from polymeric materials, is increasing. This trend arises from both the convenience of the process and concern about the toxicity of chemical sterilants. Information about the radiolysis products of natural and synthetic polymers used in the biomedical industry is required for the evaluation of the safety of the process.

UV and high-energy irradiation of polymer coatings on metals and other substrates has been developed for various industrial purposes. Irradiation induces modification of polymer properties as the basis of heat-shrinkable films and tubings, cross-linked polymers, and graft copolymers. Advanced polymeric materials are used in many aerospace industry applications because of their high strength and low-weight performance. However, these materials are subject to bombardment by oxygen atoms and to degradation in space by UV and high-energy radiation.

A thorough fundamental understanding of the chemistry involved in irradiation of polymeric materials is, thus, critical to many high-technology industries. Further technological advances require study of the interactions and processes taking place within a given polymer with the goal of designing appropriate chemistry for a specific application.

The Workshop on Radiation Effects on Polymer Materials, upon which this book is based, emphasized the fundamental aspects of polymer radiation chemistry and the technological significance of the effects of radiation on polymers. Subject areas include examination of the fundamental processes by which radiation interacts with polymers and the mechanisms leading to chemical changes in materials. Several applications of related topics are also presented. They include materials for microlithographic applications, aerospace (radiation-durable) materials, biomedical materials, and applications of radiation-modified polymers.

The authors and editors are indebted to the National Science Foundation of the United States and to the Department of Industry, Technology, and Commerce of the Australian Government for financial support to the U.S.–Australia Workshop on Radiation Effects on Polymer Materials. Additional financial resources were provided by Ethicon, Inc.; Davis and Geck, Inc., a Division of the American Cyanamid Company; OCG Microelectronics Materials, Inc.; Hitachi, Ltd.; and AT&T Bell Laboratories. Our sincerest thanks are extended to Anne Wilson and the production staff of the Books Department of the American Chemical Society.

ELSA REICHMANIS
AT&T Bell Laboratories
Murray Hill, NJ 07974

CURTIS W. FRANK
Stanford University
Stanford, CA 94305–5025

JAMES H. O'DONNELL
University of Queensland
Brisbane, Queensland, 4072 Australia

Received November 11, 1992

Chapter 1

Radiation Effects on Polymeric Materials

A Brief Overview

Elsa Reichmanis[1], Curtis W. Frank[2], James H. O'Donnell[3], and David J. T. Hill[3]

[1]AT&T Bell Laboratories, Room 1A–261, 600 Mountain Avenue, Murray Hill, NJ 07974
[2]Department of Chemical Engineering, Stanford University, Stanford, CA 94305–5025
[3]Polymer Materials and Radiation Group, Department of Chemistry, University of Queensland, Brisbane, Queensland 4072, Australia

The effects of radiation on polymer materials is an area of rapidly increasing interest. Several high technology industries require specialty polymers that exhibit a specific response upon exposure to radiation. For instance, the electronics industry requires materials that undergo radiation induced scission or crosslinking for resist applications, while aerospace and medical applications require highly radiation stable materials. The design and development of appropriate chemistry for these applications requires full understanding of the effects of radiation on polymer materials. It is through fundamental understanding of the radiation chemical processes occurring in polymers that the technological advances required by today's industries can be realized.

The study of the effect of radiation on polymer materials is an area of rapidly increasing interest (1-3). The radiation regimes of primary utility are either high energy, ionizing radiation such as from gamma or neutron sources, or ultraviolet radiation from arc lamps, excimer lasers or synchrotron sources. The major difference between the two types of radiation is the initial or primary event following absorption of the radiation. In the case of ionizing radiation, the initial absorption is typically a spatially random process and leads to free radical or ionic species production, whereas, ultraviolet absorption is molecular site specific and often leads to electronically excited states. Subsequent events in both cases can involve side group or main chain scission or crosslinking. Even small amounts of radiation can induce significant changes in the physical or mechanical properties of a polymer, with the extent of these changes being dependent upon the chemical structure of a particular polymer. In some cases, even a few crosslinks or scission sites per molecule can dramatically affect the strength or solubility of a polymer. These changes, in turn, will determine whether a particular polymer will have

0097–6156/93/0527–0001$06.00/0

application in a particular industry. Applications that are covered in this Proceedings include microelectronics, radiation sterilization, modified polymers and surface coatings.

Research topics in the area of radiation effects on polymers may be broken into areas of radiation physics and of radiation chemistry. For both the high energy ionizing radiation and the lower energy ultraviolet radiation the radiation physics area includes study of the types of radiation sources. In the case of ionizing radiation, it is then typical to consider the primary event of radical or ion formation as well as the subsequent reactions and rearrangements as all part of the topic of radiation chemistry. For ultraviolet radiation, however, it is typical to speak separately of the photophysics and photochemistry associated with the absorption of ultraviolet light. The former area includes the issues of electronic excited state formation, radiative and nonradiative deactivation processes and energy transfer and migration phenomena. The latter area is entered if the electronically excited state leads to a chemical reaction such as bond scission. Aspects of each of these fundamental areas are included in this Workshop Proceedings. However, the emphasis is placed on the ionizing radiation chemistry as well as the ultraviolet radiation physics.

FUNDAMENTAL POLYMER RADIATION CHEMISTRY ISSUES

Radiation Type

High-energy radiation may be classified into photon and particulate radiation. Gamma radiation is utilized for fundamental studies and for low-dose rate irradiations with deep penetration. The radioactive isotopes, cobalt-60, and cesium-137 are the main sources of gamma radiation. Lower energy x-radiation is produced by electron bombardment of suitable metal targets with electron beams, or in a synchrotron. Electron irradiation is normally obtained from electron accelerators to give beams with energies in the MeV range. The corresponding penetration depths are in the mm range. Much lower energy electron beams, e.g. 10-20 keV, are used in electron microscopy and in electron-beam lithography. In these cases, a large proportion of the energy is deposited in µm thick polymer films (4).

Nuclear reactors are a source of high radiation fluxes comprised mainly of neutrons and gamma rays, and large ionized particles (fission products) close to the fuel elements. The neutrons largely produce protons in hydrocarbon polymers by "knock-on" reactions, thus, the radiation chemistry of neutrons is similar to that of proton beams, which may alternatively be produced using positive-ion accelerators.

Classical sources for ultraviolet radiation (5) include mercury and xenon arc lamps, which have the advantage of broad band excitation for spectroscopic studies, but suffer from the corresponding broad monochromator controlled bandwidth. Excimer laser sources provide intense coherent radiation in the deep ultraviolet region. Finally, synchrotron sources provide a broad range of radiation energy, including the deep ultraviolet.

Radiation Chemical Processes

Absorption of high-energy radiation by polymers produces excitation and ionization and it is these excited and ionized species that are the initial chemical reactants (6-11). The ejected electron must lose energy until it reaches thermal energy. Geminate recombination with the parent cation radical may then occur and is more likely in substrates of low dielectric constant. The resultant excited molecule may undergo homolytic or heterolytic bond scission. Alternatively, the parent cation radical may undergo spontaneous decomposition, or ion-molecule reactions. The initially ejected electron may be stabilized by interaction with polar groups, as a solvated species or as an anion radical. The radiation chemistry taking place within a polymeric material is, thus, quite complicated, involving chemistry of neutral, cation and anion radicals, cations and anions, and excited species.

Although the absorption of radiation energy is dependent only on the electron density of the substrate and, therefore, occurs spatially at random on a molecular scale, the subsequent chemical changes are not random. Some chemical bonds and groups are particularly sensitive to radiation-induced reactions. They include COOH, C-X where X=halogen, $-SO_2-$, NH_2, and C=C. Spatial specificity of chemical reactions may result from intramolecular or intermolecular migration of energy or reactive species such as free radicals or ions.

Enhanced radiation sensitivity may be designed into polymer molecules by incorporation of radiation sensitive groups as in lithographic materials. Aromatic groups have long been known to give significant radiation resistance to organic molecules. There was early work on the hydrogen yields from cyclohexane (G=5) and benzene (G=0.04) in the liquid phase, and of their mixtures, which showed a pronounced protective effect upon addition of the aromatic moiety.

A substantial intramolecular protective effect by phenyl groups in polymers is demonstrated by the low G values for H_2 formation and crosslinking in polystyrene (substituent phenyl), polyarylene sulfones (backbone phenyl), as well as many other aromatic polymers. While the relative radiation resistance of different aromatic groups in polymers has not been extensively studied, most aromatic substituents afford similar protection. It has been observed, however, that biphenyl and phenolic moieties provide increased radiation resistance to polymer materials relative to other aromatic groups. In one example, it was demonstrated that the phenolic group in tyrosine protects the radiation sensitive carboxylic acid groups of glutamic acid even though tyrosine itself may undergo crosslinking to a greater extent than would phenylalanine.

The molecular changes occurring in polymers as a result of radiation-induced chemical reactions may be classified as i) chain crosslinking effecting an increase in molecular weight and formation of a macroscopic network (polymer solubility decreases with increased radiation dose); ii) chain scission effecting a decrease in molecular weight and, thus, substantially changing a polymer materials properties (strength, both tensile and flexural, decreases, and the rate of dissolution in a given solvent increases). In addition to these changes, irradiation of polymers will frequently give rise to small molecule products, resulting from bond scission followed by abstraction or combination reactions.

Information regarding such processes can give valuable insight into the mechanism of the radiation induced degradation process. Gaseous products, such as CO_2, may be trapped in the polymer, and this can lead to subsequent crazing and cracking due to accumulated local stresses. Structural changes in the polymer, which will accompany the formation of small molecule products from the polymer, or may be produced by other reactions, can cause significant changes to the material properties. Development of color, e.g. in polyacrylonitrile by ladder formation, and in poly(vinyl chloride) through conjugated unsaturation, is a common form of degradation.

APPLICATIONS OF POLYMER RADIATION CHEMISTRY

Electronics

The importance of radiation chemistry in this rapidly developing high technology industry is widely recognized (12-15). Microlithography is an essential process in the fabrication of silicon chip integrated circuits and involves modification of the solubility or volatility of thin polymer "resist" films by radiation. While almost all of today's commercial devices are made by photolithographic techniques that utilize 365-436nm UV radiation, within the next 3-8 yrs., new lithographic strategies will be required. The technological alternatives to conventional photolithography are deep-UV (240-260 nm) photolithography, scanning electron-beam and x-ray lithography. The leading candidate for the production of devices with features as small as 0.25 μm is deep-UV lithography.

No matter which technology eventually replaces photolithography, new resists and processes will be required, necessitating enormous investments in research and development. The polymer materials that are used as radiation sensitive resist films must be carefully designed to meet the specific requirements of the lithographic technology and device process. Although these requirements vary according to the radiation source and specific process, the following properties are common to all resists: sensitivity, contrast, resolution, etching resistance, shelf-life and purity.

Generally, the desired resist properties can be achieved by careful manipulation of the structure and molecular properties of the resist components. As mentioned, new resists will be required that are both sensitive to different types of radiation and compatible with advanced processing requirements. Since all the alternatives to conventional wavelength photolithography employ rather high energy radiation, resists designed for one type of radiation, e.g. e-beam, will frequently be applicable to the other lithographic technologies. Various resist chemistries, ranging from novel chemically amplified resist mechanisms for UV and x-ray applications, to surface imaging techniques and the affects film microstructure and morphology on conventional resist performance are discussed in subsequent chapters of this book.

Aerospace

The effects of radiation on polymer materials continues to be of importance for aerospace programs as the design lifetimes of satellites and other space vehicles are increased (16). Advanced polymer materials are desirable due to their high strength/low weight performance. For example, the use of graphite fiber reinforced polymer-matrix composites for structural applications in weight critical space structures has increased dramatically in recent years. The primary advantage of composites over more conventional metallic materials is their high specific strength and stiffness and low thermal expansion. As a result, composites are the leading candidate materials for nearly every space structure under development or consideration. Radiation resistant polymers also find application in the production of lightweight, high strength matrix resins and adhesives. Examples of such polymers include polyimides, aromatic polysulfones, and polyether ketones.

Polymeric materials are, however, subject to degradation in space by UV and high energy radiation, and oxygen atom bombardment. Thus, the effects of radiation on candidate materials for aerospace applications must be characterized. Such characterization would include an understanding of the immediate reactions taking place in the materials, in addition to continued post-exposure degradation processes. A full understanding of how the various forms of radiation interact with, and affect, materials chemistry and properties can be utilized in the design of improved radiation resistant polymers. Studies pertaining to the effects of radiation on polymers typically used in this industry may be found later in this volume.

Radiation Sterilization

Radiation is being used increasingly for the sterilization of medical and pharmaceutical items for convenience and, more importantly, because of concern about the toxicity of chemical sterilants (3). For instance, conventional sterilization involves the use of the highly toxic chemical, ethylene oxide. As a result, there are progressive restrictions placed on its use and it is anticipated that ethylene oxide based chemical sterilization techniques will ultimately be prohibited.

The radiation sterilization of biomedical polymeric materials, particularly implantable surgical devices, raises significant concerns. Polymers typically undergo some radiation-induced degradation leading to discoloration and associated deterioration in properties. Knowledge of the relationship between the chemical composition of polymer materials and their radiation sensitivity is necessary to enable selection of appropriate materials for radiation sterilization.

Understanding the radiation chemical processes taking place in biomedical materials and evaluating the dependence of these reactions on dose and storage time is also important. As noted in the previous section, post-irradiation reactions resulting from trapped radicals within the polymer matrix are commonly observed upon high energy radiation of polymers. Knowledge and understanding of such reactions in candidate polymers for biomedical applications is critical.

Information on the radiolysis products from both natural and synthetic polymers is further required by the food industry as radiation sterilization of products related to this industry becomes more commonplace.

The chapters by Shalaby and Bezwada specifically address the radiation chemistry associated with polymers of biomedical significance.

Modified Polymers

Irradiation of polymers causes modification of properties which is currently the basis of major industries in heat shrinkable film and tubing, crosslinked polymers and grafted copolymers (3). This area is now entering an era of new technology resulting from greater knowledge of the chemical processes. There are many applications of modified polymers including their use in the medical and health-care fields. For instance, these materials may be used as implantable materials, controlled release drug preparations, and hydrophilic wound dressings. Alternative applications include the textile, mining and construction industries. In one example detailed in this book, radiation crosslinked polyethylenes have several advantageous applications in areas as diverse as rock bolts for mining, reinforcement of concrete, manufacture of light weight high strength ropes and high performance fabrics.

Surface Coatings

UV and high energy irradiation of polymer coatings on substrates such as metals, plastics and optical fibers is being developed industrially for a wide variety of purposes (2). Adhesion of these coating is critical to ensure proper function. Processes such as radiation induced chemical grafting may enhance adhesion of a particular polymeric coating to a substrate. Applications of surface coating polymers are wide spread. They are routinely used for magnetic tapes and discs, compact discs. Often, the surfaces of medical materials are coated with a protective polymer, and surface coating of wood with polyurethane and acrylic lacquers is an area of renewed interest.

SUMMARY

The electronics and aerospace industries are major high technological endeavors that require specialty polymers. The first case requires materials that undergo radiation induced scission or crosslinking for resist applications while the latter involves the use of highly radiation stable materials. Similarly, biomedical applications require stable materials. These diverse applications have resulted in a renewed interest in fundamental radiation chemistry of polymers.

Numerous parameters must be examined and evaluated in understanding the effects of radiation on polymer materials. The rates and nature of chemical reactions in irradiated polymers are not only dependent on chemical factors, but physical factors as well. In other words, polymer morphology will play a role in the radiation chemistry of polymers. Temperature effects are equally important. In one example, formation of active sites, particularly free radicals by chain scission, which are identical to propagating radicals, can lead to polymer depropagation. The probability of depropagation will increase with temperature and can have an important role in the radiation degradation of polymers with low activation energies for propagation. Thus, poly(alpha-methyl styrene) and poly(methyl methacrylate) show increasing amounts of monomer formation during irradiation above 100 and 150°C, respectively.

Other factors to be evaluated include stress, which can lead to a decrease in the lifetime to failure of a polymer exposed to high-energy radiation; structural changes that occur upon irradiation; changes in radiation chemistry that occur as a result of the environment, i.e., air vs. vacuum, ordered vs. disordered; and post-irradiation effects that may lead to progressive strength reduction, cracking or crazing, etc.

It is only through fundamental understanding of the above processes that the technological advances required by today's industry can be realized.

Literature Cited

[1] "The Effects of Radiation on High-Technology Polymers", *ACS Symposium Series* **381**, Reichmanis, E., O'Donnell, J. H. eds., ACS, Washington, D.C., 1989.

[2] "Radiation Curing of Polymeric Materials", *ACS Symposium Series* **417**, Hoyle, C. E., Kinstle, J. F., eds., ACS, Washington, D.C., 1990.

[3] "Radiation Effects on Polymers," *ACS Symposium Series* **475**, Clough, R., Shalaby, S. W., eds., ACS Washington, D.C., 1991.

[4] Kinstle, J. F., In "Radiation Curing of Polymeric Materials", ACS Symposium Series **417**, Hoyle, C. E., Kinstle, J. F., eds., ACS, Washington, D.C., 1990, 17-23.

[5] Philps, R., "Sources and Applications of Ultraviolet Radiation", Academic Press, New York, 1983.

[6] O'Donnell, J. H., Sangster, D. F., "Principles of Radiation Chemistry", Edward Arnold, London, 1970.

[7] Swallow, A. J., "Radiation Chemistry of Organic Compounds", International Series of Monographs on Radiation Effects in Materials, Vol. 2, Charlesby, A., Ed., Permagon Press, Oxford, 1960.

[8] Charlesby, A., "Atomic Radiation and Polymers", International Series of Monographs on Radiation Effects in Materials, Vol. 1, Charlesby, A., Ed., Permagon Press, Oxford, 1960.

[9] Chapiro, A., "Radiation Chemistry of Polymeric Systems", High Polymers, Vol. 15, Interscience, New York, 1962.

[10] Dole, M., "Fundamental Processes and Theory", The Radiation Chemistry of Macromolecules, Vol. 1, Dole, M., Ed., Academic Press, New York, 1972.

[11] Dole, M., "Radiation Chemistry of Substituted Vinyl Polymers", The Radiation Chemistry of Macromolecules, Vol. 2, Dole, M., Ed., Academic Press, New York, 1973.

[12] "Introduction to Microlithography", ACS Symposium Series **219**, Thompson, L. F., Willson, C. G., Bowden, M. J., Eds., ACS, Washington, D.C., 1983.

[13] Moreau, W. M., "Semiconductor Lithography, Principles, Practices and Materials", Plenum, New York (1988).

[14] "Polymers in Microlithography, Materials and Processes", Reichmanis, E., MacDonald, S. A., Iwayanagi, T., Eds., ACS Symposium Series **412**, ACS, Washington, D.C., 1989.
[15] "Electronic and Photonic Applications of Polymers", Bowden, M. J., Turner, S. R., ACS Advances in Chemistry Series **218**, ACS, Washington, D.C., 1988.
[16] Tenny, D. R., Slemp, W. S., In "The Effects of Radiation on High-Technology Polymers", Reichmanis, E., O'Donnell, J. H., Eds., ACS Symposium Series **381**, ACS, Washington, D.C., 1989, pp. 224-251.

RECEIVED November 11, 1992

Chapter 2

Dosimetry: The Forgotten Art

Philip W. Moore

Radiation Technology, Australian Nuclear and Technology Organization, Private Mail Bag, Menai 2234, New South Wales, Australia

To do dosimetry properly we need to exercise control in three areas. First, the dosimeter must be both reliable and accurate and its calibration should be traceable to a national or international standard. Second, it must be used with some understanding to ensure that it is measuring what we want and not something else. It must also be used within recognized limits of dose and dose rate and under conditions of electronic (or charged particle) equilibrium. Corrections for irradiation temperature and measurement temperature may also have to be made. Finally, we may need to apply corrections to the dosimeter response based on cavity theory to take account of the perturbation in the medium caused by the presence of the dosimeter or we may need to use mass energy absorption coefficients or mass electron stopping powers to calculate the dose absorbed in one medium from the dose absorbed in another. The selection of the correct values is not as straightforward as you might think, and is the subject of this chapter.

One suspects that more than a few polymer chemists take the irradiated polymer or monomer as the starting point of their investigation. Possibly their materials are sent to a national nuclear laboratory for irradiation or possibly they do their own irradiations and regard the dosimetry as an easy, routine, and somewhat peripheral part of the procedure. The recipe for dosimeter solution is given in several textbooks on radiation chemistry. All that is needed is to irradiate a small volume of this solution, to measure the absorbance, and to calculate the irradiation dose by using the equation and values presented in the book. This calculation gives the dose rate of the irradiation facility, and it is a simple matter to calculate the irradiation time required for a particular dose. After starting disasterously, those people then proceed to measure changes in properties of their materials with infinite care.

0097–6156/93/0527–0009$06.00/0

The problems with this kind of approach fall into several areas and, while you might not transgress in all (or any) of the areas, I do not think that the factors involved are so trivial or so widely understood and observed that most of us would not benefit from a review of dosimetry practice.

Dosimeter Preparation and Measurement

The Fricke dosimeter (with which we shall mainly concern ourselves) was developed over sixty years ago (Fricke & Morse, 1927, 1929) and is based on the oxidation of aerated ferrous sulphate solution. It has come to be regarded as a standard since it meets most of the requirements of a dosimeter such as simplicity of preparation, sensitivity, reproducibility, stability before and after irradiation, linearity of response and independence of yield on pH, solute concentration, dissolve oxygen concentration, dose rate, quality of radiation and irradiation temperature within sufficiently wide limits. Its principle shortcomings are its limited does range (40 - 400 Gy), a dependence on LET and its unsuitability for use with high dose rate electron irradiation owing to its dose rate dependence above and about 10^6 Gy s^{-1}.

It is also widely accepted that the presence of sodium chloride removes the necessity for extreme purity of reagents and cleanliness of glassware. I believe that this has been taken too far. We have found it necessary to reserve special glassware for dosimetry solutions and to adopt certain proven procedures for washing, rinsing and drying of this glassware. We also believe that variations in water quality are a principle source of batch-to-batch variation in dosimeter response, particularly with ceric-cerous dosimetry. We do not regard double distillation from a deionised water feed as good enough with respect to trace organic impurities. We use a Millipore reverse osmosis/MilliQ system for water purification which provides ASTM Type 1 reagent water.

Water used in dosimetry should not be stored in plastic containers or in containers with plastic caps or plastic cap liners. Two useful indicators of impurity-free water are (1) a linear dose response with the Fricke dosimeter up to 350-400 Gy and (2) identical responses in the presence and absence of 0.001 M sodium chloride.

Chemical reactions between the inside wall of the dosimeter and the solution are frequently overlooked. Typical effects are illustrated in Figure 1 and 2 which compare responses in conditioned borosilicate glass A ampoules with those in conditioned polyethylene vials of 2-4 mL volumes. There is clearly some radiolytic reaction between the glass and the dosimeter solution. Marked sensitivity to changes to internal glass surfaces produced by heat, radiation and chemical treatment have been reported over the years but are still largely unexplained. Glass cells can be pretreated by either (i) ultrasonics cleaning in hot distilled water, (ii) heating the washed cells at 550°C for one hour or (iii) by preirradiating cells filled with water or 0.4 M sulphuric acid followed by rinsing and drying. Polyethylene cells can be resued and should be kept filled with either irradiated or unirradiated Fricke solution between uses and refilled with fresh solution shortly before use.

A related phenomenon is the effect of cell size. For reasons discussed later,

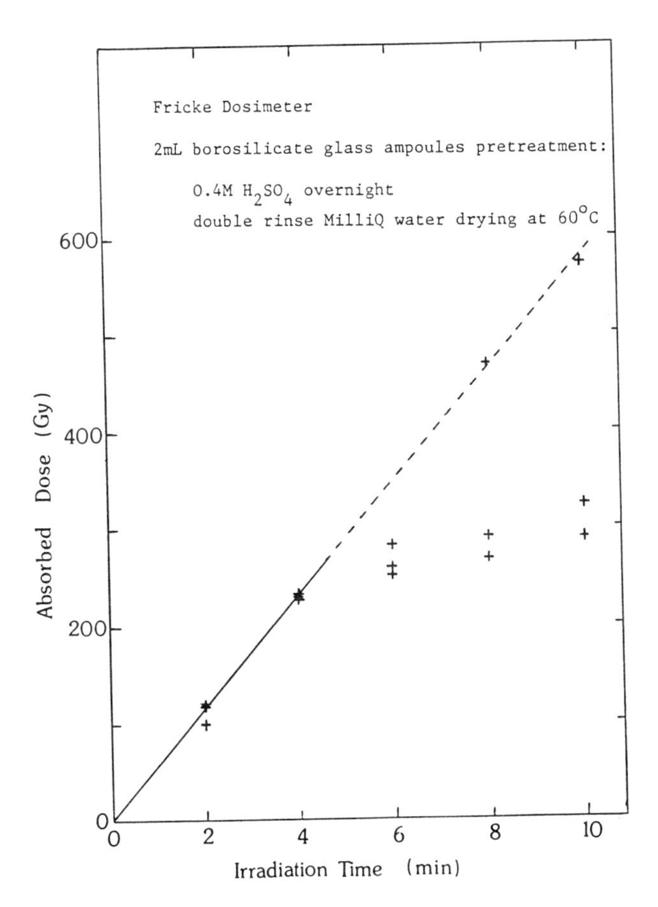

Figure 1: Wall Effects: Variation with Dose

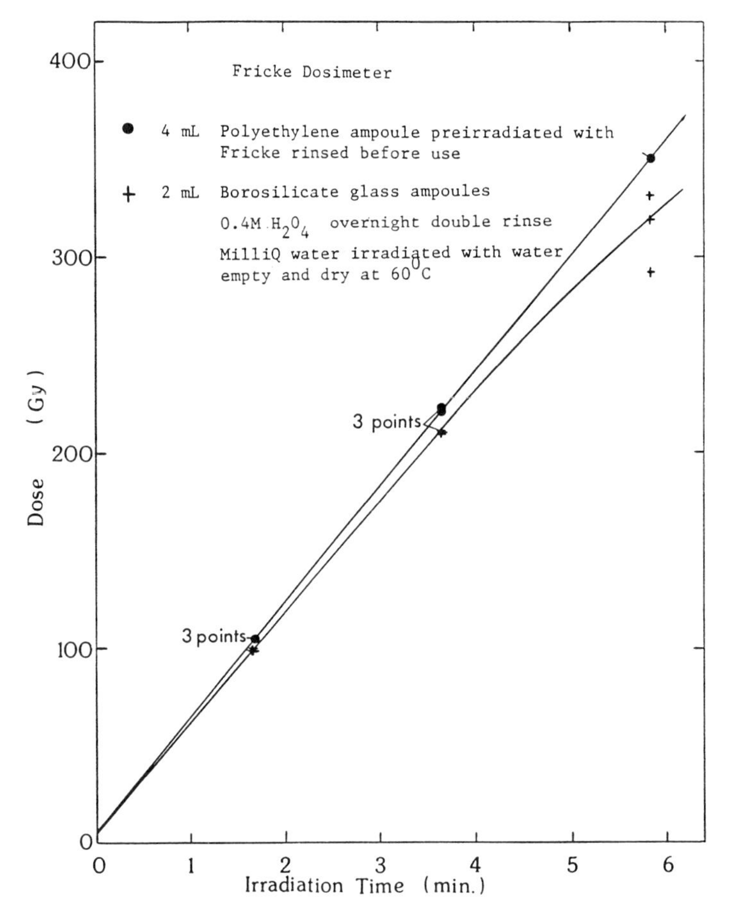

Figure 2: Wall Effects: Comparison of Glass and Polyethylene Ampoules

the dosimeter is usually not quite in electron (or charged particle) equilibrium. According to theory slightly more energy is carried from a glass wall into the solution than out of it. Consequently, the yield should become larger for smaller cell sizes. For polyethlene cells the reverse should apply. There is some evidence for this with cylindrical glass cells below about 4 mm diameter but not with polyethylene cells. The effect of thin glass walls is illustrated in Figure 3 which shows the depth/dose distribution as measured by concentric layers of thin radiochromic dye film dosimeters inside a 2 mL glass ampoule surrounded by a water-equivalent absorber. There is a sharp dose gradient within 40 μm of the wall with relatively uniform dose thereafter.

In some cases, however, and increase in oxidation rate of ferrous ion with decreasing size (and corresponding increasing ratio of cell surface area to cell volume) has been observed over a much larger range of sizes. These results are probably complicated by a contribution from chemical effects of the types mentioned above. Nevertheless, it would be wise to restrict dosimeter cell sizes to diameters above about 8 - 10 mm and to completely fill dosimeters with solution. Wall effects in gamma ray dosimetry have been discussed by Burlin and Chan (1969) and Jarret et al (1978) among others.

Then there is the matter of experimental technique. Clearly it is not good enough to use only one exposure time since the linearity of response cannot be checked and the transit dose (which could be appreciable for a short irradiation times) produces a zero error of unknown magnitude. Also, since the extinction coefficient (or molar linear absorption coefficient) for ferric ion has a large temperature coefficient it is necessary to either measure the solution at 20° or 25°C or to correct the measured absorbance to a standard temperature. Equations for correcting both G(Fe^{3+}) and the absorbance are given in ASTM, E1026. Similar equations are given in ASTM, E1205 for the ceric-cerous system.

What about the accuracy of the spectrophotometer? When was it last checked for wavelength accuracy, wavelength precision, stray light, photometric linearity, photometric precision and photometric accuracy? Is the same narrowest possible slit width always used? Is sufficient care taken with cell cleanliness, orientation and matching? And the solutions themselves - is each batch calibrated against a standard and does each have a preparation and expire date? How are the solutions stored and are they checked periodically to make sure they have not deteriorated?

While all this may seem a little finicky, experience indicates that this degree of care is necessary for accurate consistent results. ICRU Report 14 (1969) recommended a value for the molar extinction coefficient for ferric ion equivalent to 219.6 ±0.5 m^2 mol^{-1} at 25°C and 303 nm in a agreement with the value proposed by Fricke and Hart (1966) based on a mean of published values. This value was changed in ICRU Report 17 (1970) and ICRU Report 21 (1972) to 220.5 ±0.3 m^2 mol^{-1} based on 83 published values and then in ICRU Reports 34 and 35 (1982, 1984) to 216.4 ±0.8 m^2 mol^{-1} based on a very careful determination by Eggermont et al (1978). All this suggests that the value should be accurately determined for the particular spectrophotometer.

In an international comparison conducted in 1978 by the IAEA by the IAEA the measured dosimeter responses (means of 5 replicates for each dose) from eight national laboratories were compared. The responses ranged from 98% to 123% of

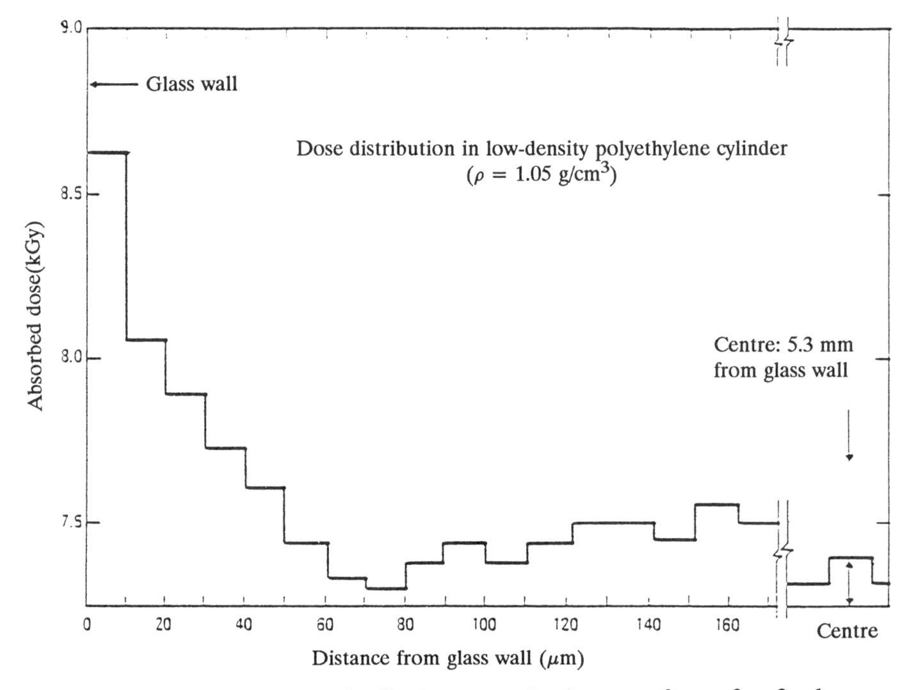

Figure 3: Depth-dose distribution near the inner surface of a 2-ml glass ampoule in a stack of thin polyvinyl pyrollidone radiochromic dye films (film thickness 10 µm) backed by a low-density polyethylene cylinder and at the center of the cylinder were three of the dye films dose meters held in a central bisecting plane. The ampoule had 0.6 mm thick glass walls and was held tightly in a polystyrene cylinder with a 4.7 mm wall thickness. The cobalt-60 gamma radiation field was coaxially symmetric around the ampoule. (W. L. McLaughlin, Center for Radiation Research, National Bureau of Standards, Washington, DC, United States of America). (Reproduced from reference 14.)

the nominal dose for ceric-cerous dosimeters and from 88% to 116% for radiochromic dye film. These are large variations considering that the participating laboratories were major laboratories specializing in dosimetry and the discrepancies were in the measurement alone. They do not include other common sources of error.

Even if such sources of discrepancy could be eliminated, there is a further difficulty in that the use of the expression

$$D = \frac{\Delta A}{\rho\ G\ \varepsilon d}$$

for the determination of absorbed dose treats the Fricke dosimeter as a primary standard dosimeter although it is only a secondary standard since the values of G or the product Gε are experimentally determined. (I am aware that the ASTM regards the Fricke dosimeter as a primary standard dosimetry system (see E170), but this seems to stretch the definition of a primary standard and is not generally accepted by other authorities). Secondary and routine dosimeters should be calibrated against reference standards traceable to a national or international primary standard.

Use of Dosimeters

I refer you to the question posed in the abstract. When you place a perspex dosimeter on the outside of a container what are you expecting to measure - the dose in air, the dose in perspex, the dose in water, the dose at the surface, or the dose to the material inside? I don't know the answer to that. But if the question is "What are you actually measuring?" then the answer is "None of the above".

The basic requirement for dosimetry is that the dosimeter must be in a condition of electron (or charged particle) equilibrium. This is defined as the condition existing in a volume within a medium uniformity exposed to gamma or X-radiation whereby the sum of the kinetic energies of the electrons liberated by the primary photons entering the volume equals the sum of the electrons leaving the volume. Usually a bare or thin-walled dosimeter does not meet this requirement since there is a region within a few millimeters of the wall or surface in which the electron fluence is either higher or lower than in the rest of the medium (Figures 4 and 5).

Figure 4 shows a build up of electron fluence and absorbed dose followed by attenuation in the medium. This pattern is typically observed for the irradiation of thick condensed material in air.

Figure 5 shows the changes in electron fluence in the 2 or 3 mm adjacent to the cell wall in an aqueous dosimetric solution in a thick-walled container of higher average atomic number such as glass. When the dosimeter is irradiated from all sides there is a slight net increase in electron fluence, and hence in absorbed dose, near the walls.

Figure 6 illustrates the situation for small, intermediate and large (compared to the secondary electron range) dosimeters in a medium of atomic number greater than that of the dosimeter. In each case the net electron fluence in the dosimeter is taken as the sum of the electron fluence generated in the dosimeter and the

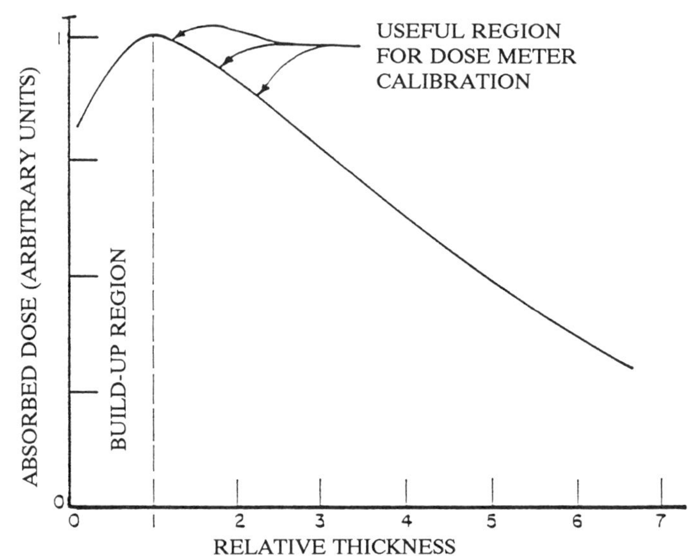

Figure 4: Radiation energy deposition as a function of thickness (expressed in multiples of thickness at maximum dose), showing build-up region, the thickness to achieve electron equilibrium (dashed vertical line), and the useful region for dose meter calibration.
(Reproduced with permission from reference 15. Copyright 1977 International Atomic Energy Agency.)

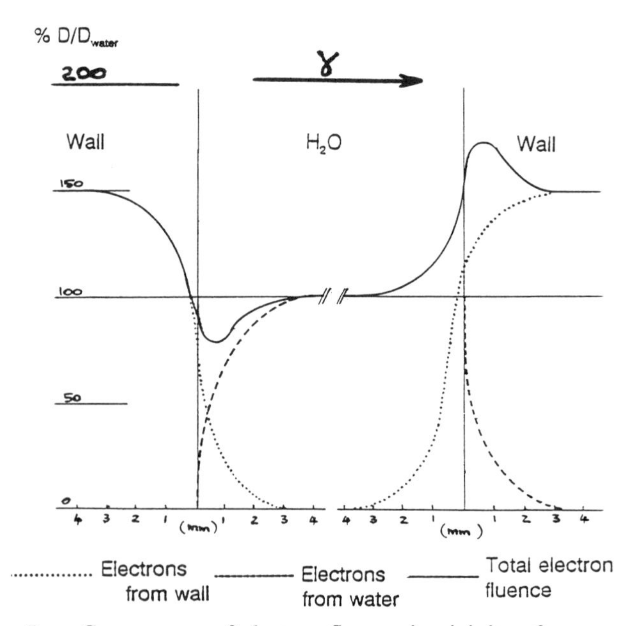

Figure 5: Components of electron fluence in vicinity of entry wall and exit wall of a dosimeter for wall of higher atomic number than water (after Dutreix & Bernard, 1966).

electron fluence generated in the surrounding medium. The absorbed dose is highest in the small dosimeter and lowest in the large dosimeter. It is intermediate and very uneven in the dosimeter of comparable size to the secondary electron range. It follows from this that correction factors will be smaller and dosimetry will be more accurate if the absorption characteristics of the dosimeter are closely matched to those of the surrounding material.

When a primary standard dosimeter is used to measure an irradiation dose, the quantity measured is the absorbed dose in water. Hence, other dosimeters calibrated against this standard also measure the absorbed does in water and should normally, therefore, be surrounded by a suitable thickness of water or water-equivalent material such as polyethylene and polystyrene to achieve electron equilibrium (there are exceptions to this such as when thin film dosimeters are embedded tightly in a homogenous medium of similar absorbing qualities to give a direct measure of the dose absorbed by the medium at that point or when they are used to provide spatial resolution in regions of high dose gradient such as at the interface between dissimilar materials (e.g., Figure 3)).

The thickness required for electron equilibrium based on maximum electron range is given by curve A in Figure 7 for different primary photon energies. It has been found experimentally that the depth of the peak in the absorption curve (given by curve B) is somewhat less than the maximum electron range. It would appear, therefore, that the minimum thickness of water-equivalent material needed to achieve electron equilibrium for Co-60 irradiation is between 2.5 and 5.5 mm. These considerations apply to irradiation by a point source or a parallel beam where the electron fluence is predominantly in the forward direction. However, for irradiation close to a large plaque source or for irradiation from all sides as in a Gammacell or an irradiation pond, the non-equilibrium region near the surface may be smaller still. The dose response should, therefore, be checked using different wall or absorber thicknesses to determine the peak of the energy deposition curve and, hence, the optimum wall thickness for dosimeters.

The three conditions illustrated in Figure 6 lead to the basic equations of cavity theory for correcting dosimeter response to determine what the absorbed dose would be at that point in the medium in the absence of the dosimeter. By this theory, a dosimeter placed in a medium during irradiation constitutes a discontinuity or cavity in the medium which alters the secondary electron fluence but not the photon fluence. The energy absorbed by the dosimeter is, therefore, not the same as that absorbed by the unperturbed medium and, hence, corrections are needed to calculate the relationship between the two doses. If the absorption characteristics of the dosimeter are well matched to those of the medium, the corrections are less than about 5%.

The three cases for irradiation by photons are:

Case 1 Large Cavities (i.e., 2mL, 5mL ampoules and vials with water-equivalent absorber

The dose absorbed in the dosimeter is due to secondary electrons produced in the dosimeter. Electrons entering from outside can be ignored. The effect of the

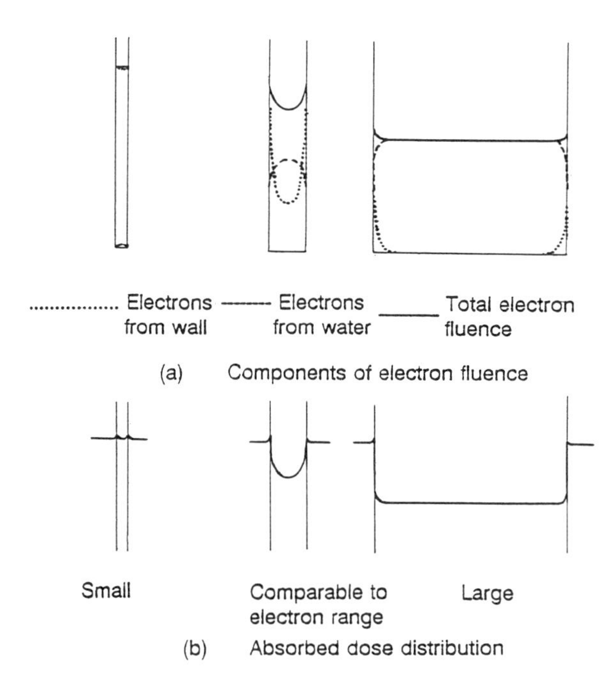

Figure 6: Electron fluence and absorbed dose distribution for small, medium and large dosimeters irradiated by gamma radiation from all sides (after Burlin, 1970).

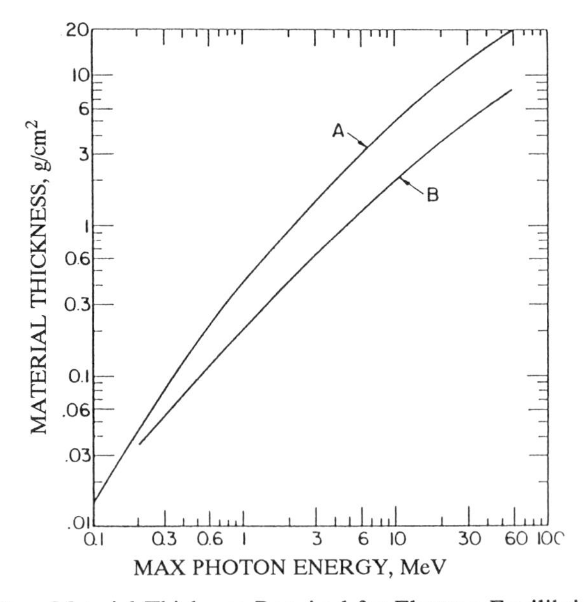

Figure 7: Material Thickness Required for Electron Equilibrium. A - Electron Range. B - Depth of Peak Dose. (Reproduced with permission from reference 16. Copyright 1989 American Society for Testing and Materials.)

dosimeter on the photon fluence can be neglected if the dosimeter is similar to the surrounding medium.

The relationship between the dose in the cavity (i.e., dosimeter) and the dose in the medium is given by

$$f = \frac{D_c}{D_m} = \left[\frac{\mu_{en}}{\rho}\right]_m^c = \frac{\int_o^{Emax} \Psi(E) \left[\frac{\mu_{en}(E)}{\rho}\right]_c dE}{\int_o^{Emax} \Psi(E) \left[\frac{\mu_{en}(E)}{\rho}\right]_m dE} \quad (1)$$

where f = cavity correction factor

D_c = dose in cavity (i.e., dosimeter) (Gy)

D_m = dose in medium (Gy)

$\frac{\mu_{en}}{\rho}$ = mass energy absorption coefficient ($m^2 kg^{-1}$)

Ψ = photon energy fluence (Jm^{-2})

dE = fraction of photon energy

and subscripts c and m refer to cavity and medium respectively

Case 2 Small Cavities (i.e., thin film dosimeters without water equivalent absorber)

The dose absorbed in the dosimeter is due to secondary electrons generated outside the dosimeter. Energy deposited by electrons generated inside the dosimeter is negligible. The dosimeter has a negligible effect on the photon fluence (and vise versa).

The relationship between the dose in the cavity (i.e., dosimeter) and the dose in the medium is given by

$$f = \frac{D_c}{D_m} = \left[\frac{S_{col}}{\rho}\right]_m^c = \frac{\int_o^{Emax} \Phi(E) \left[\frac{S_{col}(E)}{\rho}\right]_d dE}{\int_o^{Emax} \Phi(E) \left[\frac{S_{col}(E)}{\rho}\right]_m dE} \quad (2)$$

where Φ = electron fluence (m^{-2})

$\frac{S_{col}}{\rho}$ = electron collision mass stopping power ($J\ m^2\ kg^{-1}$)

If the dosimeter and surrounding medium are of similar composition, equation 2 may be calculated for a single electron energy since the ratios of stopping powers do not vary greatly with energy between 0.05 and 1.3 MeV.

Equation 2 then becomes

$$f = \frac{D_c}{D_m} = S_m^c = \frac{\left(\frac{S_{col}(E)}{\rho}\right)_C}{\left(\frac{S_{col}(E)}{\rho}\right)_M} \tag{3}$$

where $\bar{E}$ = mean energy of electron spectum at the measurement location in the medium

Case 3 **Cavities of Similar size to the Range of the Hightest Energy Secondary Electrons (i.e., perspex dosimeters, ampoules and vials)**

Both the photon fluence and the secondary electron fluence contribute to the absorbed dose in the dosimeter.

The relationship between the dose in the cavity (i.e., dosimeter) and the dose in the medium is given by

$$f = \frac{D_c}{D_m} = d \cdot \left(\frac{S}{\rho}\right)_m^c + (1 - d) \cdot \left(\frac{\mu_{en}}{\rho}\right)_m^c \tag{4}$$

where d = fraction of energy deposited in cavity by electrons from the surrounding medium

$$d = \frac{1 - e^{-\beta g}}{\beta g} \tag{5}$$

β = effective mass attenuation coefficient for electrons in the cavity

$$\beta = \frac{16}{\left(E_{max} - 0.036\right)^{1.4}} = 4.6R^{-1} \tag{6}$$

R = extrapolated range of secondary electrons

g = average path length through the dosimeter

$$g = \frac{4V}{A} = \frac{4 \text{ x volume}}{\text{surface area}} \tag{7}$$

The use of Equations 1 and 2 requires some knowledge of both the degraded photon spectrum resulting from Compton scattering under the irradiation conditions and the recoil Compton electron spectrum. These are shown in Figures 8 and 9. Energy degradation can occur through scattering in the cobalt source material, the double stainless steel encapsulation, the source holder, the irradiation can, steel or lead shielding and water in an irradiation pond as well as in the target material.

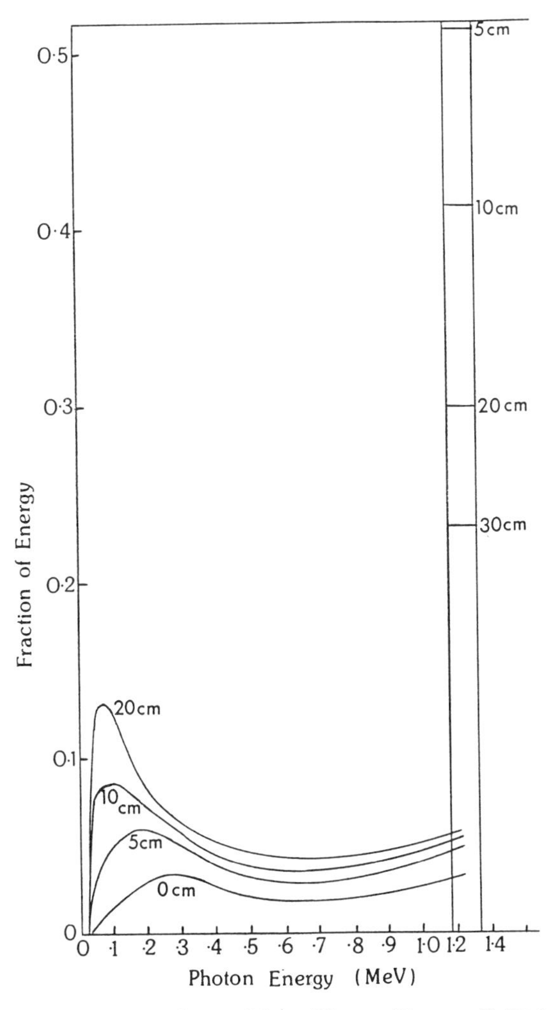

Figure 8: Fraction of Primary Mean Photon Energy (1.17-1.33 MeV) and Scattered Degraded Photon Energies (0.05-1.25 MeV) for irradiation in air (0cm) and at Various Depths in water-equivalent material with a cladded Co-60 Source Plaque (after Seltzer, 1990).

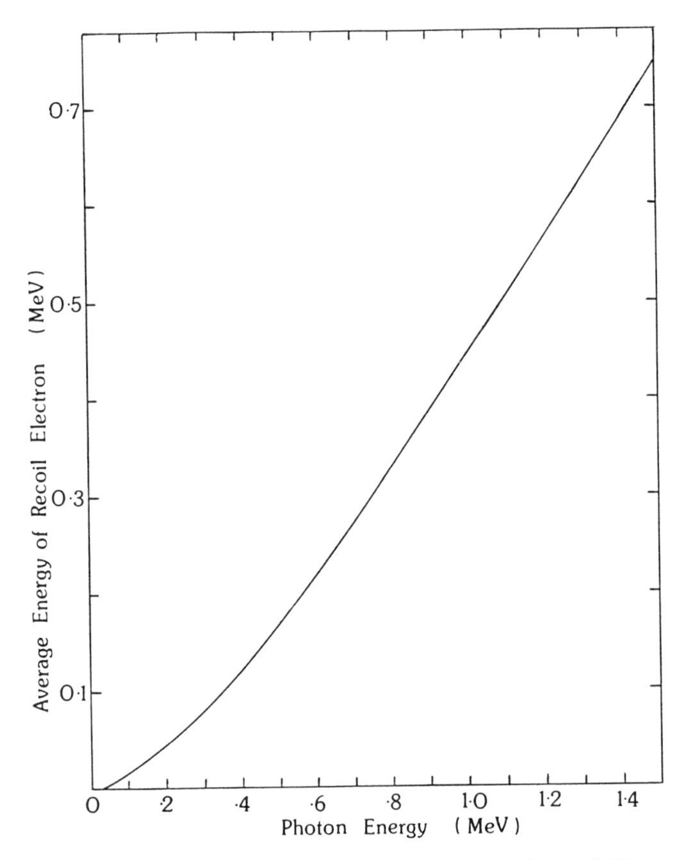

Figure 9: Relationship between photon energy and recoil Energy of Ejected Electron.

Mass energy absorption coefficients vary with energy and are generally higher around 0.05 - 0.2 MeV than for the primary undegraded photons, particularly for high atomic number materials. Electron mass collision stopping powers are also higher around 0.05 - 0.2 MeV but show similar patterns of variation, hence, the ratios of stopping powers do not vary gently over the range 0.05 - 10 MeV.

Mass energy absorption coefficients are tabulated in Hubbell (1982) and Storm and Israel (1970) for elements and a small range of materials. Those in Storm and Israel are inaccurate in the low energy region. Electron stopping powers are tabulated in ICRU Report 37 (1984) for a limited range of elements and materials or can be calculated from fundamental parameters of the elemental constituents using equations and values given in Seltzer and Berger (1984). Tabulated or calculated values for selected energy ranges are multipled by the appropriate weighting factor (dE) for the conditions and the values so obtained are summed over the energy range 0.05 - 1.25 MeV for Co-60 irradiation or up to the maximum electron energy for electron irradiation.

Energy absorption coefficients and electron stopping powers of some common dosimetric materials for use in Equations 1 and 2 are given in Tables 1 and 2. The fact that the values differ by 10% or more underlines the importance of matching the dosimeter to the material under treatment.

Energy absorption coefficients and electron stopping powers can be more widely applied to enable calculation of the absorbed dose in one medium from that in another. Although the values can be used directly, it is often convenient to express conversion factors as the ratio of absorption in the medium to that in water.

If the absorbed dose has been determined under equilibrium conditions in a material, M_1, (i.e. water) then the absorbed dose in another material of interest, M_2, irradiated under the some conditions is given by

$$D_{m_2} = \frac{(\mu_{en}/\rho)_{m_2}}{(\mu_{en}/\rho)_{m_1}} \cdot D_{m_1} \qquad \text{*c33E} \qquad (5)$$

where D_{m1} = dose absorbed in material m_1
D_{m2} = dose absorbed in material m_2

Also, for mixtures and compounds

$$(\mu_{en}/\rho)_{total} = \sum_i k_i \, (\mu_{en}/\rho)_i \qquad (6)$$

where k_i = weight fraction of constituent i
$(\mu_{en}/\rho)_i$ = mass energy absorption coefficient of constitutent i

Values for the ratio of absorption coefficients, and hence, of absorbed dose, for water and some common polymers are given in Table 3. The general pattern shows that the absorption characteristics of low atomic number materials composed of H, C, N and O do not change appreciably with radiation quality. Those such as PTFE without hydrogen have significantly lower absorption coefficients than water and hydrocarbons while those containing high atomic number materials such as chlorine show substantially increased absorption with increasing degradation of the energy spectrum. This applies to a greater or lesser degree to other materials such as stainless steel and glass.

Finally, it should be emphasised that correction factors of this kind should be applied judiciously. To illustrate the point, I leave you with a question. In the sterilization of a suture pack containing a length of suture threaded to a stainless steel needle, what is the ratio of absorbed dose in the two materials. The ratio of mass energy absorption coefficients for suture/needle is 0.947 for irradiation in air and 0.476 for irradiation in an irradiation pond. But the dimensions of the suture and the needle are small compared to the range of the secondary electrons. Hence, the dose absorbed may be better described by Equation 2 than Equation 1 and the relevant factor should be the ratio of electron stopping powers. This is 1.41 for irradiation in air and 1.37 for irradiation in the pond. Which factor is correct in the circumstances? The answer is given at the foot of the page.

Table 1. Photon Mass Energy Absorption Coefficients for Some Common Dosimetric Materials for Irradiation with a Cladded Co-60 Source Plaque in Air and at Various Depths in Water-Equivalent Material

$$\frac{\left(\mu_{en}\right)}{\rho} \qquad \text{units: } m^2\ kg^{-1} \times 10^3$$

Substance	Depth (cm)			
	0	5	10	20
water	3.0156	3.0411	3.0538	3.0680
Fricke solution	3.0067	3.0363	3.0535	3.0604
0.01 M Ce^{4+}/Ce^{3+} soln.	3.0651	3.1291	3.1641	3.2228
alanine	2.9278	2.9426	2.9464	2.9508
perspex	2.9278	2.9442	2.9456	2.9475
nylon	2.9566	2.9716	2.9714	2.9713
polyvinyl chloride	2.8542	3.0733	3.3197	2.5998
graphite	2.7099	2.7214	2.7084	2.7151

Answer: Neither is correct since the two materials are not in electronic equilibrium. The absorption characteristics of the needle differ greatly from those of the surroundings. The question does illustrate the wide variations in absorbed dose that can occur, depending upon the radiation quality and the size and atomic number of the absorbing materials.

Table 2. Electron Mass Collision Stopping Powers for Some Common Dosimetric Materials for Irradiation with a Cladded Co-60 Source Plaque in Air and at Various Depths in Water-Equivalent Material

$$\frac{\left(S_{col}\right)}{\rho} \quad \text{units: MeV cm}^2\ \text{g}^{-1}\ *$$

Substance	Depth (cm)			
	0	5	10	20
water	2.5511	3.3476	4.0233	4.7790
Fricke	2.5425	3.3328	4.0033	4.7533
0.01 M Ce^{4+}/Ce^{3+} soln.	2.5377	3.3263	3.9954	4.7440
alanine	2.4989	3.2858	3.9538	4.7009
perspex	2.4830	3.2593	3.9177	4.6542
nylon	2.5626	3.3745	4.0635	4.8343
polyvinyl chloride	2.2436	2.9281	3.5069	4.1542
graphite	2.2575	2.9721	3.5776	4.2545

* 1 MeV $cm^2\,g^{-1} = 1.602 \times 10^{-14}$ J m^2 kg^{-1}

Table 3. Ratio of Absorbed Dose in Common Polymers to Absorbed Dose in Water irradiated by Co-60 Radiation in Air and at Various Depths in Water-Equivalent Material

Substance	Depth (cm)			
	0	5	10	20
polyethylene	1.026	1.021	1.015	1.008
polypropylene	1.026	1.021	1.015	1.008
polystyrene	.967	.963	.967	.951
polychlorostyrene	.956	.980	1.008	1.043
polyester (mylar)	.937	.934	.931	.928
polycarbonate (lexan)	.948	.945	.939	.937
polyvinyl acetate	.962	.959	.956	.953
nylon	.980	.977	.973	.968
polyvinyl chloride	.946	1.011	1.087	1.173
polyvinylidene chloride	.923	1.008	1.107	1.221
PTFE	.866	.870	.874	.879

Literature Cited

[1] *Radiation Dosimetry*; Attix, F. H. and Roesch, W. C. Eds.; Academic Press, NY & Lond; 1968, Vol 1.

[2] Burlin, T. E. and Chan, F. K. *The Effect of the Wall on the Fricke Dosimeter*; Int., J. Appl. Radiat. Isot. 20, 1961, pp 767.

[3] Burlin, T. E., *Theory of Dosimeter Response*; Holm, N. W. and Berry, R. J., Eds.; Manual on Radiation Dosimetry; Marcel Dekker: NY, 1970.

[4] Dutreix, J. and Bernard, M.; *Dosimetry at Interfaces for High Energy X and Gamma Rays*; Br. J. Radiol 39, 1966, pp 205.

[5] Eggermont, G.; Buysse, J.; Janssens, A.; Thielens, G.; Jacobs, R. *Descrepancies in Molar Extinction Coefficients of* Fe^{3+} *in Fricke Dosimetry*; National and International Standardization of Radiation Dosimetry lAEA-SM-222/42; lAEA Vienna; 1978; Vol 2, pp 317.

[6] Fricke, H.; Hart, E. J. *Chemical Dosimetry* in "Radiation Chemistry"; Attix, F. H.; Roesch, W. C., Eds.; Academic Press: NY and Lond., 1966, Vol 2.

[7] Fricke, H; Morse, S. *The Chemical Action of Roentgen Rays on Dilute Ferrous Sulfate Solutions as a Measure of Dose*; Am. J. Roentgenol. Radium Therapy. Nucl. Med. 18; 1927; pp 430.

[8] Fricke, H.; Morse, S. *The Action of X-Rays on Ferrous Sulfate Solutions*; Phil. Mag. 7, 7, 1929, pp 129.

[9] Hubbell, J. H.; *Photon Mass Attenuation and Energy Absorption Coefficients from 1 keV to 20 MeV*; Int. J. Appl. Radiat. Isot. 33, 1982, pp 1269.

[10] Jarrett, R. D.; Brynjolfsson, A.; Wang, C. P. *Wall Effects in Gamma-Ray Dosimetry*; lAEA-SM-222/33 National and International Standardization of Radiation Dosimetry; lAEA Vienna, 1978; Vol 2, pp 91.

[11] Seltzer, S. M.; Berger, M. J. *Improved Procedure for Calculating the Collison Stopping Power of Elements and Compounds for Electrons and Positrons*; Int. J. Appl. Radiat. Isot. 35; 1984, pp 7.

[12] Seltzer, S. M. *Photon Fluence Spectra and Absorbed Dose in Water Irradiated by Cladded* ^{60}Co *Source Plaques*; Radiat. Phys. Chem. 35; 1990; 4-6, pp 703; Int. J. Radiat. Appl. Instrum. Part C.

[13] Storm, E.; Isreal, H. I. *Photon Cross Sections from 1 KeV to 100 MeV for Elements* $Z = 1$ *to* $Z = 100$; Nucl. Data Table A7, 1970; pp 565.

[14] *High Dose Measurements in Industrial Reduction Processing;* International Atomic Energy Agency: Vienna, Austria, 1981; Technical Report Series 205.

[15] *Manual of Food Irradiation Dosimetry;* International Atomic Energy Agency: Vienna, Austria, 1977; Technical Report Series 178.

[16] *1989 Annual Book of ASTM Standards;* American Society for Testing and Materials: Philadelphia, PA, 1989; Section 12, Vol. 12.02, Nuclear, Solar, and Geothermal Energy.

RECEIVED February 10, 1993

Chapter 3

Heterogeneous Nature of Radiation Effect on Polymers

Yoneho Tabata

Faculty of Engineering, Tokai University, 1117, Kitakaname, Hiratsuka, Kanagawa 359–112, Japan

The radiation chemical phenomena caused by either gamma ray or high energy electrons are essentially inhomogeneous, in spite of the fact that those radiations have low LET values in an order of 10^{-2} ev/A. As one typical example, radiation effects on saturated hydrocarbons including model compounds of polyolefins and ethylene-propylene rubber (EPR) is discussed. It is made clear that crosslinking takes place very inhomogeneously in those materials.

Experimental Methods

A series of model compounds of linear alkanes, squarane, and ethylene-propylene rubber were used as the examples. Those samples are listed in Table 1. Transient species were detected by pulse radiolysis at the University of Tokyo. Electron pulses with 10ps and 2ns from an electron linear accelerator of 35MeV were used. The dose per pulse was about 1 krad for 10ps and 4-5 krad for 2ns one (1). Gas chromatography (2), mass-spectrometry (3), liquid chromatography (4), were carried out for determining and measuring the radiolysis products, and crosslinking.

Results and Discussion

(1) Transient Species. In our pulse radiolysis work on saturated hydrocarbons RH including model compounds of poly-olefins, and a copolymer of ethylene-propylene rubber, the following transient species have been directly observed (5,6).

0097–6156/93/0527–0027$06.75/0

Table 1. Samples for Experiments

Sample name	Formula	
cyclohexane(c-C_6H_{12})		
metylated cyclohexane(Me-C_6H_{11})		
n-hexane(n-C_6H_{14})	$CH_3(CH_2)_4CH_3$	
n-octane(n-C_8H_{18})	$CH_3(CH_2)_6CH_3$	
n-nonane(n-C_9H_{20})	$CH_3(CH_2)_7CH_3$	
n-undecane(n-$C_{11}H_{24}$)	$CH_3(CH_2)_9CH_3$	
n-dodecane(n-$C_{12}H_{26}$)	$CH_3(CH_2)_{10}CH_3$	
n-tridecane(n-$C_{13}H_{28}$)	$CH_3(CH_2)_{11}CH_3$	
n-tetradecane(n-$C_{14}H_{30}$)	$CH_3(CH_{12})_4CH_3$	
n-pentadecane(n-$C_{15}H_{32}$)	$CH_3(CH_2)_{13}CH_3$	
n-hexadecane(n-$C_{16}H_{34}$)	$CH_3(CH_2)_{14}CH_3$	
n-eicosane(n-$C_{20}H_{42}$)	$CH_3(CH_2)_{18}CH_3$	triclinic
n-heneicosane(n-$C_{21}H_{44}$)	$CH_3(CH_2)_{19}CH_3$	orthorhombic
n-tricosane(n-$C_{23}H_{48}$)	$CH_3(CH_2)_{21}CH_3$	orthorhombic
n-tetracosane(n-$C_{24}H_{50}$)	$CH_3(CH_2)_{22}CH_3$	triclinic
squalane($C_{30}H_{62}$)	$(CH_3CH(CH_2)_3CH(CH_2)_3CH(CH_2)_2)_2$	giass
	CH_3 CH_3 CH_3	
ethylene-propylene rubber		amorphous

$RH^{\bullet +}$	cation radical
RH^*	singlet excited state
$R(-H)^{\bullet +}$	olefinic cation radical
$R^{\bullet}$	alkyl radical
e_t^-	trapped electron in alkanes and polymer EPR
$P^{\bullet +}$	polymer cation radical
P^*	polymer single excited state

Transient absorption spectra obtained by pulse radiolysis for both n-dodecane and ethylene-propylene rubber together with other compounds are shown in Figure 1 (7). For the polymer, a broad band in a range of 400-1000 nm has been observed. By subtraction of spectrum obtained in the presence of 100 mM CCl_4 from that of the neat system, absorption peaks at 700 nm due to p* and at a maximum longer than 1000 nm due to $P^{\bullet +}$ are discriminated. It is interesting to compare those peaks with maxima of model compounds of saturated hydrocarbons. The peak due to RH* is shifting toward a higher wavelength as the number of carbons increases and levels off at 700 nm. The absorption maximum due to $P^{\bullet +}$ is extrapolated to be wavelength longer than 1000 nm as an extreme higher number of hydrocarbons of model compounds.

According to the picosecond pulse radiolysis of a few alkanes including cyclohexane, the majority of alkyl radicals are formed very fast from the excited cation radical ($RH^{\bullet +*}$) and super excited state RH** (7). This has been also demonstrated by electron spin resonance method at very low temperatures (8). The absorption peak due to the alkyl radical appears around 250 nm. Although there are a few papers concerning the formation of olefinic cation radical (9), no conclusion has obtained until now.

Transient absorption around 240 nm to 350 nm was observed for all alkane compounds, as shown in Figure 2, 3 and 4 (7). From the osciloscope trace at the peak of the absorption, it is clear that two components, fast and very slow decay ones, are overlapped. The fast decay component is mainly ascribed to be olefinic cation radical and the very slow decay component alkyl radical. Absorption of the fast decay component is enhanced in the presence of electron scavengers and the slow decay component is suppressed to some extent, by the scavenger, as shown in Figure 3.

The fast decay component is subtracted by reducing optical densities at 80 ns after the pulse from that at immediately after the pulse. They are shown in Figure 4(A) and (B).

It is clear from the Figure 4(A) that the higher the electron affinity of additives, the larger the extent of enhancement. In the Figure 4(B), the effect of cation scavenger of C_2H_5OH on the formation of olefinic cation radical is shown. The concentration of the cation is reduced to a large extent by C_2H_5OH. Osciloscope traces indicating the formation and decay of the transient species are shown in the same figure. The long lifetime component which is ascribed to be alkyl radical is not affected by C_2H_5OH, while the short lifetime component, olefinic cation radical, is affected to a large extent by C_2H_5OH, that is: the initial yield is reduced and the decay is enhanced as well by adding the cation scavenger of C_2H_5OH.

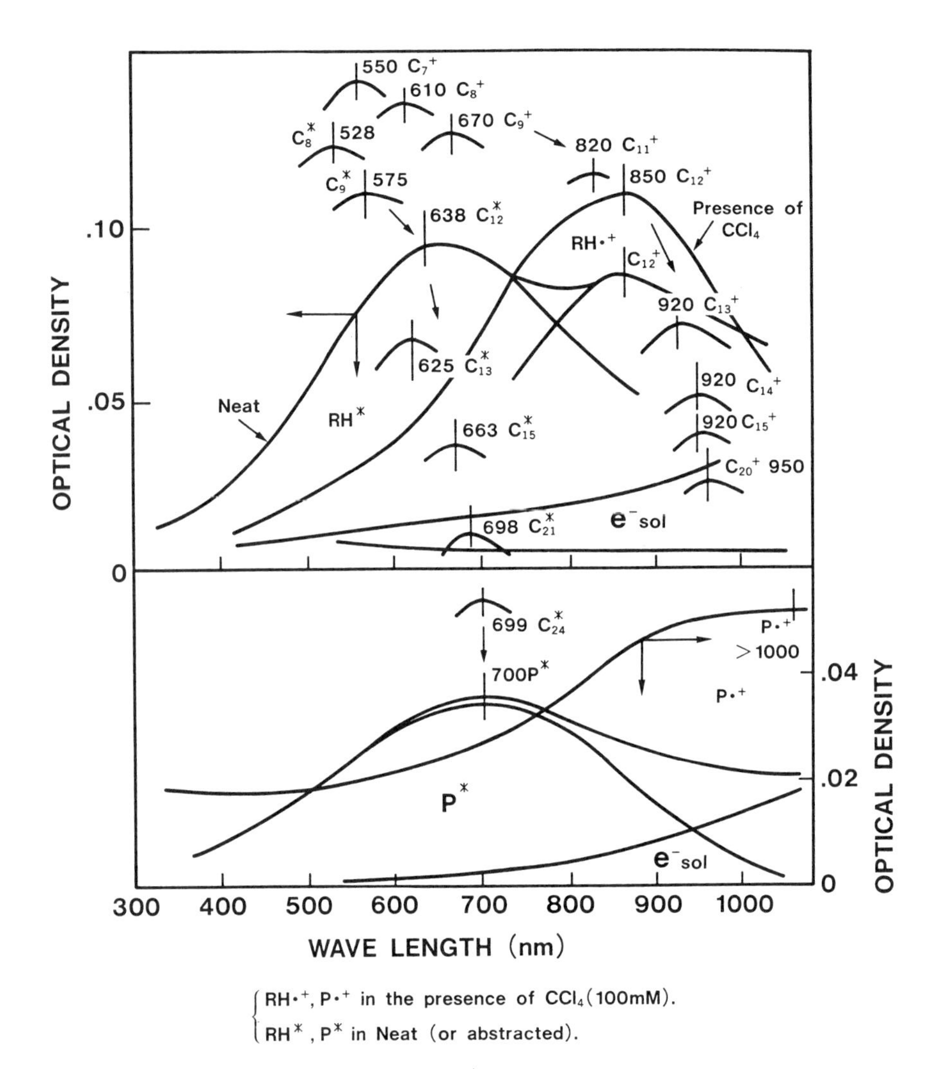

Figure 1: Transient absorption spectra of a series of n-alkanes and ethylene-propylene rubber. Absorption peaks of singlet excited state and cation radical, e^-_{sol} are shown in the figure.

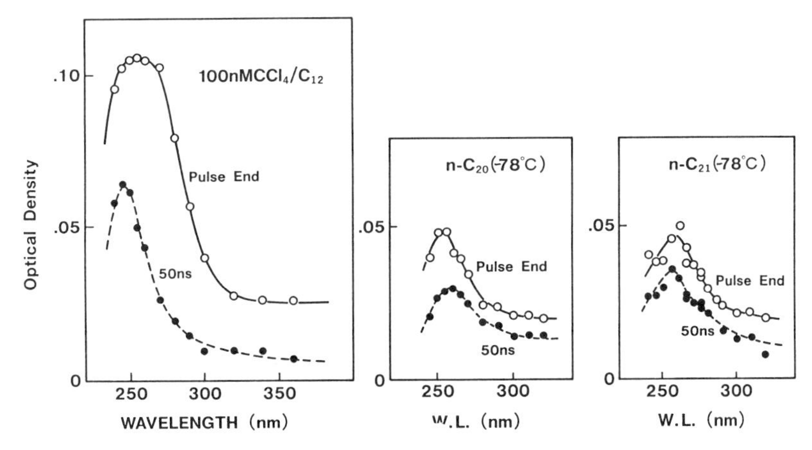

Figure 2: Transient absorption spectra of $n-C_{12}H_{26}$, $n-C_{20}H_{42}$ and $n-C_{21}H_{44}$ in the presence of 100 mM CCl_4. o Pulse end (2ns), • at 50ns after the pulse.

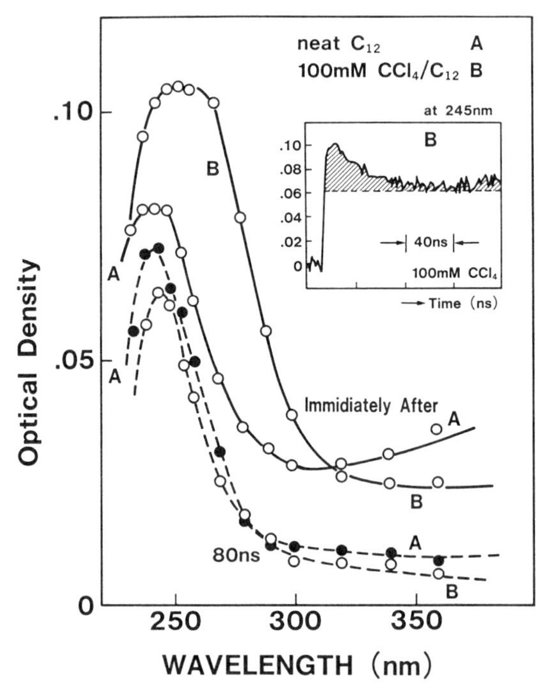

Figure 3: Transient absorption spectra of $n-C_{12}H_{26}$ in neat (A) and in the presence of CCl_4 (B).
o : immediately after the pulse (2ns).
• : 80ns after the pulse. Osciloscope trace at 245 nm is shown in the figure.

As the lifetime of olefinic cation radical in bulk is very short, pulse radiolysis with very high time resolution can only detect and observe the formation of the species. Therefore, a small absorption peak is only observed in the neat system in the case of n-dodecane, as one of the examples. In the presence of an electron scavenger with high enough concentration such as 100 mM of CCl_4, the olefinic cation radical is clearly observed.

It is interesting to note that the double bond formation is scavenged much more effectively than alkyl radical. This may suggest that the precursors for both species are different from each other. The similar trend could be postulated for higher carbon number of alkanes including hydrocarbon polymers.

From those experimental results mentioned above, the following conclusion could be reached, concerning the nature of olefinic cation radical. The precursor of olefinic cation radical, ($RH^{\bullet +}$), is scavenged to a large extent, by C_2H_5OH, while the precursor of alkyl radical RH^{+*} is not scavenged by the scavenger.

It could be said that the major process for the formation of olefinic cation radical is thermal and not hot, in contrast with the formation of alkyl radical.

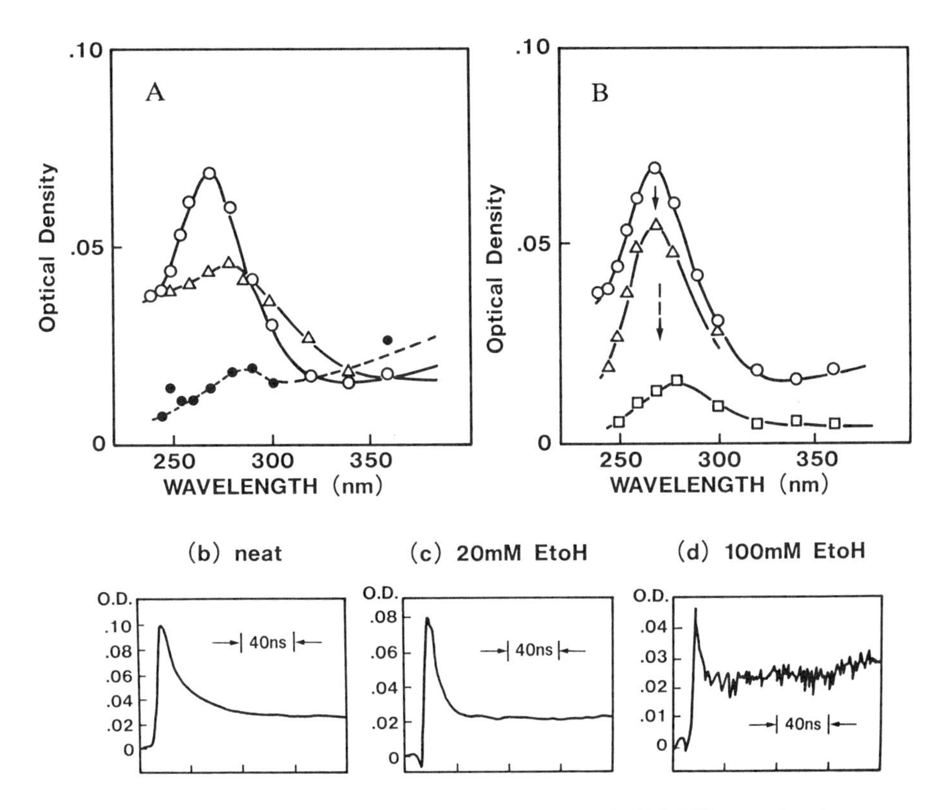

Figure 4: Transient absorption spectra around 270-280 nm in the presence of electron scavenger (CCl_4 and N_2O) A and cation scavenger B. Spectra were obtained by reducing absorption at 80ns after the pulse from that at immediately after the pulse. Osciloscope traces are shown in the same figure.

The small reduction of alkyl radical yield in the presence of CCl_4 in Figure 3 may be due to the decrease in the process "3" in Figure 5. Because the geminate recombination $RH^+ + e^- \rightarrow RH^*$ is reduced by capture of the geminate electron with CCl_4.

Therefore, the formation processes of olefinic cation radical and double bond C=C can be described as follows. $\{RH^{\bullet +}\}$ is assumed to be the major precursor of olefinic cation radical.

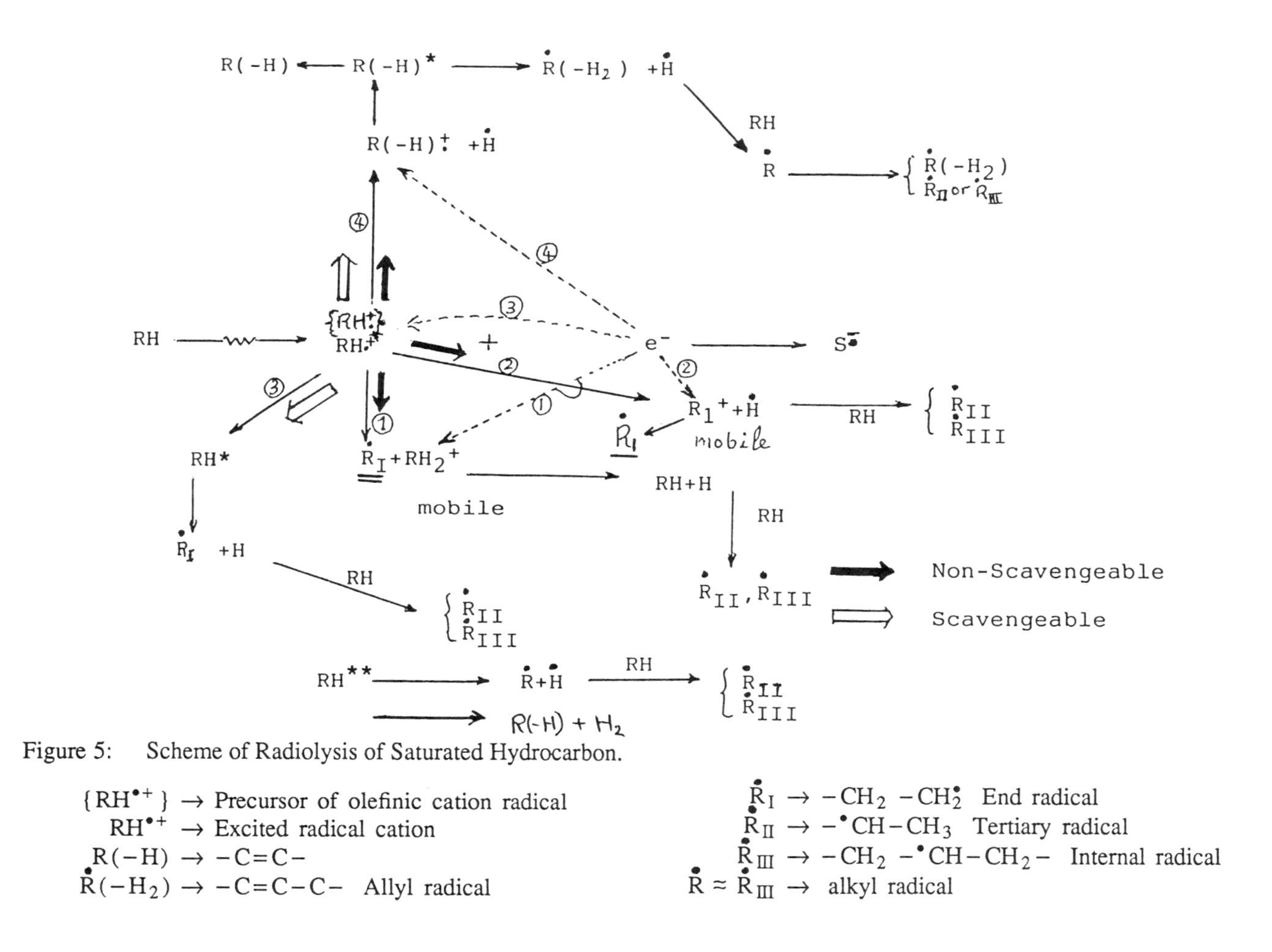

Figure 5: Scheme of Radiolysis of Saturated Hydrocarbon.

$\{RH^{\bullet+}\}$ → Precursor of olefinic cation radical
$RH^{\bullet+}$ → Excited radical cation
$R(-H)$ → $-C=C-$
$\dot{R}(-H_2)$ → $-C=C-C-$ Allyl radical

$\dot{R}_I$ → $-CH_2-CH_2^{\bullet}$ End radical
$\dot{R}_{II}$ → $-{}^{\bullet}CH-CH_3$ Tertiary radical
$\dot{R}_{III}$ → $-CH_2-{}^{\bullet}CH-CH_2-$ Internal radical
$\dot{R} \approx \dot{R}_{III}$ → alkyl radical

$$RH \longrightarrow RH^{\bullet+*},\ \{RH^{\bullet+}\},\ RH^{\bullet+} + e^-$$
$$RH \longrightarrow RH^{**}, RH^{*}$$
$$\{RH^{\bullet+}\} \longrightarrow R(-H)^{\bullet+} + H_2 \qquad \text{major process}$$
$$RH^{\bullet+} + e^- \longrightarrow RH^{*}$$
$$RH^{\bullet+*} \longrightarrow R(-H)^{\bullet+} + H_2 + e^- \qquad \text{minor process}$$
$$RH^{**} \longrightarrow R(-H) + H_2$$
$$R(-H)^{\bullet+} + e^- \longrightarrow R(-H)^{*} \qquad \text{neutralization}$$
$$R(-H)^{*} \longrightarrow R(-H_2)\bullet + H\bullet \qquad \text{dissociation}$$
$$R(-H)^{*} \longrightarrow R(-H) \qquad \text{relaxation}$$

Effect of CCl_4 (S) and C_2H_5OH (A) on the formation of olefinic cation radical can be expressed as follows.

$$\{RH^{\bullet+}\} + e^- \dashrightarrow R(-H)^{\bullet+} + e^- + H_2$$
$$S + e^- \dashrightarrow S^-$$
...........enhancement of absorption

$R(-H)^{\bullet+}$ due to e^- capture

$$\{RH^{\bullet+}\} + e^- \dashrightarrow R(-H)^{\bullet+} + e^- + H_2$$
$$\{RH^{\bullet+}\} + e^- + S + A \dashrightarrow R^{\bullet} + S^- + AH^+$$
...........scavenging of formation of $R(-H)^+$

$$R(-H)^{\bullet+} + A \dashrightarrow R(-H_2)\bullet + AH^+$$
...........enhancement of decay $R(-H)^{\bullet+}$

The scheme of the radiolysis of saturated hydrocarbons is summarized in Figure 5.

(2) G-Values of H_2, Crosslinking and C=C(10). G-values of hydrogen formation obtained from gas chromatography and those of crosslinking obtained from liquid chromatography are summarized in Table 2. The G-value of crosslinking was determined from the slope of monomer consumption. G-value of crosslinking from that of H_2.

(3) Effect of Pyrene on Irradiation Effect of Ethylene-Propylene Rubber (7). In the presence of pyrene in ethylene-propylene rubber, charge and energy transfer take place from the medium to pyrene. The pulse radiolysis results are shown in Figure 6. It is obvious from the figure that charge transfer from the polymer medium to the additive of pyrene occur very effectively. It is well known that three peaks from shorter wavelength to the longer are assigned to be triplet excited state, cation and anion radicals of pyrene, respectively. It is obvious from the osciloscope trace in the figure that charge transfer occur very fast. As the absorption coefficients are known for those species, their G-values could be easily estimated.

Emission spectra have been also observed, as shown in Figure 7. Fast formation of both the monomer singlet excited state and the excimer can be seen.

Table 2. G-Values of H_2, G(X), and G(C=C) in a Series of Model Compounds

	Irrad. Temp	$G(H_2)$	G(X)	G(C=C)	Phase
$n\text{-}C_{20}H_{42}$	-77°C	2.14	1.01± 0.05	1.13±0.05	crystal
	25	2.26	1.15± 0.05	1.11±0.05	crystal
	55	3.32	1.66± 0.07	1.66±0.07	liquid
$n\text{-}C_{21}H_{44}$	-77	2.16	1.02± 0.05	1.14±0.05	crystal
	25	2.38	1.15± 0.05	1.23±0.05	crystal
	55	3.22	1.72± 0.07	1.50±0.07	liquid
$n\text{-}C_{23}H_{48}$	-77	2.25	1.00± 0.05	1.25±0.05	crystal
	25	2.45	1.14± 0.05	1.31±0.05	crystal
	55	3.28	1.82± 0.07	1.46±0.07	liquid
$n\text{-}C_{24}H_{50}$	-77	2.16	1.01± 0.05	1.15±0.07	crystal
	25	2.52	1.15± 0.05	1.37±0.05	crystal
	55	3.22	1.70± 0.07	1.52±0.07	liquid
$C_{30}H_{62}$	-77	2.79	1.3± 0.1	1.49±0.1	glass
	25	3.27	1.6± 0.1	1.67±0.1	liquid
	55	3.26	1.6± 0.1	1.66_0.1	liquid

After the fast formation of both species, a slow process for formation of the additional excimer is seen from the figure.

Those processes mentioned above play an important role for stabilization of the rubber against radiation.

It is shown that the formation of polymer radicals and unsaturated double bonds, and the evolution of hydrogen molecules in the rubber are scavenged and inhibited significantly by pyrene. G-values of Py^+ and Py^- as a function of pyrene concentration are shown in Figure 8, together with G values of the evolution of hydrogen molecule. It is evident from the figure that G value of both ions of Py^+, Py^- approaches to 3.8 and almost all charges formed in the medium transfer to pyrene molecules at higher concentrations of pyrene more than about 3wt%. According to our experiments, it has been made clear that the evolution of hydrogen molecules is saturated at a concentration of pyrene more than 2wt%. G-value of the non-scavengeable hydrogen production is 1.70. This is shown in Figure 9.

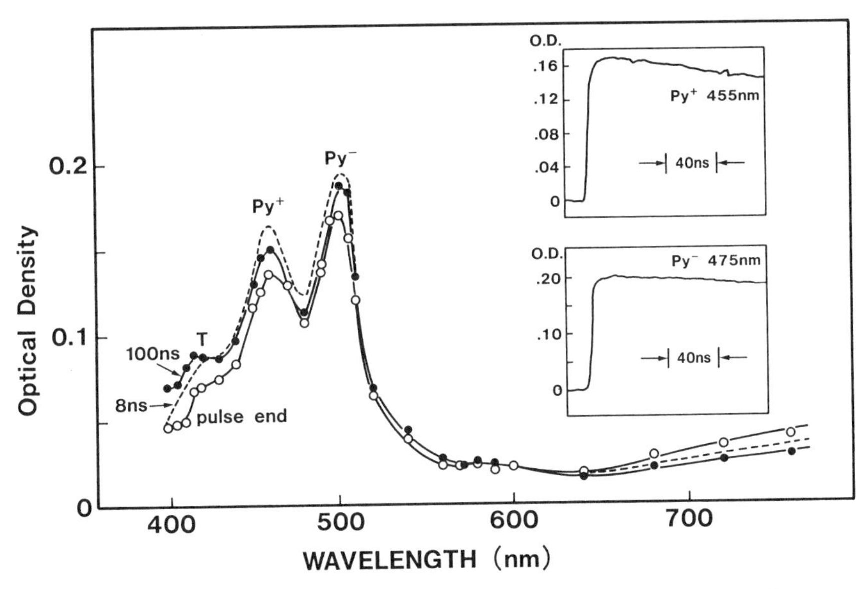

Figure 6: Transient absorption spectra of pyrene in ethylene propylene rubber and osciloscope traces of pyrene cation (455nm) and anion (475nm). 1.07 weight percent of pyrene.

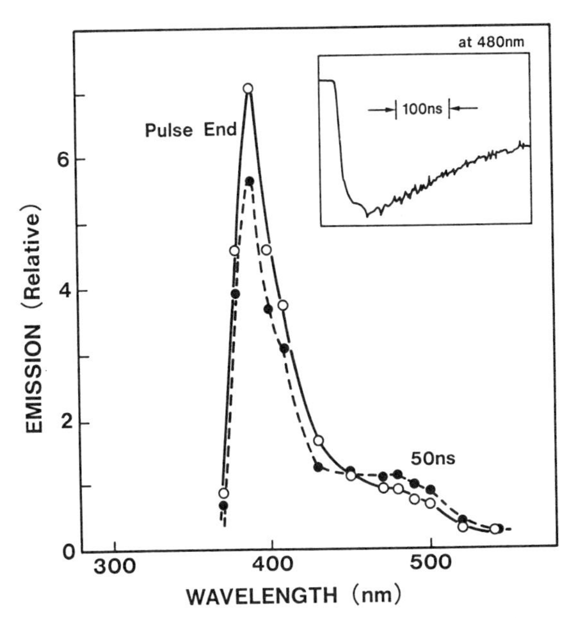

Figure 7: Transient emission spectra of pyrene in ethylene propylene rubber and osciloscope trace at the excimer peak (480nm). 1.0 weight percent of pyrene.

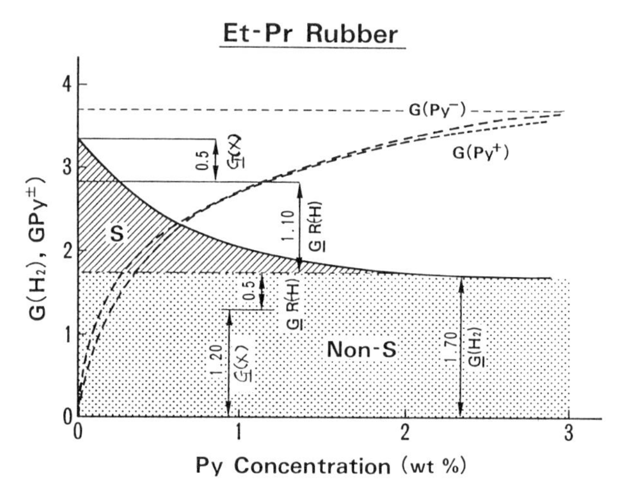

Figure 8: G-values of H_2 production (from Figure 9), and of pyrene cation Py^+ and anion Py^- in ethylene-propylene rubber as a function of pyrene concentration. Rough estimation of G-values of alkyl radical and trans-vinylene R(-H) which can be scavenged (S) and not be scavenged (Non-S). R(-H) corresponds to C=C and X to the crosslinking. Non-scavengeable crosslinking yield was roughly estimated from G-values of squalane at low temperature.

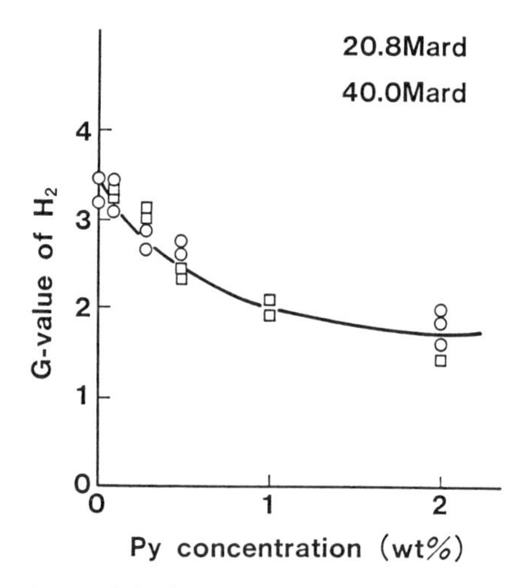

Figure 9: G-values of hydrogen evolution $G(H_2)$ from e ethylene-propylene rubber as a function of concentration of the additive, pyrene.

(4) Heterogeneous Crosslinking in a Series of Paraffins. Dimer, trimer, tetramer and higher oligomers are formed through crosslinking reaction successively and can be analyzed by means of liquid chromatography, mass-spectroscopy. The results are shown in Figure 10 and Figure 11. In the LC analysis, both infrared and ultra-violet detection methods were used. By the infrared method, a series of oligomers can be separated by the molecular weight as well as the shape of molecules. On the other hand, by the UV method, the contents of unsaturated bonds in the fraction can be estimated.

Three examples of n-$C_{20}H_{42}$,n$-C_{21}H_{44}$ and n$-C_{24}H_{58}$ are cited in Figure 10. The irradiation was made in both crystalline and molten states. It is interesting to note that a single mode of dimer formation is observed in the molten state, while two different modes of dimer formation in the crystalline state. About 50% of dimer is an extended form without double bond structure, and another 50% of the dimer is a less extended form with double bond structure. For comparison, the LC spectrum for n-$C_{40}H_{82}$ which is corresponding to the complete linear dimer of n-$C_{40}H_{42}$ is indicated in the figure. It is emphasized that two different modes of dimerization exist in the crystalline state. Based upon the selective alkyl radical formation in the edge of crystals (which is explained later), crosslinking of about 40 to 50% occurs at the end of monomer chains. Trans-vinylene and allyl radical seem to be formed randomly and distributed rather homogeneously in the crystal matrices, in contrast with alkyl radicals. Crosslinking in the crystalline phase may take place by combining of radicals formed at double bond site with radicals migrating in the matrices.

In the molten state, alkyl radicals are formed randomly along the molecular chain because of TG conformation in the medium. Crosslinking occurs by encountering of both alkyl radicals, alkyl and allyl radicals, and both allyl radicals through the molecular motion in the molten medium.

In the crystal, it is evident from the LC experiment that certain amounts of double bonds are formed in the monomer without dimerization. This is indicating that recombination of olefinic cation radical with electron occurs without further dissociation of the species. This process is much favorable in the crystalline state than in the molten state. One of our experimental results on the mass-spectrometry of irradiation alkanes is shown in Figure 11 (11). n$-C_{16}H_{34}$ was irradiated at room temperature and the irradiated sample was subjected to the mass-analysis. Distribution of double bonds in the series of oligomers from dimer to hexamer are shown in the figure. It is clear from the figure that the higher the number of monomer units in the oligomer, the higher the concentration of double bonds. Crosslinking proceeds, step by step, from dimer to trimer, trimer to tetramer and so on. It is indicating that as the crosslinking proceeds, double bonds are enriched and accumulated in the crosslinked oligomers. This phenomena could be explained by assuming that if once a double bond is formed, the site becomes the location where the next step of crosslinking starts. The scheme is discussed later.

The other factor affecting the successive crosslinking after the first step of crosslinking will be a specific structure of the crosslinking site.

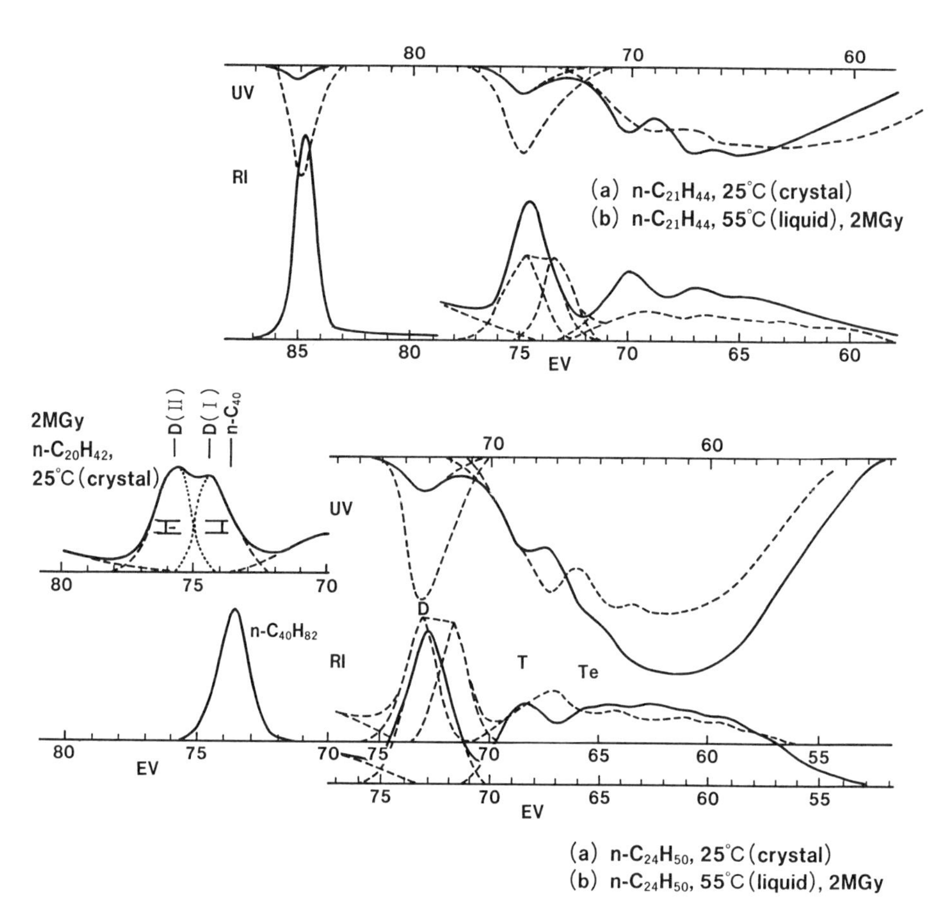

Figure 10: Product analysis by liquid chromatography (LC) of $n-C_{20}H_{44}$ and $n-C_{24}H_{50}$ at 25°C in crystalline state and at 55°C in molten state. LC spectra from the refractive indes (RI) and ultraviolet (UV) detectors. EV: elution volume (ml). LC spectrum for $n-C_{40}H_{82}$ is shown in the figure for comparison with the dimer product D: dimer, T: trimer, Te: tetramer

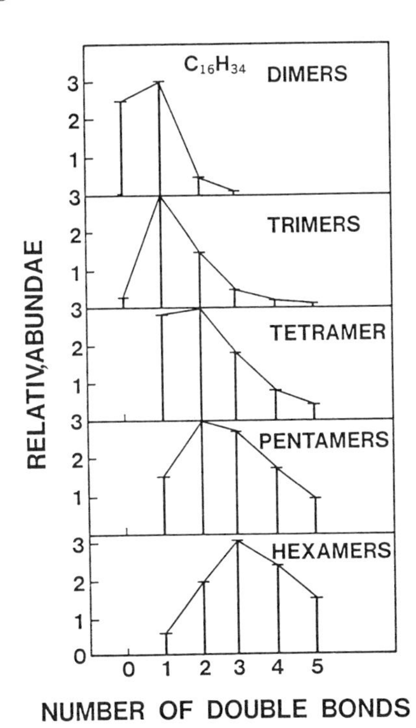

Figure 11. Analysis of mass-spectrometry of irradiated n-$C_{16}H_{34}$. Irradiation dose was 2.24MGy. (Reproduced with permission from reference 11b. Copyright 1992 Pergamon.)

$$
\begin{array}{ccccccc}
| & & | & & | & & | \\
CH_2 & & CH_2 & & CH_2 & & CH_2 \\
| & & | & & | & & | \\
CH\bullet & + & CH\bullet & \rightarrow & HC & \text{———} & CH \\
| & & | & & | & & | \\
CH_2 & & CH_2d & & CH_2 & & CH_2 \\
| & & | & & | & & | \\
CH_2 & & CH_2 & & CH_2 & & CH_2 \\
| & & | & & | & & |
\end{array}
$$

There are two interesting, important tendencies regarding the radiation chemical effect for the molecular structure in the crosslinking site. That is: one is the easier, selective formation of tertiary alkyl radical.

$$
\begin{array}{ccccccc}
 & | & & | & & | & & | & \\
 & CH_2 & & CH_2 & & CH_2 & & CH_2 & \\
 & | & & | & & | & & | & \\
H- & C & \text{———} & C-H & \rightarrow H- & C & \text{———} & C\bullet & + H\bullet \\
 & | & & | & & | & & | & \\
 & CH_2 & & CH_2 & & CH_2 & & CH_2 & \\
 & | & & | & & | & & | &
\end{array}
$$

This has been shown by spin trapping experiments (11). Another is the high possibility to form double bond through charge transfer to the site from medium.

```
              |         |                        |           |
              C         C                        C           C•+
              |         |                        |           |
RH+, RH•† + HC ─────── CH ───→ RH + H−C ─────────────────── C−H        Charge
              |         |                        |           |           transfer
              C         C                        C           C
              |         |                        |           |

              |         |                        |           |
              C         C•+                      C           C•+
              |         |        −H2             |           |
             HC ─────── C−H      ───→            C ═════════ C + H2     Dehydro−
              |         |                        |           |          genation
              C         C                        C           C
              |         |                        |           |

              |         |                        |           |
              C         C•+                      C           C*
              |         |                        |           |
              C ═══════ C+       e− ───→         C ═════════ C          Neutralization
              |         |                        |           |
              C         C                        C           C
              |         |                        |           |

              |         |                        |           |
              C         C*                       C           C•
              |         |                        |           |          Radical
              C ═══════ C        ───→            C ═════════ C + H•     formation
              |         |                        |           |
              C         C                        C           C
              |         |                        |           |
```

It is easily understood that the monomers can not diffuse or migrate in the crystalline medium. Under the condition, it is very strange that not only dimer, but also oligomers higher than trimer are produced in the rigid medium, as indicated in the series of the model compounds in crystalline state. The structure of oligomers should be a very specific one reflecting the specific condition. The structure might be a star type oligomer. Approximately 50% dimerization in the early stage of reaction occurs at the edge of crystals (at the end of monomer chain). Further crosslinking reaction will occur in the vicinity of the site on the surface of the crystal. Therefore, a large fraction of oligomers obtained in crystalline state may have structures just as star polymers.

The early phenomena of the crosslinking at the end of chain through recombination of pair radicals Because the memory of reaction can be retained during irradiation and the effect be accumulated keeping the memory until the end of high dose irradiation. The slow processes of crosslinking occurring step by step depend upon very much irradiation dose. The high dose irradiation is needed in order to get enough amounts of the product to make the analysis possible.

Spin-trapping experiments on the series of alkanes demonstrates clearly the existence of the tertiary alkyl radical based upon crosslinking (12). The radicals are preferencially formed, as far as the spin-trapping experiment is concerned.

Selective charge transfer to the crosslinking site produced initially from the surrounding molecular segments can occur, because ionization potential of the crosslinking site lower than those of surrounding molecules. Tertiary alkyl radical could be produced either from the ion directly, or from the excited state after the charge recombination (neutralization).

According to Miyazaki's experiments (13) by electron spin resonance at low temperatures including 77K, on irradiation effect of both linear and branched alkanes, olefinic cation radicals are formed much more easily from branched alkanes than from linear ones. Olefin cations are formed by the fragmentation of alkane cations in the ground state. This is based upon following two reasons; one is that ionization potential of branched alkane is lower than that of linear alkanes, another is that the dissociation reaction from alkane cation to the corresponding olefinic cation is much feasible from the view point of the heat of formation ΔH. ΔH is much lower for branched cation than for linear ones.

According to our pulse radiolysis work on a series of alkanes, it is quite clear that selective charge transfer from the medium to the double bond site occurs. One of the experimental results is shown in Figure 12. Before subjecting to the pulse radiolysis experiment, alkane such as n-$C_{24}H_{50}$, for example, was irradiated with certain amount of dose in order to introduce enough numbers of crosslinking sites and C=C bonds. It is obvious from the figure that a broad absorption peak due to formation of alkane cation radical $RH^{\bullet+}$ (n-$C_{24}H_{50}$) appears around 700 nm for the fresh, non-preirradiated sample, however, the absorption due to $RH^{\bullet+}$ is reduced drastically and a broad strong absorption due to olefinic and stable conjugated olefinic cation radicals appears in a wave length region lower than 500 nm. These results are indicating that effective charge transfer from the medium to the pre-existing sites of double bonds, takes place. As radicals are formed from those ions, crosslinking occurs successively in the vicinity of the double bonds.

Therefore, these transient phenomena support very much results obtained by both LC and mass-spectroscopy of irradiated alkanes mentioned above.

$$RH^{\bullet *}\left\{RH^{\bullet +}\right\} + -C=C-C- \rightarrow RH + -C=C^{\bullet +}-C-$$

$$RH^{+}\left\{RH^{\bullet +}\right\} + \text{Crosslinking Site} \rightarrow RH + (C.S.)^{\bullet +}$$

$$(C.S.)$$

. . . charge transfer

$$\left.\begin{array}{l} -C=C^{\bullet +}\ \ -C- + e^{-} \rightarrow -C=C^{*}-C- \\ (C.S.)^{\bullet +} + e^{-} \rightarrow (C.S.)^{*} \end{array}\right\} \text{neutralization}$$

$$\left.\begin{array}{l} -C=C^{*}-C- \rightarrow -C=C-C^{\bullet}- \ + H\bullet \\ (C.S.)^{*} \rightarrow \left\{(C.S.)\ -H\right\}\bullet + H\bullet \end{array}\right\} \text{radical formation}$$

(5) Crosslinking in Crystalline State. Very fast formation of alkyl radical in saturated hydrocarbons within 10 picosecond has been confirmed by our

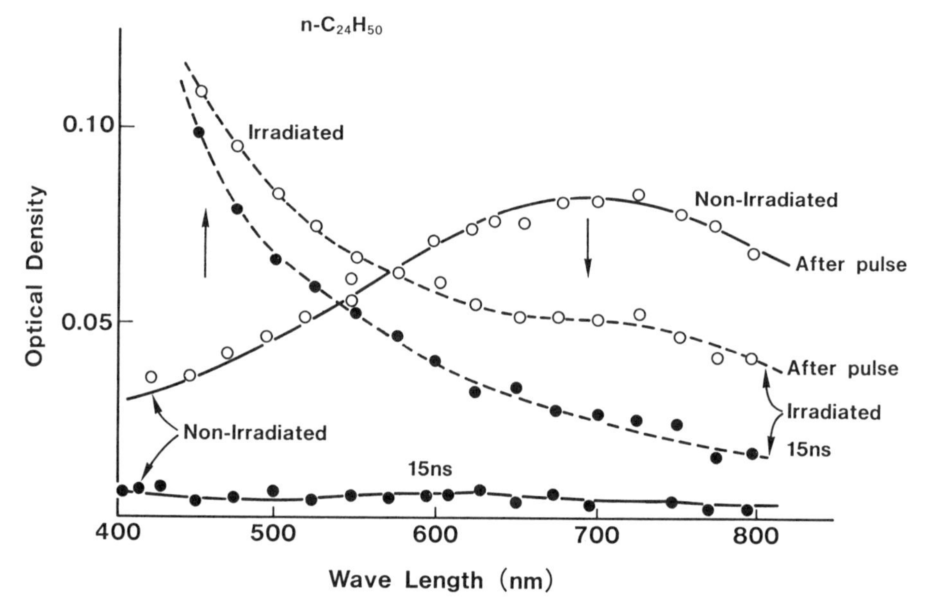

Figure 12: Pulse radiolysis of n-tetracosane $n-C_{24}H_{50}$ for fresh (Non-Irradiated) and pre-irradicated (Irradiated) samples at the end of 2ns pulse and at 15ns after the pulse.

picosecond pulse radiolysis (1,5,6). Almost all radicals are generated by this very fast process. This is indicating that almost all radicals generated by ionizing radiation in saturated hydrocarbons come from either their excited cation radical ($RH^{\bullet *}$) or their super excited state (RH**). Only a small fraction of alkyl radicals come from excited states RH* after the geminate recombination. This has been confirmed by Iwasaki, Toriyama and their collaborators by their electron spin resonance experiments on irradiated single crystals of a series of alkanes at very low temperatures including temperatures lower than 4K. The single crystals were irradiated at those temperatures, and measured at the same temperatures. They have found that almost all alkyl radicals more than 90% are generated as pair radicals at those low temperatures. They have concluded that alkyl radicals are formed from the excited cation radical $RH^{\bullet +}$ through both deprotonation and dissociation of hydrogen atom. This is not dissociation process of RH* after the recombination of $RH^{\bullet +}$ with the electron. They have demonstrated that more than 40% of pair radicals in the single crystals are formed at the edge of the crystal (at the end of paraffin molecular chain). This can be reasonably explained by spin-density distribution of $RH^{\bullet +}$. That is: spin density is the highest along C-H bond at the end of chain, as shown in Figure 13.

This is indicating that bond cleavage occurs preferenically at the end of chain on the surface of crystals through deprotonation. Therefore, crosslinking can take place by the combination of the pair of radicals at the end of the chain, giving extended linear dimers. This has been supported by our LC experiments of irradiated alkane crystals in which about 50% dimers have been found to be extended, linear ones without having any unsaturated groups. This is typical heterogeneous nature of crosslinking which is directly related to an early radiation action or radiation chemical phenomenon. G-value of crosslinking is approximately equal to G-value of C=C bond formation for almost alkanes including saturated hydrocarbon polymers. As has been shown in our LC experiments in Figure 10, 50% of dimers contain unsaturated C=C bonds and deviated from the extended linear form. This means that about 50% of crosslinking, besides at the end of chain, occur inside of the crystal matrices in the vicinity of double bonds. This can be explained by assuming that C=C bonds are formed randomly in the crystals matrices. This is based upon our pulse radiolysis experiments.

$$\{RH^{\bullet +}\} \rightarrow R(-H)^{\bullet +} + H_2$$
$$R(-H)^{\bullet +} + e^- \rightarrow R(-H)^*$$
$$R(-H)^* \rightarrow R(-H_2)\bullet + \quad H\bullet \text{....radical formation}$$
$$R(-H)^* \rightarrow R(-H) \text{.............double bond formation}$$

For further irradiation after the formation of C=C bond, charge transfer from the medium to the double-bond occurs, followed by recombination with electron to give allyl radical.

$$RH^{\bullet \ddagger}\{RH^{\bullet +}\} + R(\text{-H}) \rightarrow RH \ + \ R(\text{-H})^{\bullet +}$$
$$R(\text{-H})^{\bullet +} + e^- \rightarrow R(\text{-H})^*$$
$$R(\text{-H})^* \rightarrow R(\text{-H}_2)\bullet \ + \ H\bullet$$

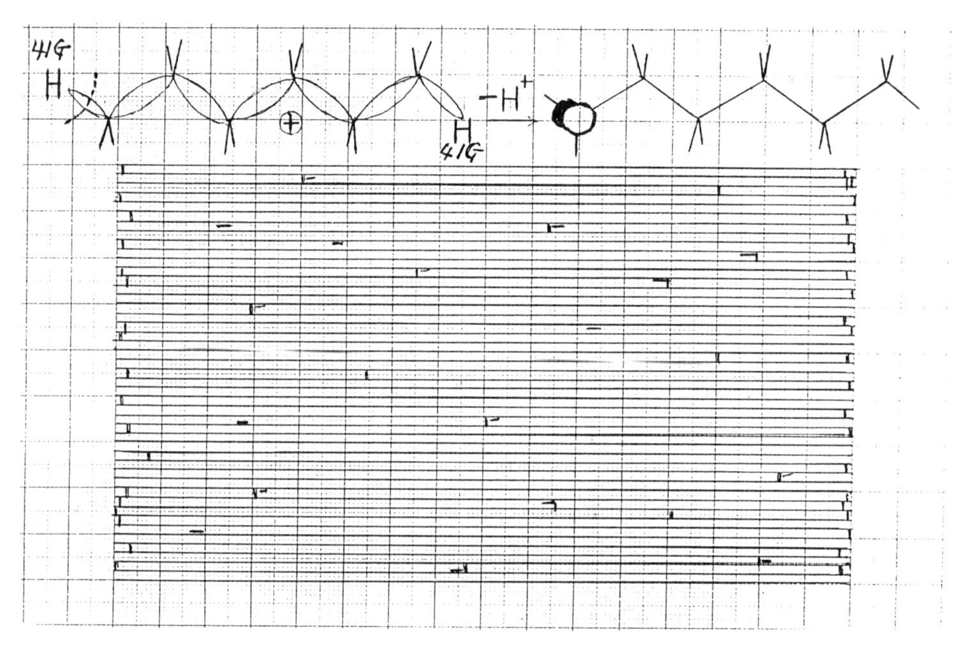

Figure 13: Delocalized σ state in extended alkane cation (trans-trans TT Conformation). High density of spin state at the end of chain of C-H(41G) where a selective bond cleavage occurs to form alkyl radical. The schematic presentation of selective crosslinking at the end of chain in crystalline state. Double bond formation and crosslinking at the site of double bond occur in the crystal matrices. Symbols of |, -, and ⊥ represent sites of crosslinking, trans-vinylene crosslinking at the double bond.

As the site of allyl radical, $R(-H_2)^*$, is not mobile and fixed in the molecular chain, crosslinking should occur by recombining with a mobile alkyl radical migrated. This is a slow process of crosslinking.
Ally radical is also well known to be formed through migration of alkyl radical to the site of C=C bond. Of course, the charge transfer process is much faster than that of radical migration.

$$-C=C-C- + \underset{\text{mobile}}{R\bullet} \rightarrow -C=C-\dot{C}- + RH$$

It is concluded from the experiments that there are two crosslinking processes existing in crystalline alkanes, that is; one is fast selective crosslinking at the edge of molecular chain followed by successive crosslinking on the surface of crystals in the vicinity of linking at the sites of C=C double bonds generated in the crystalline matrices, followed by successive crosslinking in the vicinity of the crosslinking site.

(6) Crosslinking in Molten State. According to Iwasaki's work (8) on electron spin resonance in glassy alkanes in freon matrices at low temperature, the molecular conformation take GT structure in the glassy system and spin density along C-H bond at the site of GT conformation is high and then a selective cleavage can occur with high probability at the C-H bond to produce an internal alkyl radical.

On the other hand, C=C double bonds are formed rather randomly along molecular chain and then allyl radicals are finally formed. Therefore, crosslinking of alkanes in molten state takes place randomly through those alkyl radicals and allyl radicals. The site of C=C plays an important role for the successive crosslinking, as in crystalline state. The qualitative expression of heterogeneous crosslinking in the molten state is shown in Figure 14 together with spin density for GT conformation of molecular chains.

In our experiments, a single mode of dimer formation has been observed in the liquid alkanes, as shown in Figure 10 which is very different from crosslinking in the crystalline alkanes. As is shown from the figure, unsaturated bonds are distributed rather homogeneously in the dimers. This result is consistent with one obtained by the electron spin resonance method.

Conclusion

A series of model compounds of polyethylene and ethylene-propylene rubber, and the copolymer have been studied by various methods in order to make the heterogeneous nature of crosslinking in those systems clear. It has been found from the studies that the heterogeneous crosslinking occurs clearly in both their crystalline and amorphous (or molten) states, but in different ways. In the crystalline state of the series of linear alkanes, alkyl radicals responsible for crosslinking are formed in heterogeneous spacial distribution due to the specific early excitation processes. In the amorphous (molten) state of those materials, both alkyl and olefinic radicals are formed rather homogeneously. After the first step of crosslinking initiated by the early action of radiation, the proceeding

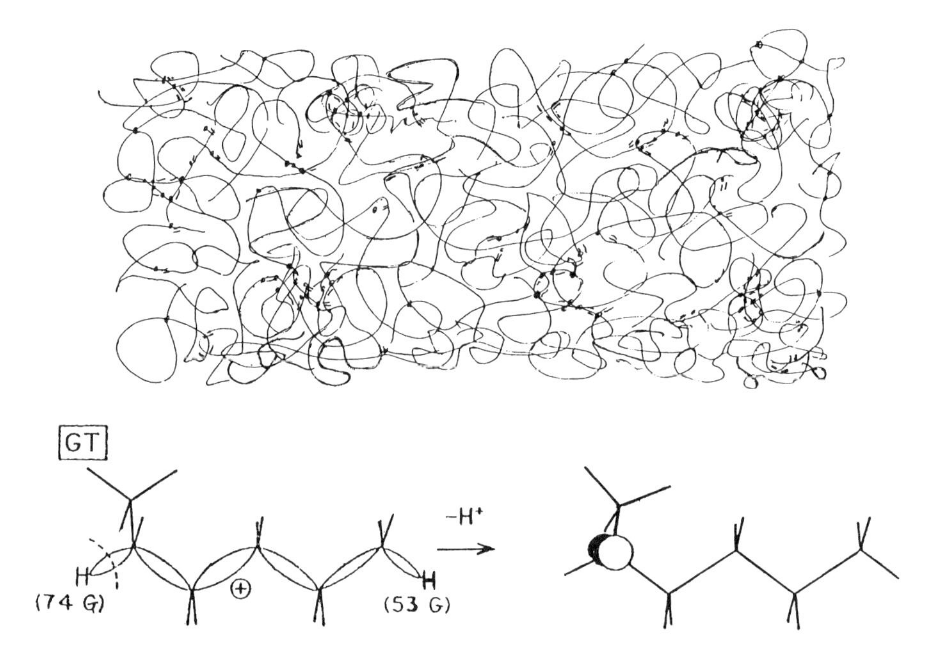

Figure 14: Delocalized σ state in bended alkane cation (gauch-trans GT conformation). Higher densities of spin state at sites of the GT conformation and the end of chain. Selective formation of alkyl radical at those sites. The schematic presentation of heterogeneous crosslinking in the molten state. Random(less selective than crystal line state) crosslinking in the early stage and then the heterogeneity increase with the accumulation of crosslinking sites and double bonds

crosslinking reaction is very much affected by the formation of both crosslinking and double bond sites formed initially. Because, selective charge transfer from the medium to those sites play an important role for the next stage of crosslinking. Those phenomena enhance the heterogeneity of crosslinking. This is based upon experimental results on pulse radiolysis of linear alkanes and ethylene-propylene-rubber, electron spin resonance of linear and branched alkanes at low temperatures, and spin trapping in alkanes. In addition, the heterogeneity has been proved clearly by the mass-spectroscopic and LC experiments.

Those results could be extended to polymeric systems, even though they are rather complicated. Our results surely help us to understand the phenomena occurring in polymeric systems; the mechanism of crosslinking and the heterogeneous nature of the process.

Literature Cited

[1] a. Tabata, Y.; Tanaka, J.; Tagawa, S.; Katsumura, Y.; Ueda, T. and Hasegawa, K.; *J. Fac. Eng. Univ. of Tokyo*, 1978, **34**, 12; Phys. Lett., 1979, **54**, 258; c. Tagawa, S.; Washio, M. and Tabata, Y., *Chem. Phys. Lett.*, 1979, **68**, 276.

[2] Seguchi, T.; Hayakawa, N.; Tamura, N.; Hayashi, N.; Katsumura, Y. and Tabata, Y., (1988), *Radiat. Phys. Chem.* **32**, 753.

[3] Seguchi, T.; Katsumura, Y.; Hayashi, N.; Hayakawa, N.; Tamura, N. and Tabata, Y., (1991), *Radiat. Phys. Chem.* **37**, 229.

[4] Seguchi, T.; Hayakawa, N.; Tamura, N.; Hayashi, N.; Katsumura, Y., Tabata, Y.; (1989), *Radiat. Phys. Chem.* **33**, 119.

[5] Tagawa, S., ACS Symposium Series *475*, 2 (1991).

[6] Tabata, Y., ACS Symposium Series *475*, 31(1991).

[7] Hayashi, N., Doctor Thesis, Univ. of Tokyo (1985).

[8] a. Iwasaki, M.; Toriyama, K., *J. Am. Chem. Soc.* **103**, 3591 (1981); b. Toriyama, K.; Numome, K.; Iwasaki, M., *J. Chem. Phys.* **77**, 5891 (1982); c. Toniyama, K., "Radical Ionic Systems", p.99 Kluer Acad. Pub. Netherlands (1979); d. Iwasaki, M.; Toriyama, K.; Fukuya, M.; Muto, H. and Nunome, K., *J. Phys. Chem.* **89**, 5278 (1985).

[9] a. Trifunac, A. D.; Werst, D. W.; Percy, L. T., *Radiat. Phys. Chem.*, **1989**, 34, 547; *Radiochem. Radionanal. Lett.* **1978**, 35, 85; c. Jonah, C. D., *Radiat. Phys. Chem.* **1983**, 21, 53; d. Klassen, N. V.; Teather, G. G., *J. Phys. Chem.* **1985**, 89, 2048; e. Mehnert, R.; Brede, O.; Cserep, G., *Radiat. Phys. Chem.*, 1985, 26, 353.

[10] Seguchi, T.; Katsumura, Y.; Hayashi, N.; Hayakawa, N.; Tamura, N.; Tabata, Y., *Radiat. Phys. Chem.* **1991**, 37, 29.

[11] a. Yamaguchi, T.: Thesis, Tokai University (1990) b. Soebianto, Y. S., Yamaguchi, T.; Katsumura, Y.; Ishigure, K.; Kudo, J.; Koizumi, T., *Radiat. Phys. Chem* **1992**, 39, 251.

[12] Tabata, M., Private communication, Katsumura, Y., Private communication.

[13] Miyazaki, T., *Radiat. Phys. Chem.* **1991**, 37, 11.

RECEIVED January 28, 1993

Chapter 4

Temperature Effects on the Radiation Degradation of Poly(isobutene) and Poly(α-methylstyrene)

David J. T. Hill, T. T. Le, James H. O'Donnell, M. C. Senake Perera, and Peter J. Pomery

Polymer Materials and Radiation Group, Department of Chemistry, The University of Queensland, Brisbane, Queensland 4072, Australia

The effect of temperature on the radiation chemistry of poly(isobutene) and poly(α–methyl styrene) has been studied over the temperature range at which the polymers begin to degrade thermally. The radical intermediates have been investigated using ESR spectroscopy and their chemical yields and reactivities have been examined using thermal annealing techniques. The low molecular weight products of radiolysis at 300 K have been investigated and their yields measured. Hydrogen was the major low molecular weight product of radiolysis together with monomer, but no significant depropagation of the chain radical was found to occur at 300 K. Molecular weight studies at 300 K showed that the polymers undergo predominantly chain scission at room temperature, and the yields of scission and crosslinking have been determined at this temperature. At higher temperatures both polymers were found to undergo radiation induced depolymerization, and in these temperature regions, the values of G(S) and G(–M) have been determined.

High energy radiation causes chemical and physical changes in polymers due to the formation of reactive intermediate species such as radicals, ions and excited states. Small volatile molecules may also be formed as the result of the cleavage of polymer side chains. For example, hydrogen and low molecular weight olefins are formed on the radiolysis of polyethylene[1]. Side chain cleavage can also result in the formation of chain radicals which may undergo chain crosslinking reactions, as occurs in polyethylene[2] and polymethylacrylate[3]. Main chain scission may also occur, either as a direct consequence of the energy absorption or as a subsequent step in a sequence of reactions involving the loss of the side chain, as in poly(methyl methacrylate)[4]. The chain scission and crosslinking reactions have obvious consequences for the molecular weight changes in the polymer. Scission would be expected to predominate over crosslinking in polymers which contain a tetrasubstituted carbon atom in each monomer unit in the chain, such as in poly(methyl methacrylate) for example. In this polymer it has been suggested that the polymer main chain may undergo β–scission following the removal of the ester side chain, with concomitant

0097–6156/93/0527–0050$06.00/0

formation of a side chain scission radical[4]. In cases where this chain scission mechanism occurs, the general sequences of reactions would be:

$$-CH_2-\overset{\overset{R_1}{|}}{\underset{\underset{R_2}{|}}{C}}-CH_2-\overset{\overset{R_1}{|}}{\underset{\underset{R_2}{|}}{C}}- \xrightarrow{\gamma} -CH_2-\overset{\overset{R_1}{|}}{\underset{\underset{R_2}{|}}{C}}-CH_2-\overset{\overset{R_1}{|}}{\underset{\cdot}{C}}- + R_2\cdot$$

$$\downarrow \beta\text{-scission}$$

$$-CH_2-\overset{\overset{R_1}{|}}{\underset{\underset{R_2}{|}}{C}}\cdot + CH_2=\overset{\overset{R_1}{|}}{C}- \qquad (1)$$

The process shown in reaction scheme (1) results in the formation of a radical which is identical to the chain propagation radical for vinyl polymerization. Because free radical vinyl polymerizations are reversible, spontaneous depolymerization of this chain radical may take place if the prevailing conditions of temperature and pressure favour the reaction.

In this paper some aspects of the effects of temperature on the radiation chemistry of poly(isobutene), PIB, and poly(α-methyl styrene), PαMS, under vacuum will be discussed, with particular attention being directed to the radiation induced depolymerization of these materials at elevated temperatures.

Experimental

Polyisobutene was obtained from Aldrich Chemical Co and was found to have an $M_w = 4.33 \times 10^5$. The polymer was purified by reprecipitation from chloroform/methanol. Poly(α-methyl styrene) was prepared by cationic polymerization using BF_3-ether in dichloromethane as the initiator. The polymer was reprecipitated using dichloromethane/methanol, and found to have $M_w = 1.44 \times 10^5$.

Irradiations were performed using a ^{60}Co gamma source with a dose rate of 3–5 kGy hr^{-1}. The studies were carried out in sealed tubes under vacuum at temperatures where depolymerization does not occur, and under vacuum with continuous pumping where depolymerization does take place.

ESR studies were performed using a Bruker ER–200D X–Band spectrometer fitted with a variable temperature cavity. During the annealing experiments, samples were allowed to equilibrate at each temperature for ten minutes before the radical spectra were obtained. Gas analyses were performed as described previously using a Hewlett Packard 5730A Chromatograph fitted with a chromosorb column. Molecular weight measurements were made using a Waters Associates Chromatograph fitted with five ultrastyragel columns (10^6, 10^5, 10^4, 10^3, 10^2 A) and a refractive index detector. Samples for GPC analyses were not exposed to air before any radicals present were allowed to decay by thermal annealing.

Results and Discussion

For any vinyl polymerization of the type shown in reaction (2) below, the free energy change can be determined from the enthalpy and entropy changes for the polymerization reaction.

$$-CH_2-\overset{\displaystyle R_1}{\underset{\displaystyle R_2}{\overset{|}{\underset{|}{C}}}}\cdot + CH_2=\overset{\displaystyle R_1}{\underset{\displaystyle R_2}{\overset{|}{\underset{|}{C}}}} \longrightarrow -CH_2-\overset{\displaystyle R_1}{\underset{\displaystyle R_2}{\overset{|}{\underset{|}{C}}}}-CH_2-\overset{\displaystyle R_1}{\underset{\displaystyle R_2}{\overset{|}{\underset{|}{C}}}}\cdot \quad (2)$$

Free radical vinyl polymerizations characteristically have a large negative entropy change associated with the chain propagation reaction. Thus, as the temperature rises, the propagation process becomes less thermodynamically favourable, and at a certain critical temperature called the ceiling temperature, T_c, the free energy change will be zero[5]. Above the ceiling temperature, depolymerization is the thermodynamically favoured reaction.

For a given monomer, the ceiling temperature will depend on the polymerization conditions, such as the polymer concentration and pressure.

The ceiling temperatures for some well–known vinyl polymerizations are given in Table 1, together with the glass transition temperatures for the polymers.

The data in Table 1 provide information about the relative propensity for the various polymers to undergo depolymerization. For example, while PIB has a much lower glass transition temperature than PαMS, PαMS has a lower ceiling temperature, and thus will undergo depolymerization at temperatures at which PIB remains stable, even though the PIB chains have greater mobility. At very low pressures of monomer, the ceiling temperatures would be lower than those given in Table 1.

Table 1. The ceiling temperature, T_c, and glass transition temperatures, T_g, for representative vinyl polymerizations

Monomer	[a]T_c/°C	T_g/°C
isobutene	175	–70
methacrylonitrile	145	108
methyl methacrylate	198	114
α methyl styrene	66	180

[a] The ceiling temperatures are for polymerization of pure liquid monomer at 1 atmosphere pressure.

It is important to note that for depolymerization to occur, the chain–end propagation radical must be present. Thus polymers may be stable above their ceiling temperature when no propagation radicals are present. Indeed, the thermal cleavage of main–chain bonds in carbon based polymers to form propagation radicals typically occurs at temperatures in the range 200–300°C, which is considerably greater than T_c for many polymers.

Because high energy radiation can result in the cleavage of main–chain bonds and result in the formation of propagation radicals, high energy radiation can induce the depolymerization reaction at temperatures at which a polymer may be thermally stable.

Low Temperature Studies. Investigations of the radiolysis of polymers at 77 K allows some of the important chemical intermediates to be identified and subsequent thermal annealing of these radicals allows their reactivities to be ascertained.

ESR studies of the radiolysis of PIB[6,7] and PαMS[8] at 77 K have been reported and the radical intermediates corresponding to radicals I and II were found to be present in both polymers.

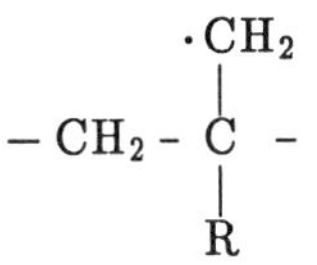

$$\begin{array}{c} \cdot CH_2 \\ | \\ -CH_2-C- \\ | \\ R \end{array} \qquad \begin{array}{c} CH_3 \\ | \\ -CH-C- \\ | \\ R \end{array}$$

I 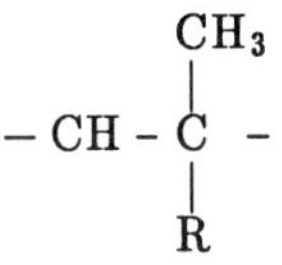 II

In addition to these two radicals, the radical anion III and the cyclohexadienyl radical IV was found in PαMS.

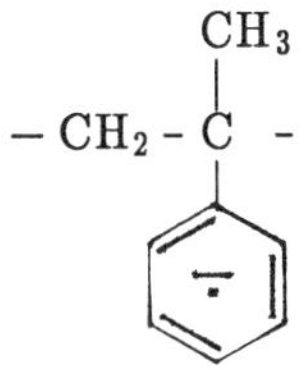

III 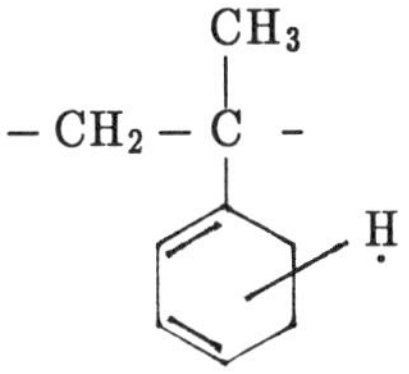 IV

It is notable that there was no evidence for the formation of any methyl radicals in these polymers at 77 K, which suggests that there is little if any cleavage of the methyl side–chains in the polymers. In addition, there was no evidence for the formation of a significant proportion of propagation radicals at 77 K in either of these polymers.

The total radical yields (G–values) at 77 K are 2.1 for PIB[7] and 0.105 for PαMS[8]. The lower G–value for PαMS is a consequence of the protective influence of the aromatic group.

Annealing of PIB after irradiation at 77 K results in a slow decay of the radicals over the temperature range 77 – 180 K, as shown in Figure 1. Both of the radicals present were found to decay simultaneously[7], because the profile of the spectrum did not change over this region. However, as the glass transition temperature is approached, the rate of decay of both radicals increases rapidly, and at 300 K the radical concentration has dropped to zero. There was no evidence for the formation of propagation radicals during the annealing process.

Annealing of PαMS[8] after irradiation at 77 K results in the decay of the anion radicals in the temperature range 100 – 200 K and of the cyclohexadienyl radicals in the region 200 – 300 K, with a consequential decrease in the total radical concentration, as shown in Figure 2. Annealing of the irradiated polymer to 300 K on irradiation of the polymer at this temperature yields an

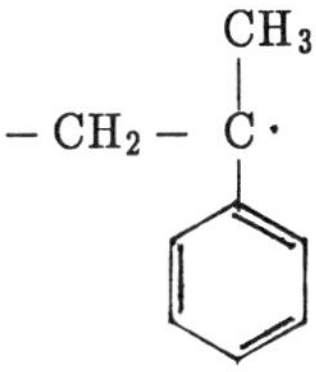

V

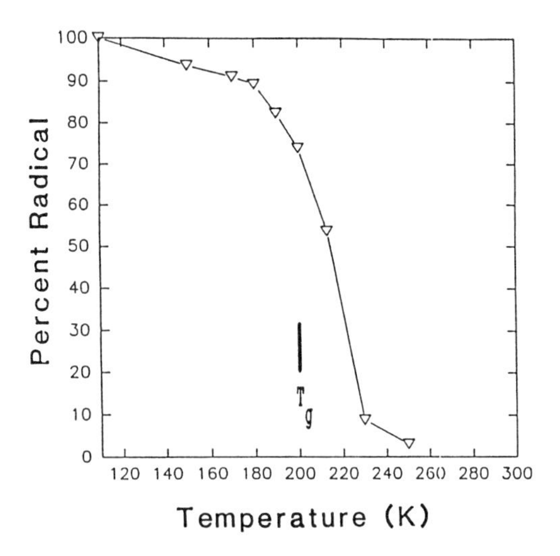

Figure 1. The variation in the radical yield with temperature for PIB after irradiation at 77 K.

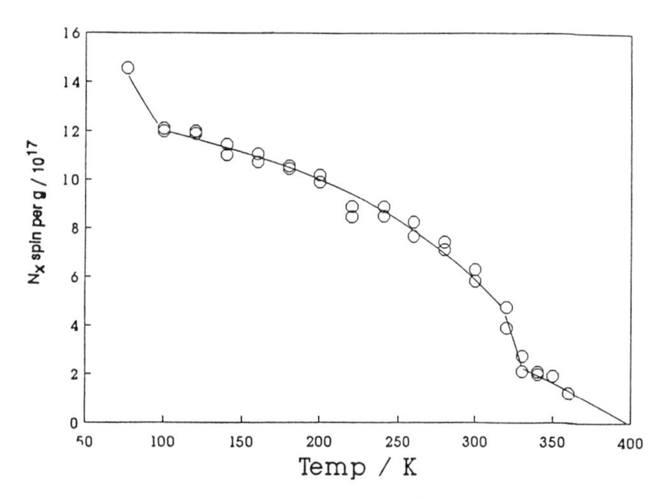

Figure 2. The variation in the radical yield with temperature for PαMS after irradiation at 77 K.

ESR spectrum with clear evidence for the presence of a significant proportion (≃ 10%) of propagation radicals[8], V, as shown in Figure 3.

Annealing to above 330 K results in a rapid decrease in the concentration of radicals, which falls to zero at approximately 400 K, as demonstrated in Figure 2.

The G–value for radical formation for irradiation at 300 K has been reported to be 0.05[8], which is approximately half of that for irradiation at 77 K, and is close to the value expected on the basis of the annealing studies summarized in Figure 2.

The low molecular weight, volatile products of the radiolysis of PIB and PαMS at 300 K are summarized in Table 2. In both polymers, the major gaseous product is hydrogen, which suggests that scission of carbon–hydrogen bonds is a major degradation pathway. This observation is consistent with the observed radicals being carbon centred radicals formed by the loss of hydrogen. In PαMS some hydrogen atoms are scavenged by the aromatic ring to form cyclohexadienyl radicals. The yield of methane isrelatively small for both polymers, and no benzene was found on radiolysis of PαMS, which indicates that little side–chain cleavage occurs in these systems.

Monomer was also formed on radiolysis of both polymers, but the observed yields of monomer are relatively small and not consistent with depolymerization playing a significant role in the radiation chemistry of the polymers at 300 K.

Table 2. G values of low molecular weight volatile products formed on radiolysis of PIB and PαMS at 300 K

Product	PIB [a]	PαMS [b]
hydrogen	1.68	0.063
methane	0.67	0.003
ethane	0.08	—
monomer	0.62	0.30

[a] From reference 7
[b] From reference 8

In polymers which do not undergo depolymerization, the G–values for scission and crosslinking can be determined from molecular weight studies in which the number average (M_n) and weight average (M_w) molecular weights are measured over a range of absorbed doses. The relationships which describe the dose dependence of the two molecular weight averages are[11]:

$$\left(\frac{1}{M_n}\right)_D = \left(\frac{1}{M_n}\right)_0 + 1.037 \times 10^{-7}\,[G(S) - G(X)]D \qquad (1)$$

$$\left(\frac{1}{M_w}\right)_D = \left(\frac{1}{M_w}\right)_0 + 1.037 \times 10^{-7}\,[0.5\,G(S) - 2\,G(X)]D \qquad (2)$$

where the dose D is expressed in kGy.

G values for scission and crosslinking obtained from molecular weight studies over a range of temperatures for which there is no evidence for significant depolymerization are given in Table 3 for PIB[7] and Table 4 for PαMS[9].

Table 3. G–values for scission and crosslinking of PIB for γ–radiolysis at a range of temperatures

Temperature	G(S)	G(X)
200 K	1.75 ± 0.05	0.05 ± 0.01
303 K	3.8 ± 0.10	0.10 ± 0.05
348 K	4.7 ± 0.10	0.2 ± 0.05

Table 4. G–values for scission and crosslinking of PαMS for γ–radiolysis at a range of temperatures

Temperature	G(S)	G(X)
300 K	0.29 ± 0.01	0.00 ± 0.01
353 K	0.48 ± 0.06	0.00 ± 0.06

For both polymers, the G–value of scission is much greater than that for crosslinking and G(S) for PαMS is much smaller than that for PIB, again reflecting the protective effect of the aromatic group in PαMS. The values of G(S) increase with increasing temperature for both of the polymers. Although significant scission takes place in both polymers, no propagation radicals were observed in PIB and only a small proportion of propagation radicals were observed in PαMS.

Thus, at low temperatures PIB and PαMS do not undergo significant depolymerization, and hence there is no significant polymer weight loss when they are irradiated under vacuum. The major bond cleavage reactions which have been observed in these polymers can be summarized by the following diagram:

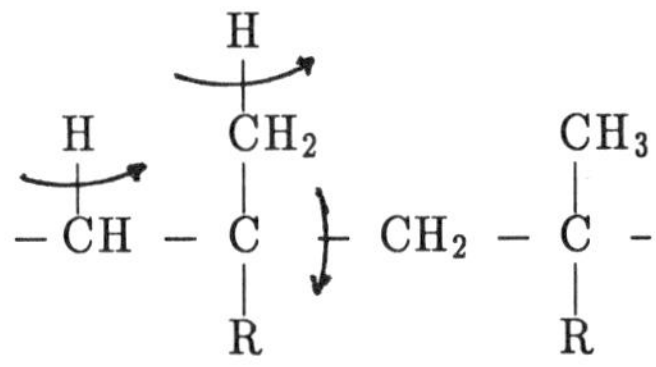

```
            H
         ⌣+↗
  H       CH2          CH3
⌣+↗       |            |
–CH  –   C   } CH2  –  C  –
          |  ↓          |
          R             R
```

High Temperature Studies At higher temperatures both polymers have been observed to undergo significant radiation induced thermal depolymerization. In PαMS this depolymerization has been observed at temperatures above 373 K and in PIB it has been observed at temperatures above 438 K. These temperatures are approximately 100°K lower than the temperatures at which significant thermal depolymerization begins to occur in the polymers. The important sequence of reactions in the radiation induced depolymerization process are:

Initiation:

$$P_n \xrightarrow{\gamma} R_{\dot{n}} + R_{\dot{m}}$$

Depropagation:

$$R_{\dot{n}} \xrightarrow{k_d} R_{\dot{n}-1} + M$$

$$\vdots$$

$$R_{\dot{n}-z+1} \xrightarrow{k_d} R_{\dot{n}-z} + M$$

Termination:

$$R_{\dot{n}-z} \xrightarrow{k_t} P_{n-z}$$

where z is the average zip length.

The unzipping reactions result in the formation of monomer with a resultant decrease in the mass of polymer. Because the irradiations in this work were carried out under vacuum and the monomer was continuously removed by pumping, the decrease in the mass of polymer can be obtained gravimetrically. For PIB and PαMS the mass of polymer was found to decrease with irradiation time (at a constant dose rate) according to first order kinetics, as shown in Figure 4 for PIB.

Thus, the weight loss can be described by the equation:

$$W_t = W_0 e^{-kt} \tag{3}$$

where W_0 and W_t are the weights of polymer initially and at time t and k is the rate constant. Hence, radiation induced thermal depolymerization at a constant dose rate follows similar kinetics to that observed for the thermally induces process[10], which takes place at a higher temperature in the absence of radiation.

The G–values for the loss of monomer G(–M), can be obtained from the first order rate constants using the relationship[9]:

$$G(-M) = \frac{k}{1.037 \times 10^{-7} \times R \times M} \tag{4}$$

where k is the rate constant in s^{-1}, R is the dose rate in kGy s^{-1} and M is the molecular weight of the monomer.

The values of G(–M) for a range of temperatures are given in Table 5 for PIB[7] and in Table 6 for PαMS[9]. The G values for both polymers increase as the temperature increases, and for PαMS at 453 K the value of G(–M) is approximately equal to the degree of polymerization (510) of the polymer used in the study.

Table 5. G(–M) for PIB over a range of temperatures

T/K	G(–M)
438	276
493	1280
523	2884

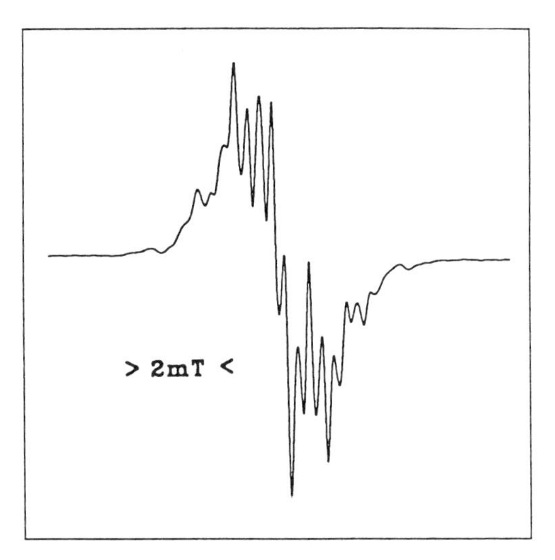

Figure 3. The ESR spectrum of PαMS after irradiation at 300 K.

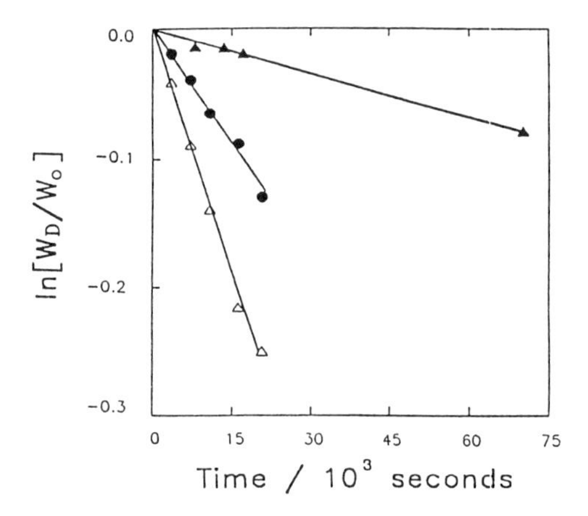

Figure 4. First order kinetic plots of the variation in the weight of PIB with irradiation time at various temperatures, 438 K (▲); 493 K (●); 523 K (△),for a dose rate of 2.64 kGy hr^{-1}.

Table 6. G(–M) for PαMS over a range of temperatures

T/K	G(–M)
373	95
383	153
393	183
453	529

Molecular weight studies can again yield information about the value of G(S) provided that the polymer weight loss and any chain end effects are taken into account. In cases where G(X) is very small compared with G(S) and where chain end effects can be ignored (ie. the average zip length, z, is small compared with the DP for the polymer), the dose dependence of the number average molecular weight is given by the equation:

$$\left(\frac{1}{M_n}\right)_D = -\frac{G(S)\cdot R}{k} + \left[\left(\frac{1}{M_n}\right)_0 + \frac{G(S)\cdot R}{k}\right] e^{-kt} \qquad (5)$$

where k is the rate constant for depolymerization, R is the dose rate and t is the radiolysis time.

The radiation induced depolymerization of PIB reported in this work satisfy the requirements of equation 5, and it has been used to calculate best values for G(S) by regression analysis of the variation of M_n with radiolysis time.

The values of G(S) obtained over a range of temperatures for PIB are given in Figure 5, together with the values obtained at lower temperatures where depolymerization does not occur. There is a clear change in the temperature dependence of G(S) above 438 K, where depolymerization becomes significant.

If the value of the average zip length for depolymerization is comparable with the average degree of polymerization of the polymer under study, equation (5) does not apply. This arises because in the derivation of equation (5) it is assumed that each chain scission results in the formation of a new chain. However, this may not be true if z is comparable with the DP of the polymer. In these cases it is necessary to use a Monte Carlo simulation to obtain G(S) from an analysis of the relationships between M_n (or M_w) and the weight loss for various values of the zip length, z[9].

The PαMS used in the study reported here had a relatively small DP of 510, so the Monte Carlo method was used to determine G(S)[9]. The values of G(S) obtained over a range of temperatures are given in Figure 6, together with the values obtained at the lower temperatures where depolymerization was not observed. Above 353 K, G(S) was found to increase substantially with increasing temperature, as was observed for PIB above 438 K.

The average zip lengths for depolymerization of PαMS over the temperature range 300 – 453 K are also shown in Figure 6. They increase from zero at 353 K to 600 at 453 K.

Figures 5 and 6 show that there is a change in the activation energy for chain scission when depolymerization becomes important. The reason for this change remains unknown, but the observations suggest that the polymer morphology is probably not responsible, because while the temperature at which

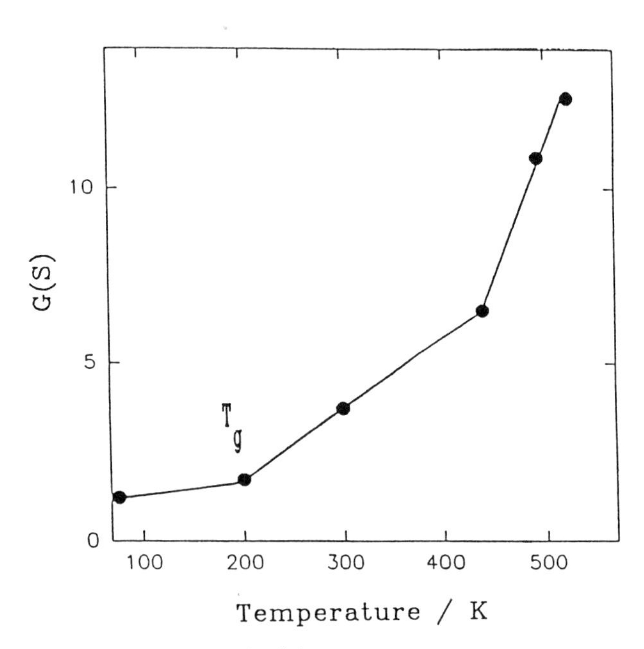

Figure 5. The variation of G(S) with irradiation temperature for PIB.

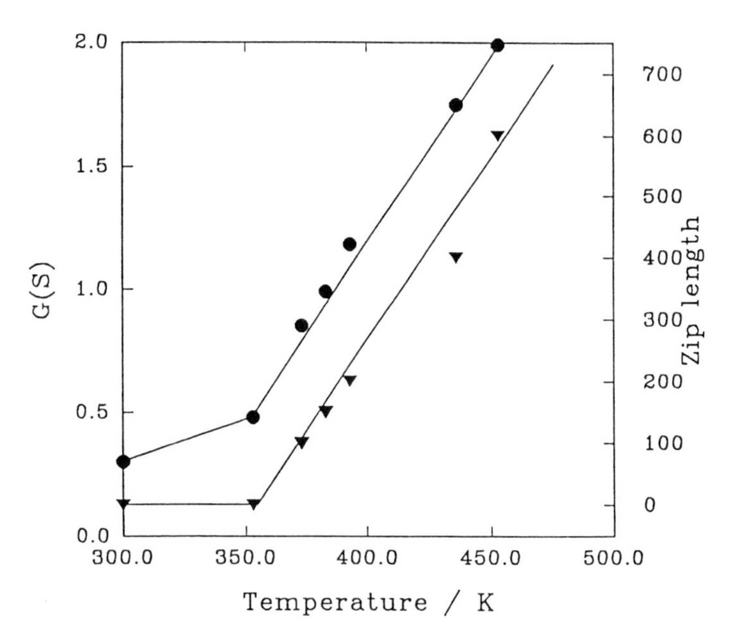

Figure 6. The variation of G(S) (●) and z (▼) with irradiation temperature for PαMS.

depolymerization becomes significant in PIB is well above T_g, for PαMS this temperature is well below the T_g of the polymer.

In summary, this work has shown that polyisobutene, which is an amorphous polymer with a low T_g (–70°C) and poly(α–methyl styrene) which is an amorphous polymer with a relatively high T_g (180°C) undergo radiation degradation involving similar radiation chemistry. The annealing studies showed that the radicals produced in PIB at low temperature are unstable above T_g, while most of the radicals formed in PαMS at 77 K undergo reaction well below T_g. In both polymers, the G–values for chain scission are much lager than those for crosslinking. Both polymers also undergo radiation induced thermal degradation, with PαMS being more susceptible to unzipping than PIB at elevated temperatures. This is in accord with the reported values of T_c for these polymers. However, the G–value for chain scission in PIB is greater than that for PαMS at the same temperature. This is due in part to the protective effect of the aromatic group in PαMS.

References

1. Bowmer, T.N.; Ho, S.Y.; O'Donnell, J.H.; O'Sullivan, P.W. *J.Macromol.Sci.,Chem.*,**1984**,*A21*,745.
2. Hill, D.J.T.; O'Donnell, J.H.; Pomery, P.J.; Sangster, D.F. In *Polymer Handbook*; Brandrup, J., Immergut, E.H., Eds.; 3rd Edn; John Wiley & Sons: New York,1989,II/387.
3. Busfield, W.K.; O'Donnell, J.H.; Smith, C.A. *J.Macromol.Sci.Chem.* **1982**,*A17*,1263.
4. Busfield, W.K.; O'Donnell, J.H.; Smith, C.A. *Polymer* **1982**,*23*,431.
5. Busfield, W.K. In *Aspects of Degradation and Stabilization of Polymers*; Jellinek, H.H., Ed.; Elsevier, Amsterdam;1978.
6. Hori, J.; Kashiwabara, H. *J.Polym.Sci.,Polym.Phys.Ed.* **1981**,*19*,1141.
7. Hill, D.J.T.; O'Donnell, J.H.; Perera, M.C.S.; Pomery, P.J. In Preparation.
8. Garrett, R.W.; Hill, D.J.T.; Le, T.T.; O'Donnell, J.H.; Pomery, P.J. *Radiat.Phys.Chem.* Accepted for publication.
9. Hill, D.J.T.; Le, T.T.; O'Donnell, J.H.; Pomery, P.J. In Preparation.
10. Boyd, R.H. *J.Chem.Phys.* **1959**,*31*,321.
11. Dole, M. *The Radiation Chemistry of Macromolecules*, Academic Press, New York, 1973, Vol.1.

RECEIVED August 14, 1992

Chapter 5

Radiolytic Formation and Decay of *trans*-Vinylene Unsaturation in Polyethylene

B. J. Lyons[1] and W. C. Johnson[2]

[1]22 Hallmark Circle, Menlo Park, CA 94025–6683 (retired)
[2]Raychem Corporation, 300 Constitution Drive, Menlo Park, CA 94025

This paper is the first of several in which relations, derived from various mechanisms of growth and decay of *trans*-vinylene unsaturation, are compared with a selection of experimental measurements, some already published and here re-examined and some newly made.

Understanding the reaction mechanism and kinetics of, and the possibility of interfering reactions with, the growth and decay of *trans*-vinylene unsaturation in polyethylene exposed to ionizing radiation is of interest for both practical and theoretical reasons. For practical reasons because, if the dependence on dose can be accurately predicted, *trans*-vinylene concentration measurements can be used as a method of dosimetry. Theoretically because, although there is general agreement on the *trans*-vinylene formation process, there are several suggested mechanisms of decay and no clear indication why any particular mechanism should be preferred. One of us (WCJ) discovered some years ago that the variation of *trans*-vinylene unsaturation with dose could be represented very accurately in the dose range up to about 1000 kilograys by an empirical equation, binomial in form. This paper is the first in a series in which we will examine the calibration and use of this relation in dosimetry, derive this binomial as a special case of a polynomial relation from consideration of scavenging of free radicals by radiolytically formed *trans*-vinylene groups, compare the binomial and polynomial relations with previously published and new measurements and comment generally on the use of measurements of *trans*-vinylene unsaturation concentration in radiation dosimetry.

A number of investigations of the growth of *trans*-vinylene unsaturation have been carried out over the years {Dole, Milner and Williams (*1*), Lyons and Crook (*2*) and others summarized in (*2*)}. There seems to be little if any disagreement as to the formation reaction but widely different opinions as to the decay reaction. These authors agree that the formation reaction is a primary process possibly involving ionization followed by the detachment of a molecule of hydrogen in a one step process.

0097–6156/93/0527–0062$06.00/0

We concur, noting however that there may also be a contribution, most likely amounting to 5% or less, from the recombination of adjacent alkyl free radicals on the same polymer chain. This contribution is more important in the crystalline phase and therefore larger in high density polyethylenes. Thus, this contribution may largely account for the difference in the G(*trans*-vinylene) (G(Vl)) values noted between high (HDPE) and low (LDPE) density polyethylenes.

As mentioned above, the growth of *trans*-vinylene unsaturation in low density polyethylene over a dose range up to about 1000 kGy can be fitted accurately to a binomial of the form:

$$Ar + D = B[Vl] + C[Vl]^2 \quad (1)$$

where A, B and C are proportionality constants, D allows for an initial non-zero *trans*-vinylene molar concentration ($[Vl]_0$) at zero dose, and [Vl] and $[Vl]_\infty$ represent molar *trans*-vinylene concentrations after r kilograys and at infinite dose. As the relations derived in references (*1*) and (*2*) can be reduced to this simple equation for values of [Vl] less than, say, $0.3[Vl]_\infty$, that is doses up to approximately 500 kGy, measurements in this range do not in themselves enable us to discriminate between the various mechanisms which have been suggested.

However, in the course of deriving a theoretical basis for this empirical equation one of us (BJL) noted that expansion of the published relations gave polynomials of quite different character to each other. Thus comparison of such expansions with the experimental results over a much wider dose range might serve to distinguish between them.

Theory

Dole and others considered the *trans*-vinylene decay reaction to involve excitation of the *trans*-vinylene group followed by further reaction with a neighboring polymer chain to form a crosslink. This analysis leads to the relation (*1*):

$$\frac{d[Vl]}{dt} = k'I - k''[Vl] \quad (2)$$

where k' and k" are the reaction constants for formation and decay of trans-vinylene groups and I is the radiation intensity. This equation, after integration and insertion of limits, becomes:

$$\frac{[Vl] - [Vl]_0}{[Vl]_\infty - [Vl]_0} = 1 - e^{-k''r} \quad (3)$$

They note that this equation is similar to equations deduced by Pearson (*3*),(*4*) and by Simha and Wall (*5*).

Equation 3 may be rewritten:

$$e^{-k''r} = \frac{1 - [Vl].[Vl]_{\infty}^{-1}}{1 - [Vl]_0.[Vl]_{\infty}^{-1}} \tag{4}$$

or

$$k''r = \ln\left(1 - \frac{[Vl]_0}{[Vl]_{\infty}}\right) - \ln\left(1 - \frac{[Vl]}{[Vl]_{\infty}}\right) \tag{5}$$

As, for most polyethylenes $[Vl]_0$ is very small, this expression reduces to:

$$k''r = -\ln\left(1 - \frac{[Vl]}{[Vl]_{\infty}}\right) \tag{6}$$

which, using the expansion:

$$\ln(1 + x) = x - \frac{x^2}{2} + \frac{x^3}{3} - \frac{x^4}{4} + \ldots \tag{7}$$

may be rewritten:

$$k''r = \frac{[Vl]}{[Vl]_{\infty}} - \frac{1}{2}\left(\frac{[Vl]}{[Vl]_{\infty}}\right)^2 + \frac{1}{3}\left(\frac{[Vl]}{[Vl]_{\infty}}\right)^3 - \frac{1}{4}\left(\frac{[Vl]}{[Vl]_{\infty}}\right)^4 + \ldots \tag{8}$$

Lyons and Crook (*2*) suggested that the *trans*-vinylene group was acting as a scavenger for mobile radicals through a reaction leading to the formation of allyl stable radicals which are trapped at room temperature. At that time agreement between estimates of allyl radical concentrations and estimates of the amount of *trans*-vinylene decay was not good. However, since then it has been found by Dole and Waterman (*6*) that estimates of the two agree quite well. Lyons and Crooke pointed out that a form of the general scavenging equation of Charlesby and Lloyd (*7*) (modified to allow for the *trans*-vinylene formation reaction) should apply. These reactions are considered to be:

$$\text{-CH}_2\text{-CH}_2\text{-} \rightsquigarrow \text{-CH=CH-} \qquad (\text{Vl}) \quad k'$$

$$\text{-CH}_2\text{-CH}_2\text{-} \rightsquigarrow \text{-CH}\cdot\text{-CH}_2\text{-} + \text{H}\cdot \qquad k_1$$

followed by

$$\text{H}\cdot + \text{-CH}_2\text{-CH}_2\text{-} \longrightarrow \text{-CH}\cdot\text{-CH}_2\text{-} + \text{H}_2 \quad (\text{R}\cdot)$$

$$2\text{R}\cdot \longrightarrow \text{CROSSLINK} \qquad k_2$$

$$\text{R}\cdot + \text{Vl} \longrightarrow \text{-CH=CH-CH}\cdot\text{-} \quad (\text{Al}\cdot) \quad k_3$$

These authors did not derive a general expression for *trans*-vinylene growth and decay from analysis of this reaction scheme, but pointed out that a plot of reciprocal *trans*-vinylene concentration against reciprocal dose was linear over a wide dose range and that this corresponded to the empirical expression:

$$K_3r = \frac{[Vl]}{(1 - [Vl].[Vl]_{\infty}^{-1})} - C \tag{9}$$

where k_3 and C are proportionality constants, which expression, by analogy to

$$(1 - x)^{-1} = 1 + x + x^2 + x^3 + x^4 + ... \tag{10}$$

may be expanded to:

$$K_3r = [Vl]\left\{1 + \frac{[Vl]}{[Vl_{\infty}} + \left(\frac{[Vl]}{[Vl]_{\infty}}\right)^2 + \left(\frac{[Vl]}{[Vl]_{\infty}}\right)^3 + \left(\frac{[Vl]}{[Vl]_{\infty}}\right)^4 + ...\right\} \tag{11}$$

The general expression for *trans*-vinylene growth and decay, assuming a free radical scavenging process, is derived in Appendix A. We here summarize the main features. From consideration of the above reactions, the rate of change of *trans*-vinylene unsaturation with dose is:

$$\frac{d[Vl]}{dt} = k'I - k_3[R\cdot][Vl] \tag{12}$$

now

$$\frac{d[R\cdot]}{dt} = k_1I - k_2[R\cdot] - k_3[Vl][R\cdot] \tag{13}$$

and, assuming steady state conditions:

$$[R\cdot] = \frac{k_1I}{k_2 + k_3[Vl]} \tag{14}$$

whence:

$$\frac{d[Vl]}{dt} = k'I - \frac{k_3[Vl].k_1I}{k_2 + k_3[Vl]} \tag{15}$$

Now if k_1 and k' are similar in value, that is $k_1/k' \approx 1$, then

$$k'r = [Vl] + \frac{k_3}{2k_2}[Vl]^2 \tag{16}$$

which is identical to the empirical equation 1 above.

For the derivation of the general expression, let $a = k_3/k_2$ and $k_1/k' = 1 - b$, i. e. $b = 1 - k_1/k'$; then, as shown in Appendix A:

$$k'r = [Vl] + \frac{k_1}{k'ab^2}\left[\frac{(ab[Vl])^2}{2} - \frac{(ab[Vl])^3}{3} + \frac{(ab[Vl])^4}{4} - \frac{(ab[Vl])^5}{5} + ...\right] \quad (17)$$

which reduces to equation 2 above for $k_1/k' \approx 1$ and/or low concentrations of *trans*-vinylene unsaturation. It will be noted that this expression differs from that of Dole and coworkers mainly in that the alternation of negative and positive coefficients starts with the third term on the right hand side whereas the alternation in equation 8, which is derived from their expression, commences with the second term on the right hand side.

Inspection of the successive coefficients of equation 17 shows that:

$$\frac{3a_3}{2a_2} = \frac{4a_4}{3a_3} = \frac{5a_5}{4a_4} = ... = ab \quad \text{and:} \quad \frac{a_1}{2a_2} = \frac{k'}{ak_1}$$

where a_1, a_2, a_3, ... are the coefficients of the terms in [Vl], $[Vl]^2$, $[Vl]^3$, ...

Lyons and Crook (*2*) also proposed that freely migrating excited species formed during radiolysis could be trapped by unsaturation to form species of much more limited migratory ability. One of us has since proposed (*8*), (*9*) that the excited species involved are hydrogen atoms. Thus the migration of free radical sites in polyethylene can occur through the migration of alkyl free radical sites, which involves a stepwise exchange:

$$\text{-CH}\cdot\text{-CH}_2\text{- + -CH}_2\text{-CH}_2\text{-} \longrightarrow \text{-CH}_2\text{-CH}_2\text{- + -CH}\cdot\text{-CH}_2\text{-}$$

in which each step, obviously, involves only a very small displacement of the free radical site, or through the infrequent reaction:

$$\text{-CH}\cdot\text{-CH}_2\text{- + H}_2 \longrightarrow \text{-CH}_2\text{-CH}_2\text{- + H}\cdot$$

followed by

$$\text{H}\cdot\text{ + -CH}_2\text{-CH}_2\text{-} \longrightarrow \text{-CH}\cdot\text{-CH}_2\text{- + H}_2$$

Because, as noted by Libby (*10*), this latter reaction of a thermalized hydrogen atom is slow and the hydrogen atom is very pervasive, hydrogen exchange can result in wide ranging migration of free radical sites whose diffusion is otherwise extremely limited. The net effect of this qualitative difference between these two migration reactions is that, when unsaturation is present initially, there is a preferential transfer of free radical sites (mainly from the crystalline regions) to those (amorphous) regions rich in vinyl and vinylidene groups which become converted to alkyl free radicals. A possible consequence of such a conversion is that while significant amounts of vinyl groups remain, the endlinking reaction occurs in place of rather than in addition to the crosslinking reaction. Indeed Randall, Zoepfl and Silverman (*11*) have shown that endlinking is by far the most predominant process in a broad molecular weight distribution HDPE at radiation doses in the pre-gel range. Scavenging of hydrogen atoms by vinyl or vinylidene groups is manifested as an additional term in the relation relating unsaturation decay to the radiation dose. Lyons and Crooke (*2*) thought this factor was manifested in the *trans*-vinylene decay reaction as well. We are inclined to doubt this because there is no reason to believe that, under normal conditions of radiolysis, the rate of formation of *trans*-vinylene groups in the crystalline phase differs significantly from that in the amorphous phase. Moreover, if, as seems likely, allyl free radicals are produced, then whether an alkyl free radical or a hydrogen atom is involved in this reaction with *trans*-vinylene, scavenging of a free radical effectively removes it from significant further participation in radiolytic reactions involving migration. We underline the preceding remark to stress that the assumption here is not that the allyl radical does not react further, rather it is that the radical site remains associated with the *trans*-vinylene group itself, because the reverse reaction (to reform a secondary alkyl free radical) as Bodily notes (*12*) is endothermic. Further reaction then becomes limited by the chain segments containing these allyl radicals. We must add here that discerning the fate of the allyl radical itself is not a trivial problem. These radicals in LDPE decay quite rapidly at room temperature, so that none are detectable after irradiation, but are more persistent in HDPE, which must be heated to at least about 80°C for the decay to occur. However, the characteristics of *trans*-vinylene growth in both LDPE and HDPE as we shall see below are identical. Interestingly, if the HDPE contains significant ($>0.2\%$ or more) amounts of a free radical scavenger (for example, an antioxidant) no allyl radicals are observed after irradiation at ambient room temperature and a larger amount of *trans*-vinylene unsaturation is found (presumably because the allyl radicals have been "repaired"). The G value for radiation crosslinking in LDPE has been shown to be constant with dose over the dose range 0-800 kGy by Lyons (*13*), a range in which the G(*trans*-vinylene decay) value increases from essentially zero to about one third of the G(*trans*-vinylene formation) value, that is to about 0.7. It seems, therefore, that the allyl radical decay reaction must involve crosslinking without reforming the *trans*-vinylene group (or, at least, without reforming an unsaturated group absorbing at 964 cm^{-1}). Finally we should consider the temperature range in which the derived relation is considered to be valid: as we show below the relation applies equally well to LDPE in which we have seen the allyl radical does not persist at ambient room temperature and to HDPE in which this radical does persist at temperatures below about 80°C. We conclude that this relation should be valid for irradiations up to and even above the melting point range of polyethylene.

Results

Lyons and Crooke (*2*) pointed out that their vinyl decay measurements and those of Dole and coworkers are very closely similar despite the large difference in the dose rate provided by the radiation sources they used. However, for *trans*-vinylene growth there is, especially at higher doses, as Figure 1 shows, a considerable difference between the results of these two groups. It might be concluded that this difference arises from differences in dose rate. However unpublished measurements by former colleagues of the author showed an even larger difference in growth rates for *trans*-vinylene unsaturation (Figure 2) between electron irradiation in air (Stivers, E. C., unpublished results (1963) reproduced by permission of Raychem Corporation) and that using Co-60 gamma radiation (as did Dole and co-workers) and a low density polyethylene irradiated in a nitrogen filled sealed cell (Lambert, A., unpublished results (1963) reproduced by permission of Raychem Corporation). However, the crosslinking rate found by Lambert, for gamma irradiation, measured by elastic

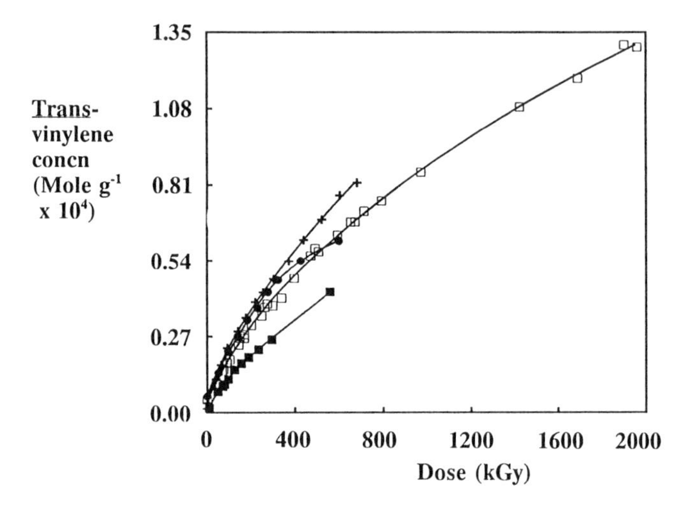

Figure 1. A comparison of measurements of radiolytic *trans*-vinylene growth in HDPE and LDPE by Dole (*1*) using Co-60 γ radiation and by Lyons (*2*) using electron radiation. ●: Dole, HDPE (Marlex 50); +: Lyons, HDPE (Rigidex 50); ■: Dole, LDPE (B3125); □: Lyons, LDPE (Alkathene 2).

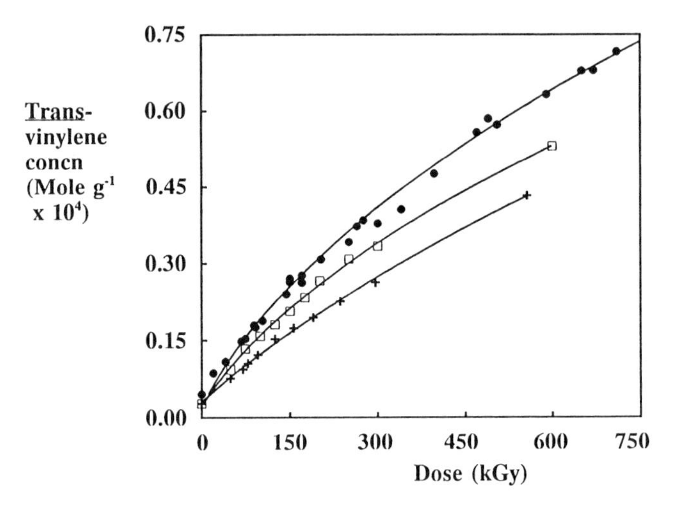

Figure 2. A comparison of measurements of radiolytic *trans*-vinylene growth in LDPE by Stivers and by Lyons (*2*) using electron radiation and by Lambert using Co-60 γ radiation. □: Stivers, (DYNK); +: Lambert (DYNK); ●: Lyons (Alkathene 2).

modulus measurements at 150°C, was identical to that found by Stivers to result from electron irradiation in air, indicating that there was no difference in their dosimetry. Later experiments by one of the present authors (*14*) showed that trans-vinylene growth in .5 mm thick films at doses above 1500 kGy was about 30% higher, for a given dose, than that in 0.75 mm films. We have recently found that low density polyethylene samples irradiated continuously with electrons exhibit lower *trans*-vinylene concentrations at higher doses than do similar samples irradiated intermittently with periodic measurement of the infra-red spectra as has been the practice of one of us (BJL) in previous work. As these irradiation methods only differ in that intermittent irradiation allows time for radiolytic hydrogen to diffuse out of the sample, this observation emphasizes the pronounced sensitivity of the trans-vinylene decay reaction to the presence of hydrogen. Indeed most if not all the differences between the various groups considered herein appears to be due to the influence of radiolytic hydrogen on *trans*-vinylene decay. Schumacher (*15*) found many years ago that polyethylene irradiated under 1 atmosphere of hydrogen to a high dose contained 30% less *trans*-vinylene unsaturation than did the same polymer irradiated in vacuum. One of us (BJL, *14*) found that the *trans*-vinylene content of a 0.75 mm HDPE film, irradiated to 1500 kGy, was about 30% less than that of a 0.25 mm film.

The results of regressions of these data on binomial and polynomial relations are shown in Table I. In these regressions, trans-vinylene concentrations were expressed as moles/kg. In the case of the polynomials, the number of terms used in the regression is that number that gave the highest F ratio. This Table shows that for all data sets, when a binomial is selected, the coefficients of the first two terms are positive. The results of Dole (*1*) do not give anything like as good a fit to the binomial as they do to a polynomial and this is in contrast to the other results where the fit to the binomial is almost as good or in one case better than the fit to the polynomial. The coefficients of the polynomial regression on Dole's results have alternating signs commencing with a_2 as required for his equation (8 above). However with the results of Lyons and Crooke (*2*) for a essentially identical polymer, the polynomial with 6 terms have alternating signs commencing with a_3 as required for equation 17 derived in this paper. With a polynomial regression of the results of Lyons and Crooke (*2*) for LDPE, the third coefficient is positive (as is required by their semi-empirical equation 11) but the F ratio is markedly less than that for the binomial; the third term is also so small that it may be neglected. With the results of Lambert although the third term is negative as required by equation 17 above and significant at the 4% level, the improvement due to inclusion of this term in the regression is very small. With a polynomial regression of the results of Stivers, the third term is positive again but the size of this coefficient indicates it is not significant. We find consistently in these regressions that the term in $[Vl]^3$ is, at best, marginally significant and further terms are not statistically significant (they are included for the first two data sets only to point out the alternations in sign of the coefficients). It is well known that polynomial expressions can be adjusted to fit any type of data pattern so the analysis attempted in this paper can only be expected to work with very precise measurements. Although the accuracy of these published data is not sufficient of itself to support a firm conclusion as to the mechanism, the surprising success of the binomial in representing trans-vinylene growth over a wide range in concentrations in four out of the five data sets, in one case over a range of 1000, in another over a range of 2000 kGy, strongly suggests free radical scavenging is the correct mechanism. The sensitivity of trans-vinylene growth to the presence of hydrogen during radiolysis further supports this conclusion.

Table I: Regressions of Various Data on Binomials and Polynomials

Source	a_0	a_1	a_2	a_3	a_4	a_5	a_6	R^2	F Ratio
Dole (*1*)	.58	7.2	138	-	-	-	-	0.9914	403
HDPE	-11.1	2581	-21034	96150	-195700	151600	-50403	0.9996	3730
Lyons (*2*)	-13	389	552	-	-	-	-	0.9982	3930
HDPE	-3.7	215	1530	-2570	11910	-10800	3680	0.9996	5129
Lyons (*2*)	-7.3	415	840	-	-	-	-	0.9954	4120
LDPE	-36	680	314	270	-	-	-	0.9957	2944
Stivers	-11	503	1220	-	-	-	-	0.9984	3140
LDPE	-16	618	690	640	-	-	-	0.9984	2026
Lambert	-29	942	780	-	-	-	-	0.9976	2055
LDPE	-11.7	528	3340	-3490	-	-	-	0.9985	2233

It follows that the generally good fits obtained with the binomial are a strong indication that G(trans-vinylene formation) and G(free radicals) values are very similar if not identical in the case of both LDPE and HDPE. Although early estimates of G(free radicals) in polyethylene were close to 4, estimates have tended downwards over the years. If 2G(X) = G(alkyl radicals), then the estimates by Lyons and Fox (*13*) that the value for G(X) in a low density polyethylene is 1.0 and more recently by Lyons (*16*) that the value for G(X) in a large number of low, linear low and high density polyethylenes is 1.0 ± 0.1 would indicate a G(free radicals) of 2.0 ± 0.2 which is square in the middle of the range of G(trans-vinylene) values found by a number of workers (see *1, 2*).

In recent years Fourier Transform Infra-Red (FTIR) machines have become readily accessible and these machines are capable of much more accurate infra-red absorption measurement than was previously possible. In the next paper in this series we report on FTIR measurements of radiolytic trans-vinylene growth in a low density polyethylene and their use in dosimetry.

Acknowledgements

The authors express their gratitude to the late Professor Malcolm Dole for kindly providing us with the original results from reference *1* and to Raychem Corporation for permission to cite unpublished work by A. Lambert and E. C. Stivers and for permission to publish this paper.

Literature Cited

1. Dole, M.; Milner, D. C.; Williams, T. F. *J. Amer. Chem. Soc.* **1958,** *80,* 1580.
2. Lyons, B. J.; Crook, M. A. *Trans. Faraday Soc.* **1963,** *59,* 2334.
3. Pearson, R. W. *Chem. and Ind.* **1956,** 903; **1957,** 209.
4. Pearson, R. W. *J. Polymer Sci.* **1957,** *25,* 189.
5. Simha, R.; Wall, L. A. *J. Phys. Chem.* **1957,** *61,* 425.
6. Waterman, D. C.; Dole M. *J. Phys. Chem.,* **1970,** *74,* 1913.
7. Charlesby, A.; Lloyd, D. G. *Proc. Royal Soc. (Lond.)* **1960,** *A254,* 207.
8. Lyons B. J. *Radiation Processing for Plastics and Rubber* **1981,** Plastics and Rubber Institute, London, England, pp. 5-1 to 5-13.
9. Lyons, B. J. *Radiat. Phys. Chem.* **1983,** *22,* 135.
10. Libby, W. F. *J. Chem. Phys.* **1961,** *35,* 1714.
11. Randall, J. C.; Zoepfl, F. J.; Silverman, J. *Radiat. Phys. Chem.* **1983,** *22,* 183.
12. Bodily, D. M.; Dole, M. *J. Chem. Phys.* **1961,** *35,* 1714.
13. Lyons B. J.; Fox, A. S., *J. Polymer Sci.,* **C21**, 159 (1968).
14. Lyons, B. J. *J. Polymer Sci.* **1965,** *A3,* 777.
15. Schumacher, K., *Kolloid Z.* **1958,** *157,* 16.
16. Lyons, B. J. *Radiat. Phys. Chem.* **1983,** *22,* 135.

Appendix A

The general expression for trans-vinylene growth and decay which we derive from consideration of the free radical reactions in the theory section is:

$$\frac{d[Vl]}{dt} = k'I - k_3[R\cdot][Vl] \qquad (19)$$

now

$$\frac{d[R\cdot]}{dt} = k_1I - k_2[R\cdot] - k_3[Vl][R\cdot] \qquad (20)$$

and, assuming steady state conditions:

$$[R\cdot] = \frac{k_1I}{k_2 + k_3[Vl]} \qquad (21)$$

whence:

$$\frac{d[Vl]}{dt} = k'I - \frac{k_3[Vl].k_1I}{k_2 + k_3[Vl]} \tag{22}$$

or

$$\frac{d[Vl]}{dt} = \frac{k'I(k_2 + k_3[Vl]) - k_3[Vl].k_1I}{k_2 + k_3[Vl]} = \frac{k'I\{1 + \frac{k_3}{k_2}(1 - \frac{k_1}{k'})[Vl]\}}{1 + \frac{k_3}{k_2}[Vl]} \tag{23}$$

Now if k_1 and k' are similar in value, that is $k_1/k' \approx 1$, then

$$\frac{d[Vl]}{dt} = \frac{k_1I}{1 + \frac{k_3}{k_2}[Vl]} \tag{24}$$

whence equation (16) may be derived:

$$k'\tau = [Vl] + \frac{k_3}{2k_2}[Vl]^2 \tag{16}$$

which, as we have seen is identical to the empirical equation (1) above.

To facilitate derivation of the general expression, let $a = k_3/k_2$ and $k_1/k' = 1 - b$, i. e. $b = 1 - k_1/k'$; then:

$$k'I.dt = \frac{(1 + a[Vl])d[Vl]}{1 + ab[Vl]} \tag{25}$$

Integrating:

$$k'\tau = \frac{[Vl] - [Vl]_0}{b} + \frac{(1 - b^{-1})}{ab} \,\mathrm{Ln}\left[\frac{(1 + ab[Vl])}{(1 + ab[Vl]_0)}\right] \tag{26}$$

This is a rather intractable expression to develop as it has two incompatible terms in the dependent variable but (as $1 - b^{-1} = (b-1)/b$) the logarithmic term can be expanded to a polynomial expression:

$$k'\tau = \frac{[Vl]}{b} + \frac{(b-1)}{ab^2}\left[ab[Vl] - \frac{(ab[Vl])^2}{2} + \frac{(ab[Vl])^3}{3} - \frac{(ab[Vl])^4}{4} + \frac{(ab[Vl])^5}{5} - ..\right] \tag{27}$$

or

$$k'r = \frac{[Vl]}{b} + \frac{(b-1)}{ab^2}ab[Vl] - \frac{(b-1)}{ab^2}\left[\frac{(ab[Vl])^2}{2} - \frac{(ab[Vl])^3}{3} + \frac{(ab[Vl])^4}{4} - \frac{(ab[Vl])^5}{5} + ..\right] \quad (28)$$

that is,

$$k'r = \frac{ab[Vl] + ab^2[Vl] - ab[Vl]}{ab^2} - \frac{(b-1)}{ab^2}\left[\frac{(ab[Vl])^2}{2} - \frac{(ab[Vl])^3}{3} + \frac{(ab[Vl])^4}{4} - \frac{(ab[Vl])^5}{5} + .\right] \quad (29)$$

or

$$k'r = [Vl] - \frac{(b-1)}{ab^2}\left[\frac{(ab[Vl])^2}{2} - \frac{(ab[Vl])^3}{3} + \frac{(ab[Vl])^4}{4} - \frac{(ab[Vl])^5}{5} ...\right] \quad (30)$$

and substituting for b-1 in the second term on the right hand side yields equation 17:

$$k'r = [Vl] + \frac{k_1}{k'ab^2}\left[\frac{(ab[Vl])^2}{2} - \frac{(ab[Vl])^3}{3} + \frac{(ab[Vl])^4}{4} - \frac{(ab[Vl])^5}{5} + ..\right] \quad (17)$$

RECEIVED August 14, 1992

Chapter 6

Radiation-Induced Structural Changes in Polychloroprene

David J. T. Hill, James H. O'Donnell, M. C. Senake Perera, and Peter J. Pomery

Polymer Materials and Radiation Group, Department of Chemistry, University of Queensland, Brisbane, Queensland 4072, Australia

γ-Irradiation of polychloroprene (98% trans 1,4) resulted in crosslinking with G(X) = 3.9 (NMR), 4.8 (swelling ratio), 3.2 (soluble fraction). A resonance at 45 ppm in the ^{13}C NMR spectra was attributed to the methine carbon in a crosslink, and resonances at 64, 80 and 95 ppm to chlorinated end groups and main chain structures. The ESR spectra after irradiation at 77K gave G(R) = 3.0, comprising radical anions and a variety of chlorinated allyl and polyenyl radicals. In contrast to polybutadiene and polyisoprene, NMR spectra did not give higher G(X) values than other methods, suggesting that crosslinking did not proceed through a kinetic chain reaction producing clustered crosslinks.

Halogen substitution is known to increase the sensitivity of aliphatic hydrocarbons to high energy radiation(*1*). This sensitization is attributed to the low dissociation energy of the C-Cl bond and the occurrence of dissociative electron capture to form the halide ion.

Low molecular weight compounds containing halogen atoms, such as CCl_4 and alkyl halides, have been used to sensitize polymers to irradiation for enhancement of crosslinking and other reactions(*2*). The radiation yield of crosslinking in cis-polybutadiene was increased from 3.6 to 14 by adding 7.4% of chloroparaffin. Incorporation of a halogen atom in the molecular structure of polymers also increases their radiation sensitivity. Substantial differences have been reported(*3*) in the effect of high energy radiation on halogenated and unhalogenated polymers. Chlorination of polyethylene caused an increase in radiation stability at $<30\%$ chlorine content but stability decreased at $>30\%$. Chlorination and hydrochlorination of 1,4-cis-polyisoprene rubber reduced the radiation stability.

The radiation chemical yields of several different chlorine-containing styrene-based resists were studied by Hartney(*4*). Increased sensitivity was observed up to one chlorine atom per monomer unit but then decreased. Butyl rubber has been shown(*5*) to undergo predominantly main-chain scission, whereas halogenated butyl rubbers crosslink with high yields at very low radiation doses.

0097–6156/93/0527–0074$06.25/0

Polychloroprene (chloroprene rubber), I, is widely used in applications which require an elastomer with resistance to organic solvents and oil. The applications include hoses, seals, chemical plant coatings, conveyor belts, surgical and industrial gloves. It has been classified as having a relatively low resistance to radiation on the basis of the deterioration in mechanical properties. The half-value dose(the dose required to reduce elongation at break to half of its original value) in air for polychloroprene was 0.5 MGy compared to 1-1.5MGy for natural rubber(*6*). There have apparently been no studies of the chemical changes produced in polychloroprene by ionizing radiation.

```
H  H      H
-C - C = C - C -
H        Cl  H
```

trans-1,4

(I)

In the present study, radiation induced changes in the molecular structure of polychloroprene were identified and determined quantitatively by NMR spectroscopy. The mechanisms of the chemical reactions resulting from irradiation were investigated by ESR spectroscopy.

Experimental

Materials and Methods. Polychloroprene (Aldrich Chemical Co.) was precipitated twice from solution in chloroform with methanol and dried in a vacuum oven at 30 ^{o}C (until no trace of solvents could be detected in the NMR spectrum). Samples were sealed in glass tubes at 10-3 Pa and irradiated using a 60Co source with a dose rate of 2.5 kGy/h. The samples were annealed at different temperatures before opening the glass tube in order to remove free radicals prior to opening to air.

Electron Spin Resonance Spectroscopy. The ESR spectra were obtained using a Bruker ER 200D X-Band spectrometer with a variable temperature unit at a frequency of 9.26 GHz with 332.4 mT field centre and 25 mT sweep width. A standard of pitch in KCl (Varian) was used as the reference for absolute radical concentrations. The data were processed using the SIMOPR software package(Garrett R.W., Personal Comunication, 1991). The spectra were observed at 77K and then warmed in progressive stages in order to study the reactions of the radicals. The samples were maintained for 10 minutes at each temperature and then cooled down to 110 K for measurement in order to obtain comparable spectra and to avoid errors in concentration measurements. Photobleaching was performed with an Oriel 1000 W Hg/Xe lamp using either 495 or 530 nm cutoff filters.

Nuclear Magnetic Resonance Spectroscopy. ^{1}H and ^{13}C NMR spectra of the unirradiated polymers were obtained in $CDCl_3$ with a JEOL GX 400 spectrometer operating at 400 MHz for hydrogen and 100 MHz for carbon. The free induction decays (FID) for ^{1}H were accumulated in 8 K data points, spectral width of 4400 Hz, 7.0 μs (90°) pulse and a recycle time of 4 s. For ^{13}C NMR, the FID were accumulated in 32 K data points, spectral width of 22000 Hz, 9.1 μs (900) pulse and a recycle time of 10 s, with gated decoupling.

Solid state ^{13}C NMR spectra were obtained with a Bruker CXP300 or MSL300 spectrometer at 75 MHz. The polymers were crushed under N_2 at 77 K using a cryogenic grinder and packed into a Zirconia rotor with a Kel-F cap and spun at the magic angle at 3-10 kHz. ^{13}C NMR spectra were obtained in the solid state from both ^{1}H dipolar decoupling and ^{1}H-^{13}C cross-polarisation experiments. The dipolar decoupling experiment consisted of a single 900 rf pulse (duration 5 μs) and a recycle time of 5 s with high-power proton decoupling during data acquisition. The cross polarisation experiment was performed with a range of contact times (0.01-15 ms) and a recycle time of 5 s. The Dixon(*7*) pulse sequence(TOSS) was used in some cases to suppress spinning sidebands.

Soluble fraction and swelling ratio measurements. The soluble fractions of samples of polychloroprene after various doses of radiation were determined by stirring each irradiated sample (0.1 g) in chloroform (10 ml) for 24 h. Then the solution was filtered and an aliquot of the filtrate dried to constant weight. The swelling ratio was measured by swelling an irradiated sample (0.1 g) in benzene for 24 h and then weighing the swollen sample. The swelling ratio was calculated using a polymer density of 1.23 g/ml, solvent density of 0.879, solvent molar volume of 89.4 ml/g mol and polymer solvent interaction parameter of 0.47 (*8,9,10,*).

Results and Discussion

Structure of polychloroprene. The solution state ^{1}H and ^{13}C NMR spectra of the purified polychloroprene sample were analysed using the assignments reported by Coleman and Brame(*11*). Chloroprene, like butadiene and isoprene, can polymerize to give different structural units in the polymer, i.e. 1-2 vinyl, 3-4 vinyl, cis 1-4 and trans 1-4. Moreover, the linkages can be head-head (HH), tail-head (TH) or tail-tail (TT). The ^{1}H NMR spectrum of the polychloroprene used in the present study (Figure 1) shows that the structure is predominantly (98.7%) 1,4 and 1.3% 1,2. Out of 98.7% 1,4 structures, 94% are trans and the rest cis. The linkages are mainly TH (65% TH, 17% HH, 12% TT) as shown in the ^{13}C NMR spectrum in Figure 2.

Radical Intermediates. The ESR spectrum recorded at 77K after irradiation at 77 K, is a broad, poorly resolved multiplet of width 15 mT containing a number of component peaks, as shown in Figure 3a. Measurements of the radical yields over the dose range of 0-15 kGy gave a value of G(R)=3.0. The spectrum obtained at higher microwave power, e.g. 1.99 mW compared with 20 μW used in obtaining the spectra shown in Figure 3a, indicated that the spectrum is due to more than one radical.

Photobleaching at 77 K was used to identify the presence of radical ions. Filters with a transmission cut-off of 530 nm or 495 nm were used to eliminate low wave lengths which might result in initiating radical reaction. These photobleaching experiments caused a decrease in the total radical concentration of about 20%. Substraction of the spectra obtained before and after photobleaching (Figure 3) showed that the decrease in concentration resulting from the photobleaching is due to loss of a broad singlet of Hpp 1.6 mT. There was no evidence for the formation of new radicals during photobleaching. The photobleachable singlet spectrum shown in Figure 3c can be attributed to a radical ion and has been assigned to RCl- since the halides have a significant electron affinity(*12*).

The radical concentration decreased linearly to approximately 70 % on warming from 110K to 225K and then decreased more rapidly to 10% on warming to 275K. The radical stability in polychloroprene can be compared with that obtained for polybutadiene over the same temperature range(figure 4). It can clearly be seen that the radicals associated with irradiated polychloroprene are inherently more stable than those associated with polybutadiene. At low temperatures the radical stability can be associated with the difference in Tg,of the polymer which is, higher for polychloroprene. Above Tg there is a marked

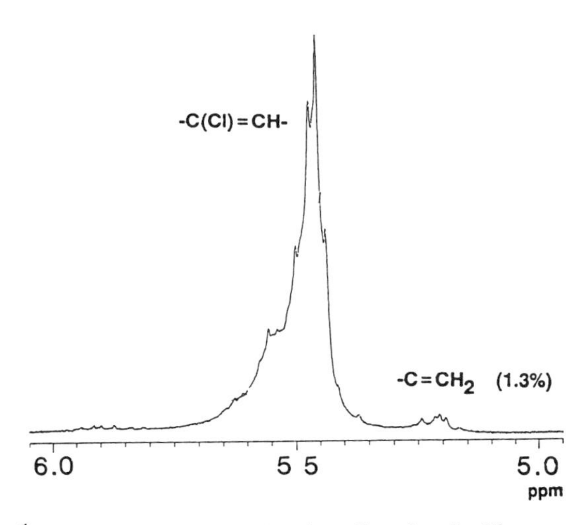

FIGURE 1: ^{1}H NMR spectrum of unirradiated polychloroprene (5-6 ppm region).

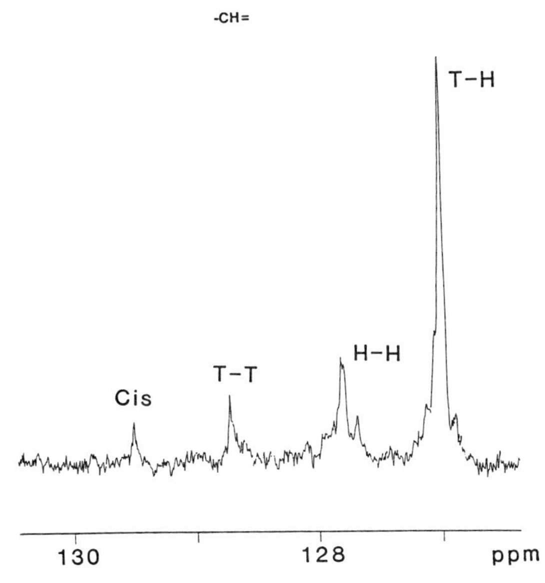

FIGURE 2: ^{13}C NMR spectra of unirradiated polychloroprene (126-130 ppm region) T = tail, H = head.

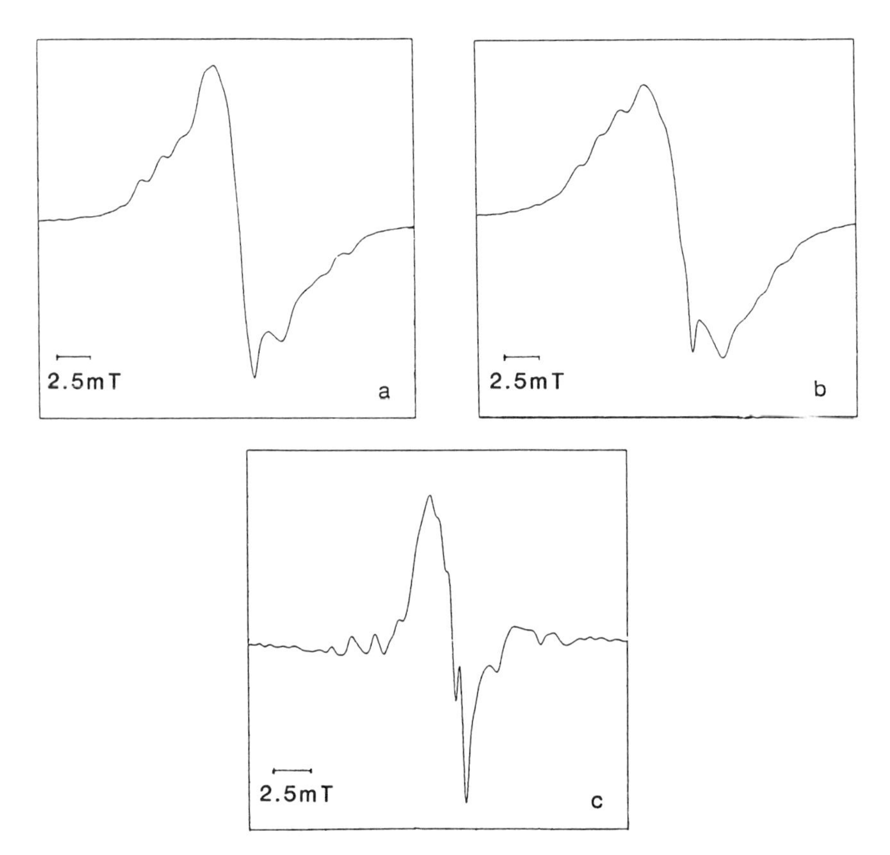

FIGURE 3: ESR spectrum (a) before photobleaching (b) after photobleaching (c) difference between a and b, of polychloroprene irradiated and measured at 77K.

decrease in the stability of radicals associated with the polychloroprene, which we believe results from the greater crosslink density in the irradiated polybutadiene and the associated decrease in radical mobility.

The shape of the ESR spectrum also changed during warming above 77 K. There was an increase in the relative height of the central line from 110 K to 205 K, as shown in Figure 5. Analysis of the spectra obtained on warming of the irradiated polychloroprene indicate that a number of radicals are present at 110K and these radicals decay or are transformed to other radicals on warming from 110K. At all temperatures the ESR spectra are broad due to the presence of several radical species exhibiting some anisotropy. However the identification of the radicals which decay on warming can be made from substraction of ESR spectra observed after warming from 110-160K, 160-225K, 225-275K and above 275K (figure 6). At 110K the ESR spectrum can be analysed in terms of three major radicals, 1) an allyl radical R1 (see scheme 1) associated with a seven line spectrum a_H=1.25mT (g=2.0027) resulting from the abstraction of a chlorine atom from a 1,2 structure. 2) A radical resulting from a hydrogen abstraction reaction giving a $-CH_2-\dot{C}(Cl)-CH-$ radical R2 (see Scheme 1) a_{H1}=4.06mT and a_{H2}=1.56mT (g=2.01) and 3) a radical $-CH_2-\dot{C}=CH-CH_2-$ R3 associated with cleavage of the C-Cl bond in the 1,4 structure a_{H1}=2.5 mT and a_{H2}=0.94 mT (g=2.0027).

On warming from 110-160K the change in the ESR spectrum can be analysed in terms of the decay of radical R1 accompanied by a rearrangement to form R2, which result from the side chain resonance form which can be expected to react at relatively low temperatures.

On warming from 160-225K the radical decay occurs as Tg is reached. The analysis of the subtracted ESR spectrum shown in figure 6b can be made in terms of loss of R2. A similar radical has been proposed in the thermal degradation of polychloroprene(*13*,*14*).

On subsequent warming to 275K the ESR spectrum undergoes a further change, the subtracted spectra in figure 6c being analysed in terms of loss of radical R3 and an allyl radical $-\dot{C}H-CR=CH-CH_2-$ a_H=1.7mT (g=2.0027) which has been observed in polybutadiene and has been shown to be stable above Tg(*15*). On heating above 275K stability of the remaining radicals is greatly enhanced and the singlet (Hpp=1.6 mT,g=2.0027) is assigned to a polyenyl radical.

Structural changes. Crosslinking is the predominant reaction on irradiation of polychloroprene and NMR spectra of irradiated samples cannot be obtained in solution. The solid-state NMR spectra of unirradiated polychloroprene recorded at 298 K with single pulse dipolar decoupling (DD) and cross polarisation (CP) are shown in Figure 7. The resolution in the NMR spectra is much less in the solid state than in solution and details of the microstructure are not observable. The line width in the solid state spectrum (50 Hz), compared with the solution spectra (4 Hz) reflects the decreased motional modulation of the C-H dipolar coupling in the solid state.

A good signal:noise ratio was observed in the CP spectrum of unirradiated polychloroprene in contrast to the spectrum of polybutadiene(*16*). This indicates that the polychloroprene molecules are sufficiently rigid to give efficient cross-polarisation even without crosslinking. This rigidity is due to the high 1,4-trans content of polychloroprene, and is reflected in its higher Tg.

The ratio of the intensity of the peaks due to aliphatic and olefinic carbons in the DD/MAS ^{13}C NMR spectra of polychloroprene was 1.005, which is in good agreement with the 1:1 ratio of these carbons in the polymer. This ratio indicates that the ^{13}C spin relaxation times of the carbons are short and the conditions for quantitative spectra are fulfilled. However, for CP/MAS spectra obtained with a contact time of 0.5 ms, the ratio was 0.85, indicating different

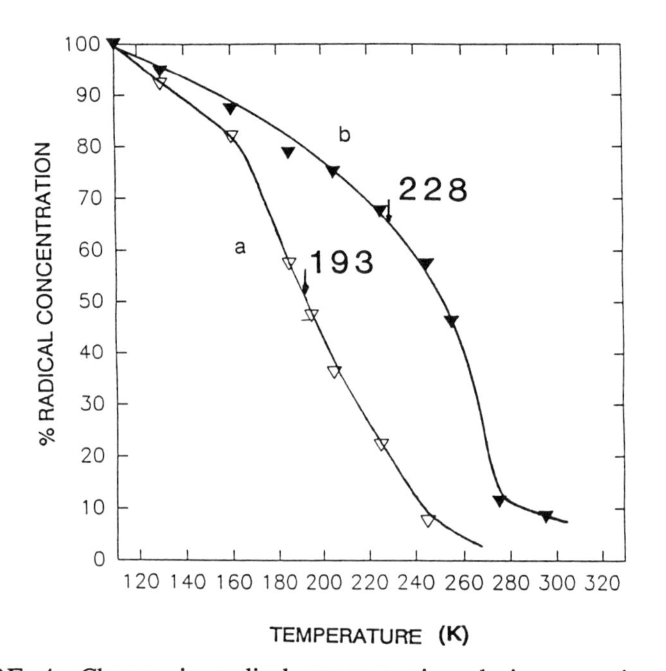

FIGURE 4: Change in radical concentration during warming of (a) polybutadiene (b) polychloroprene.

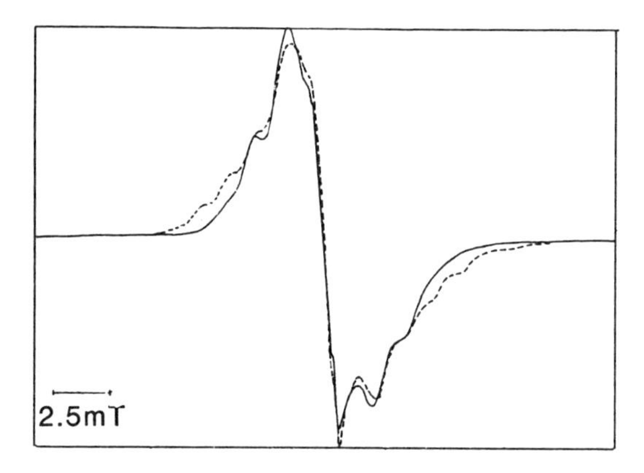

FIGURE 5: Change in the shape of the ESR spectrum of polychloroprene during warming up from (- -) 110 to (—) 225 K.

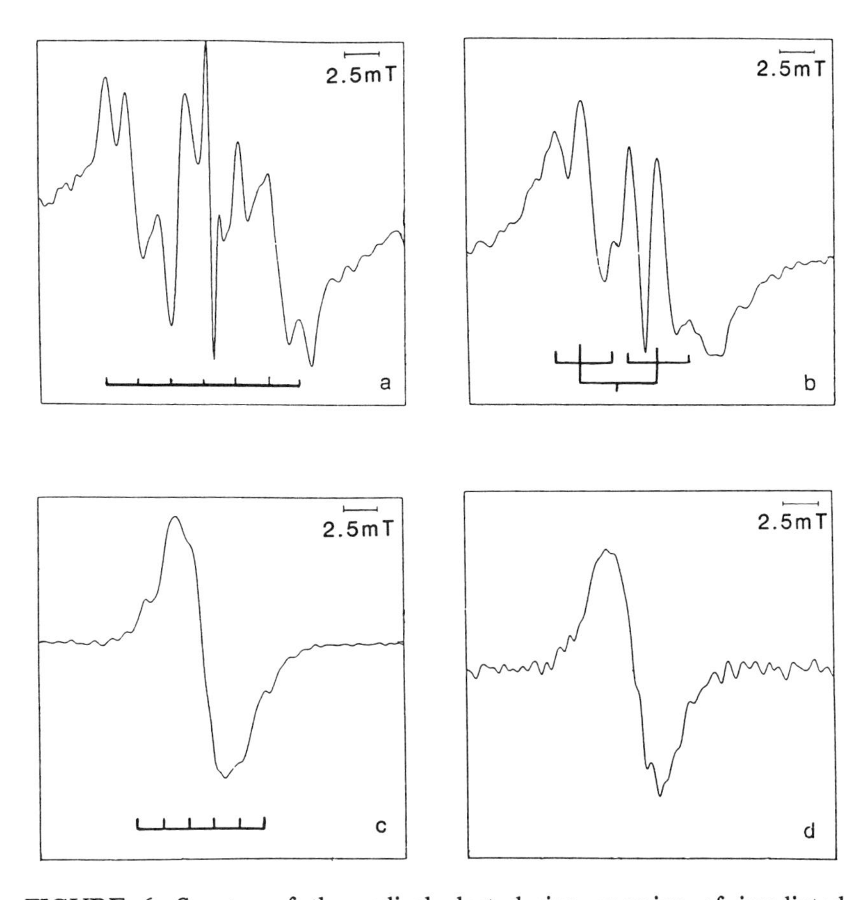

FIGURE 6: Spectra of the radicals lost during warming of irradiated polychloroprene (a) 110-160 K (b) 160-225 K (c) 225-275 K (d) above 275 K.

```
                                                            •
    CH2                        CH2                         CH2
    ||                         ||                          |
    CH         ------->        CH        <------->         CH
    |                          |                           ||
-CH2-CCl-CH2-              -CH2-C-CH2-                -CH2-C-CH2
                                •
                                            R1

    Cl                         Cl
    |                          |
-CH2-C=CH-CH2              -CH2-C-CH-CH2
                \               •/     \
  •              CH2  ------->  CH2     CH2
 CH2-CH=C /                       \    /
                                   CH=C

                                         R2

    Cl
    |                            •
-CH2-C=CH-CH2-   ------->  -CH2-C=CH-CH2-

                                 R3
```

SCHEME 1

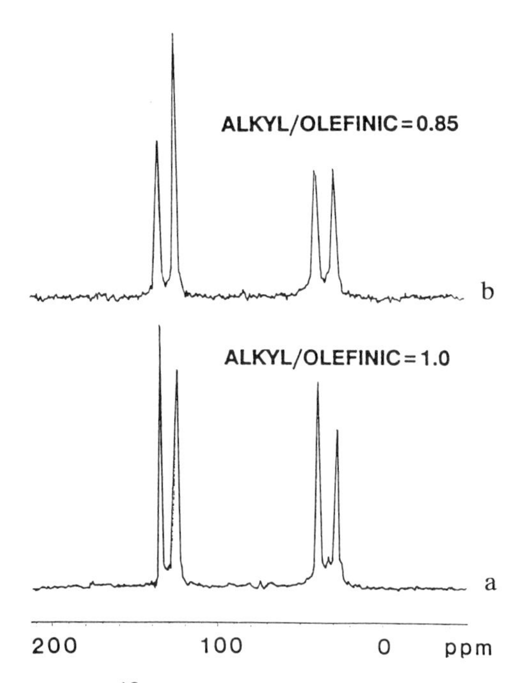

FIGURE 7: Solid-state ^{13}C NMR spectra of unirradiated polychloroprene (a) DD/MAS (b) CP/MAS.

cross-polarisation efficiencies of the olefinic and aliphatic carbons under these conditions.

After irradiation to high doses (up to 9 MGy), the DD ^{13}C NMR spectra showed a decrease in the signal:noise ratio (compare Figures 7a and 8a). It is apparent that the solution and single-pulse solid-state NMR techniques are more sensitive to mobile and uncrosslinked regions of the polymer with relatively short ^{13}C spin-lattice relaxation times, and not to more rigid structures associated with crosslinks. It was necessary to use solid-state NMR with cross polarisation in order to obtain an adequate signal:noise ratio for the irradiated polymer and to observe crosslink structures.

The line width increased from 50 Hz to 200 Hz with increasing radiation dose, due to the increased motional modulation in crosslinked samples. The CP/MAS spectra of polychloroprene irradiated to 1.3, 2.9 and 9.0 MGy shown in Figure 9 illustrates this increase in line width.

With increasing dose, several new peaks appeared in the spectrum, centred at 45, 64, 77,85 and 95 ppm, as shown in Figure 9c. This spectrum was obtained at a spinning rate of 3.0 kHz. However, with increased spinning rates and by using the TOSS pulse sequence (Figure 10) it was shown that the peaks at 77 and 85 ppm in Figure 9 are spinning side bands. A new broad peak does appear at 80 ppm. The possible carbon structures which would give resonances in this region of the spectrum are listed in Table 1.

The peak centred at 45 ppm can be assigned to the methine carbon in a crosslink (*16*). The peak at 80 ppm may be due to a $-CHCl_2$ group or a -C(C,Cl)- group. This peak is assigned to the -C(C,Cl)- group, since it has disappeared in the DD/MAS spectrum. The other possible structure, $-CHCl_2$ (an end group) should show up in the DD/MAS spectra, which is more sensitive to mobile regions of the polymer. The peak at 95 ppm is assigned to the $-CCl_3$ end group since it is also seen in the DD/MAS spectra.

The peaks at 80 and 95 ppm have been confirmed as quaternary carbons by the dipolar dephasing experiment (*20*) shown in Figure 11. Carbon-proton dipolar interactions depend on the inter-nuclear distance, hence coupling of carbons with attached protons is much greater than for carbons adjacent to proton-bearing carbons. Therefore, it is possible to differentiate between protonated and non-protonated carbons by introducing a variable proton decoupling delay. This delay can be selected so that only non-protonated carbons appear in the spectrum (Figure 11b).

Quantitative Measurements. The intensities of the peaks in a cross-polarisation NMR spectrum depend on (1) the rate of cross-polarisation from the ^1H nuclei to near-neighbour ^{13}C nuclei, described by the relaxation time T_{CH}, and (2) the relaxation of the proton magnetisation during cross-polarisation, described by the relaxation time $T_{1\rho(1H)}$. The rate of cross-polarisation (proportional to $1/T_{CH}$) depends on the interaction of static ^1H and ^{13}C dipolar moments. In an amorphous, relatively mobile polymer, such as polybutadiene, molecular motion reduces the interaction of the dipolar moments and cross-polarisation is a slow process. The overall intensities, for a given cross-polarisation contact time, are therefore reduced, as indicated by the low signal:noise ratio in the CP spectra compared to the DD spectra (*16*).

The regular trans-1,4 structure in polychloroprene gives it a higher rigidity than cis or cis/trans polybutadiene, and cross-polarisation is more efficient. This results in CP/MAS spectra with reasonably good signal:noise ratio, even for an unirradiated sample. The high rigidity of polychloroprene molecules compared to polybutadiene is also indicated in the intensity versus contact time plots for the aliphatic carbons in polymers irradiated to low dose (0.44 MGy) shown in Figure 12. During the thermal contact, the ^{13}C magnetisation increases rapidly in an exponential growth characterised by the

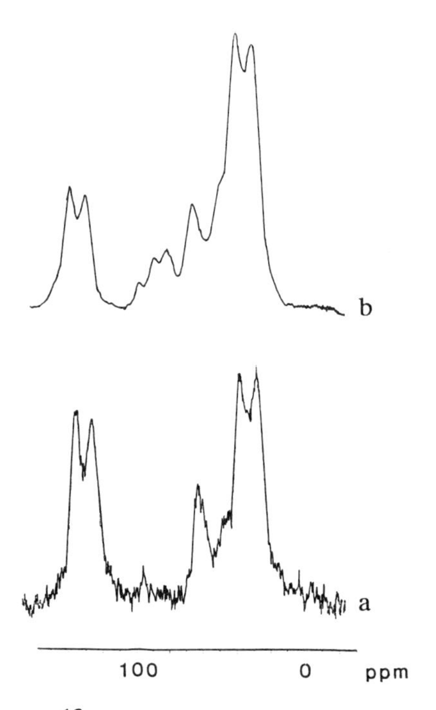

FIGURE 8: Solid-state ^{13}C NMR spectra of polychloroprene irradiated to 9.0 MGy (a) DD/MAS (b) CP/MAS.

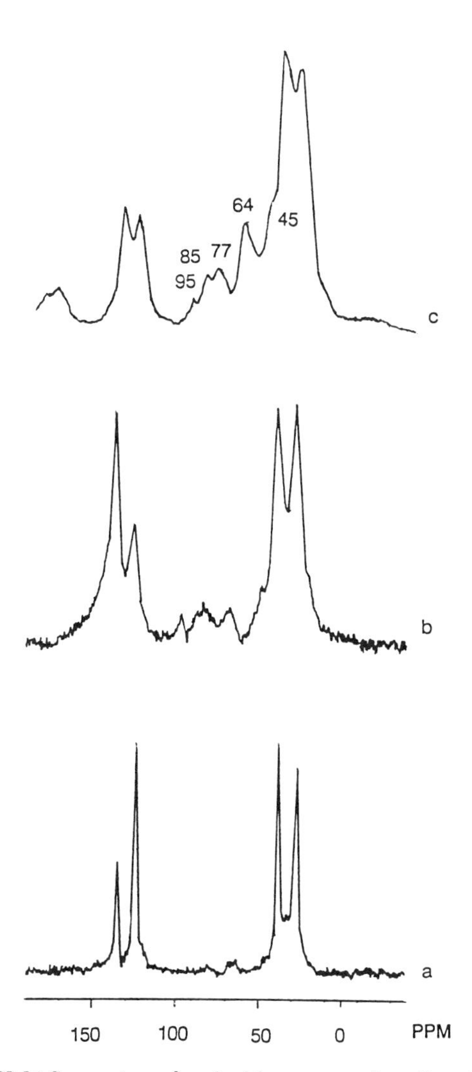

FIGURE 9: CPMAS spectra of polychloroprene irradiated to (a) 1.3 MGy (b) 2.9 MGy (c) 9.0 MGy.

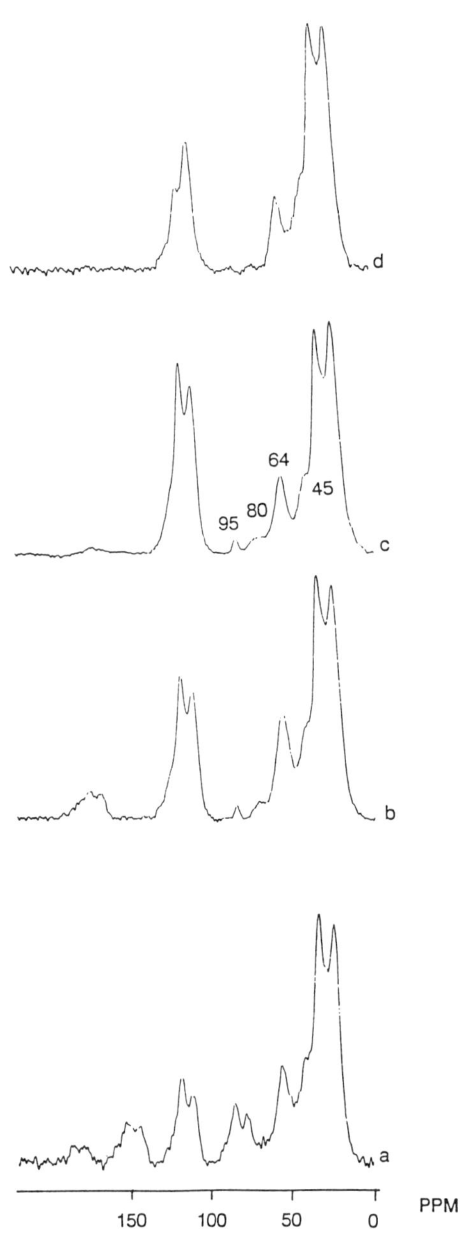

FIGURE 10: Effect of sample spinning rate and the TOSS pulse sequence on the CPMAS spectrum of polychloroprene irradiated to 9.0MGy (a) 3 kHz, (b) 5 kHz, (c) 5 kHz/TOSS and (d) 10 kHz.

Table 1: Assignments for the ^{13}C NMR spectrum of irradiated polychloroprene

Chemical shift (ppm)	Structure	Ref
45	crosslink methine carbon	16
64	-C-(C)(H)(Cl)-C-	17
80	-C-(C)(C)(Cl)-C-	18
	$-\underline{C}HCl_2$	19
95	-C-(C)(Cl)(Cl)-Cl	19

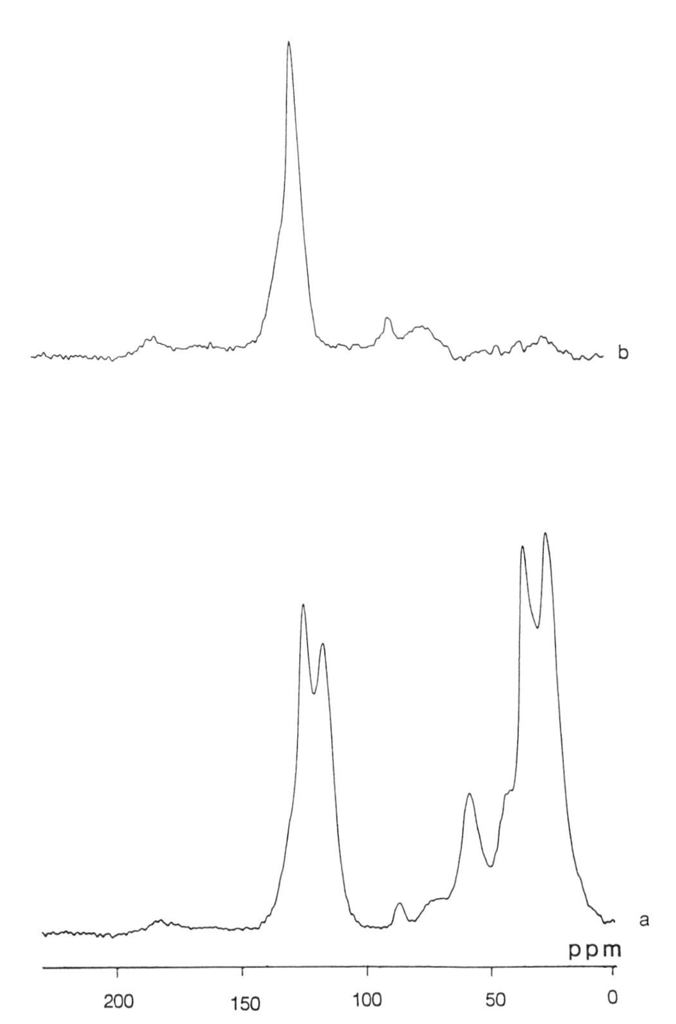

FIGURE 11: Solid-state ^{13}C NMR spectra of polychloroprene irradiated to 9.0 MGy (a) CP/TOSS/MAS and (b) Dipolar dephasing/CP/MAS.

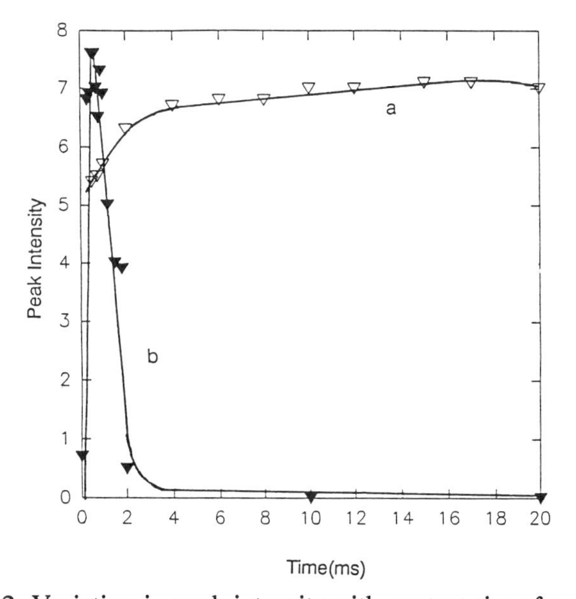

FIGURE 12: Variation in peak intensity with contact time for the olefinic peaks in the CPMAS spectra of (a) unirradiated polybutadiene (ref 16) (b) polychloroprene irradiated to 0.4MGy.

cross-polarisation time T_{CH}, which is governed by the strength of the proton-carbon dipolar coupling. Simultaneously, the proton magnetisation decreases according to the proton spin-lattice relaxation time $T_{1\rho(1H)}$. The intensity-contact time plot shows that the rates of both the initial development of the cross-polarisation and of the decay of the proton magnetisation are higher in polychloroprene than in polybutadiene.

T_{CH} is also dependent on the local concentration of the ^{1}H nuclei around the ^{13}C nuclei being examined. Therefore, the rate of cross-polarisation for olefinic methine carbons is slower than the rate for methylene carbons, and this is the explanation of the observed ratio of the peak intensities of 0.85 compared with the expected value of 1.0. The variation in the intensities of the four peaks with contact time are shown in Figure 13. Both methylene carbons show the same behaviour, and therefore only one is shown.

The exponential rise in intensity is determined by T_{CH} and the decay by the relaxation time $T_{1\rho(1H)}$. The rise in intensity is slowest for the olefinic carbon with an attached Cl atom, due to the absence of hydrogen atoms required for the efficient carbon-hydrogen interaction, determined by T_{CH}. $T_{1\rho}$ relaxation is also dependent on the molecular structure and is most efficient for carbons in rigid environments.

The broad lines in the solid-state ^{13}C NMR spectra of the irradiated polymers were analysed by computer simulation. The position, height, width and line shape parameter of each component peak was varied until an optimum fit was obtained between the simulated and experimental spectra, defined by a minimum sum of the squares of the differences. The simulation of the region between 0 and 100 ppm is shown in Figure 14. Contact time plots, as shown in Figure 13, were fitted to time constants using a Simplex routine (*21*), and the peak intensities were determined by extrapolating the curves to zero contact time. These intensities (concentrations) were used to calculate the G values for crosslinking.

Reaction mechanism. The radiation yields for the different structural changes in polychloroprene are compared with the corresponding values for polybutadiene in Table 2. The yields of radicals are much higher for polychloroprene than polybutadiene. This result is in contrast to previous reports(*22,23*) which indicated similar radical yields for the two polymers.

We have calculated the G values for total volatile products (mainly HCl and H2) from the results reported by Arakawa et al.(*26*). The yield of HCl is similar to the value for G(R) obtained in the present work, suggesting that the radicals are mainly formed by C-Cl bond scission.

Therefore, we propose that the allyl radical is formed by scission of the C-Cl bond in 1,4 structures followed by a hydrogen atom transfer from a neighbouring allylic carbon. An allylic radical, similar to that observed in polybutadiene, should be formed and this is supported by the similar G(-Cl) value obtained by elemental analysis and of the HCl yield by Arakawa et al.(*26*). We have also observed that electron irradiation of polychloroprene in an XPS spectrometer causes a considerable decrease in chlorine content. The G(H2) value for polychloroprene was similar to polybutadiene, hence the formation of allylic radicals by direct C-H bond scission must be much smaller in polychloroprene.

An interesting feature of the radiation yields shown in Table 2 is the low value for G(crosslink) determined by NMR for polychloroprene compared to polybutadiene. The values of G(X) calculated from the soluble fractions and from the swelling ratios are in good agreement with the values derived from the NMR spectra for polychloroprene, whereas the NMR-derived values are much higher than the values derived from soluble fractions for polybutadiene. The difference between the G(crosslink) values from NMR and soluble fractions for

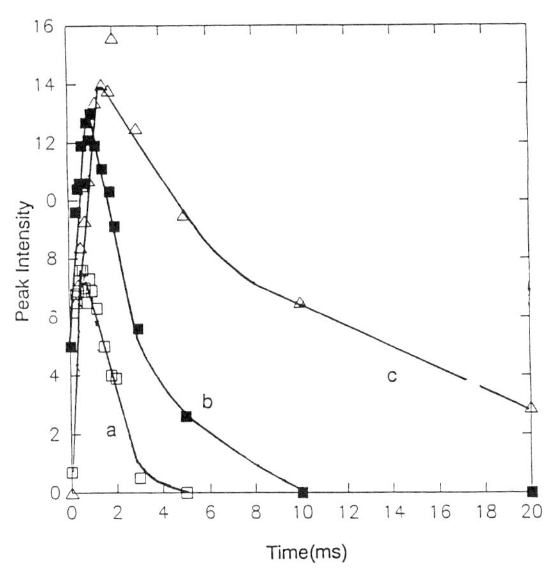

FIGURE 13: Variation in peak intensity with contact time for different peaks in polychloroprene irradiated to 0.4 MGy (a) $-CH_2-$,(b) -CH= and (c) -C(Cl)=.

Table 2: Radiation yields for polychloroprene and polybutadiene

	polybutadiene	polychloroprene
G(R) at 77K	0.43 [a]	3.0
G(total gas)	0.23 [b]	3.5 [d]
G(HCl)	-	3.3 [d]
$G(H_2)$	0.23 [b]	0.2 [d]
G(-Cl)	-	2.8
G(X)		
NMR	41 [c,e]	3.9 [f]
Swelling	4-6 [b]	4.8 [g]
soluble fraction	-	3.2 [g]

a ref 24
b ref 25
c ref 16
d ref 26
e Dose range of (0-1) MGy
f Dose range of (0-3) MGy
g Dose range of (0-0.2) MGy

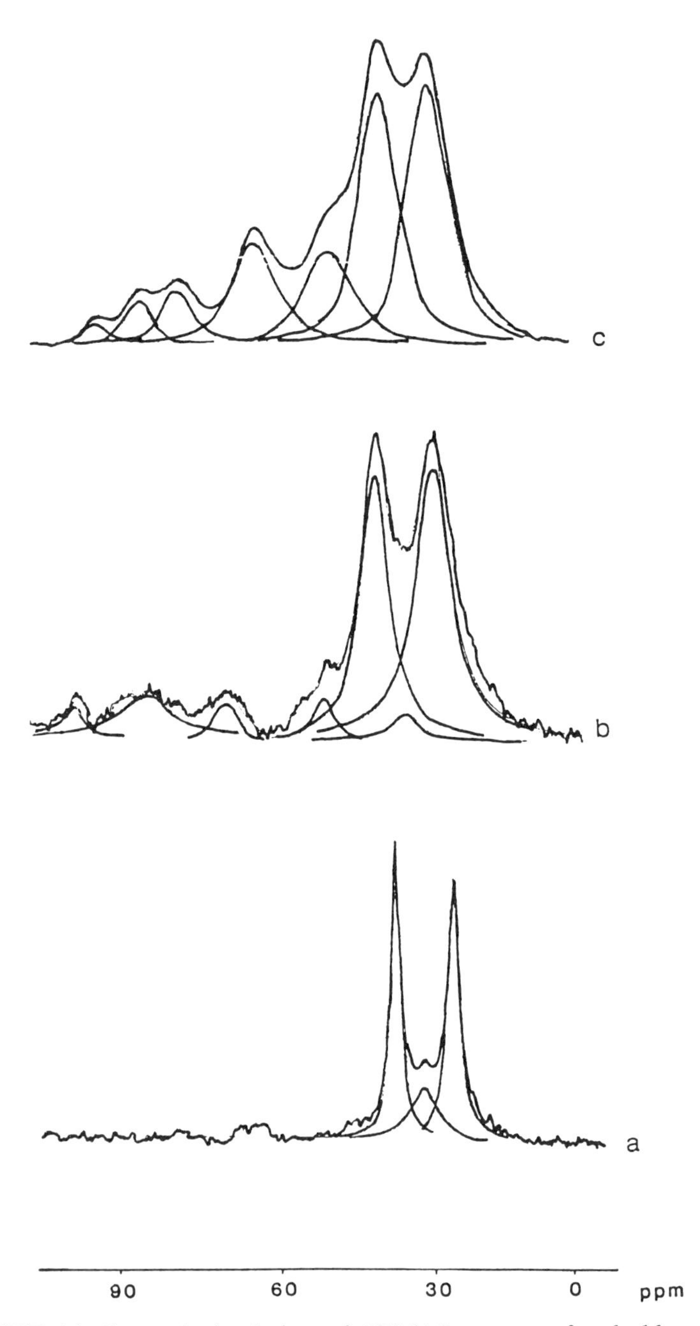

FIGURE 14: Spectral simulation of CPMAS spectra of polychloroprene irradiated to (a) 1.3 MGy (b) 2.9 MGy (c) 9.0 MGy.

polybutadiene have been attributed to clustering of the crosslinks, arising from a kinetic chain reaction for crosslinking (*16*).

The soluble fraction and swelling methods for determining G(crosslink) underestimate the numbers of chemical crosslinks when they are clustered and do not have a random spatial distribution. The agreement between the G(crosslink) values for polychloroprene obtained by the three methods in the present study indicates that the crosslinks must be randomly distributed, and hence that the crosslinking does not occur by a kinetic chain reaction. This difference in behaviour between polybutadiene and polychloroprene may be due to the rigidity of the polychloroprene molecules and their unfavourable spatial arrangement and ability to rearrange for the kinetic reaction.

Conclusions

1. Scission of the C-Cl bond is the main primary reaction occurring in polychloroprene when it is subjected to ionizing radiation.
2. The alkyl radical that is formed during this primary reaction of C-Cl scission undergoes a hydrogen transfer reaction to yield an allylic radical similar to that observed in polybutadiene.
3. The radiation yields for crosslinking of polychloroprene can be measured directly from the solid-state ^{13}C NMR spectra.
4. The agreement between the G(crosslink) values obtained from ^{13}C NMR spectra, soluble fractions and swelling ratios and the low values obtained by NMR compared with polybutadiene, indicate that the crosslinks formed in polychloroprene have a random spatial distribution, and are not formed by a kinetic chain reaction, in contrast to polybutadiene.

Acknowledgments

The authors thank the Australian Research Council (ARC) and the Australian Institute of Nuclear Science and Engineering (AINSE) for supporting their research, the Australian Nuclear Science and Technology Organization (ANSTO) for irradiation facilities and Dr A. K. Whittaker for advice on the NMR spectra. The high-speed NMR spectra were recorded at the Centre for Magnetic Resonance.

Literature Cited

1. Spinks J.W.T.;Woods R.J., *An Introduction to Radiation Chemistry*, John Wiley and Sons, New York,1964,pp 396.
2. Jankowski B.; Kroh J., *J.Appl.Polym.Sci.*,**1969**,*13*,1795.
3. Yamada S.; Hamaya T., *Kobunshi Ronbunshu*,**1982**,*39*,133.
4. Hartney M.A., *J.Appl.Polym.Sci.*,**1989**, *37*,695,.
5. Hill D.J.T., O'Donnell J.H., Perera M.C.S.;Pomery P.J., in preparation,**1992**.
6. Wundrich K., *Radiat. Phys. Chem.*,**1985**, *24*, 503.
7. Dixon W.T., *J.Magn.Reson.*,**1981**, *44*,220.
8. Brandrup J.; Immergut E.H., *Polymer Handbook*, 3rd ed, John Wiley and Sons, **1989**.
9. Barton A.F.M., *CRC Handbook of Polymer Liquid Interaction Parameters and Solubility Parameters*, CRC press, **1990**.
10. Lide D.R., *CRC Handbook of Chemistry and Physics*, 72nd ed., CRC press, **1991**.
11. Coleman M.M.; Brame E.G., *Rubber Chem.Technol.*,**1978**,*51*,668.
12. Symons M.C.R.; Smith I.G., *J.Chem.Soc.,Faraday Trans.1*,**1985**,*81*,1095.
13. Petcavich R.J.,Painter P.C.; Coleman M.M., *Polymer*,**1978**, *19*,1249.

14. Miyata Y.;Atsumi M., *J.Polym.Sci.,Polym.Chem.Ed.*,**1988**,,*26*,2561.
15. Von Raven A.;Heusinger H., *J.Polym Sci., Polym Chem. ed.*, **1974**,*12*,2255.
16.O'Donnell J.H.;Whittaker A.K., *J.Polym.Sci.,Polym.Chem.Ed.*,**1992**,*30*,185
17.Eskina M.V.; Khachaturov A.S.;Krentsel L.B.;Yutudzhyan K.K ;Litmanovich A.D., *Polym Sci., USSR*,**1988**, *30*,159.
18.Makani S., Brigodiot M., Marechal E., Dawans F.; Durand J.P., *J.Appl.Polym.Sci.*,**1984**,*29*,4081.
19.Velichko F.K.,Chukovskaya E.C.,Dostovalova V.I., Kuzmina N.A.; Freidlina R.Kh., *Org. Mag. Reson*,**1975**,*7*,361.
20. Opella S.J.;Frey M.H.,*J.Am.Chem.Soc.*,**1979**,*101*,5854.
21. Nelder, J.A.; Mead R., *Comput. J.*,**1965**, *7*, 308.
22.Kozlov V.T., *Vyskomol Soedin*,**1967**, *9*,515.
23.Kozlov V.T.; Tarasova Z.N., *Vyskomol Soedin*,**1966**,*8*,943.
24. Kozlov V.T., Gurev M.V., Yevseyev A., Keshevskaja N.; Zubov P.I., *Vyskmol Soedin*,**1970**, *12*, 592.
25. Bohm G.G.A.; Tveekram J.O., *Rubber Chem.Technol.*,**1982**, *55*,575.
26. Arakawa K.,Seguchi T.;Yoshida K., *Radiat.Phys.Chem.*, **1986**,*27*,157.

RECEIVED August 14, 1992

Chapter 7

Applications of Radiation Treatment of Ultradrawn Polyethylene

David F. Sangster

Chemistry School, University of Sydney, 253 Connels Point Road, Connels, New South Wales 2221, Australia

High density polyethylene drawn at high draw ratios to orient the fibres and molecular chains possesses such desirable properties as high tensile strength and low creep rates under load. These properties can be enhanced as much as three fold by gamma irradiation in the presence of the bifunctional crosslinking agent acetylene. Acetylene penetrates readily into polyethylene fibres of diameter 1mm but takes time to be effective with more massive rods of material. The reaction mechanisms leading to the improvement are discussed.
Suggested uses for this material include applications where the low density of polyethylene would be an advantage such as rock bolts used in mining, reinforcement of concrete, light weight high strength ropes and high performance fabrics.

For many years the crosslinking of polymers – and polyethylene has been the important one – has been the major industrial use of high energy ionising radiation. There are hundreds of electron accelerators located in several countries throughout the world being utilised commercially to treat the insulation of electrical cable or to produce foam or shrinkable material or many other products. The annual output amounts to several million tonnes.

It should be noted that no detectable radioactivity is produced in the material being irradiated by electron beams or by x–rays at the energies being used or by gamma radiation from radioactive sources and the means of detection are very sensitive indeed.

This presentation is concerned with the combination of the use of high density polyethylene, drawn to give a high orientation and gamma irradiated in the presence of acetylene in order to improve the mechanical properties. It encompasses a project (*1*) to determine the feasibility of processing large cross section (25 mm diameter) rods in this way.

0097–6156/93/0527–0095$06.00/0

Drawn Polymers

Drawing and the related technique, extrusion, are important processes. Many, perhaps most, synthetic polymer products are processed by being either drawn or extruded or treated by such extensions of these operations as blowing, rolling, pressing or spinning or extrusion into a mold. Examples range from the barrels of cheap ball point pens to tubing and piping and from plastic bags and sheets to fibres. In many cases the processing confers properties superior to that of the raw material. Tensile testing itself is an example of drawing. When the testing extends to the break point in order to determine the limits which are allowable in use, the draw ratio exceeds the optimum which would be required in industrial manufacture. A considerable improvement in the tensile strength of polyethylene can be achieved by drawing the material.

Because it can be made with a high degree of crystallinity with fewer imperfections within the crystals, high density linear polyethylenes can achieve a much higher strength than can branched material. The undrawn material is spherulitic and there is amorphous material between the crystalline regions. Drawn polyethylene is morphologically a different material from undrawn polyethylene. Under the stress of drawing, the molecular strands in the amorphous regions are pulled along the axis of draw and later the cyrstals become aligned in the same direction resulting in a microfibrillar structure. The tie molecules – those bridging two crystalline region and having one end in each – straighten out, unfold from the lamellar structure or pull out and may take some of the load. Finally the tie molecules may break, the amorphous regions may neck down and intracrystalline slippage may occur and breakage results.

Long before the breakpoint is reached the maximum stress/strain ratio (modulus) is achieved. This is dependent on the rate of drawing and the temperature and is usually characterised by the draw ratio which can be equated to the ratio of initial cross–section of the material to its final cross–section. Since extrusion or casting from the melt requires an elevated temperature and there is a temperature rise due to the work of the drawing, antioxidant is usually added because oxidation has a deleterious effect on the properties. The antioxidant also reduces the oxidation of the radicals which are formed during drawing and others which remain after irradiation. Drawing at temperatures above ambient is an advantage in promoting crystallinity and in annealing defects. The resulting material has greatly increased tensile strengths along the axis of draw without greatly affecting the properties in the perpendicular directions.

Extensive studies on the drawing of high density polyethylene have been carried out by Ward and his co–worker (*2–8*) at Leeds University. They have investigated the mechanical properties of the product as a function of draw ratio, drawing speed, temperature and the material being drawn. They have carried out a mathematical analysis of the process and the mechanism of failure and have been able to distinguish the independent processes operating.

A different approach to making high modulus polyethylene was taken by Keller (*9*) at Bristol University and Pennings (*10*) and Lemstra (*11*) at the Dutch State Mines and their co–workers. They took either a dilute solution of high molecular weight material or one so concentrated as to form a gel and slowly removed a thread allowing time for the chains to line up and crystals to be formed. The thread is subsequently drawn. A commercial process based on similar principles is used by Allied Chemical in the United States of America. In this way very high strength material can be made. The strength of polyethylene as a single crystal can be calculated from the bond strength. The modulus should be 250–350 Giga pascals (*12*). Using great care a figure of 220 GPa has been achieved by spinning from a gel but ~ 100 Gpa is more usual. Hydrostatic extrusion has been reported to give material with a modulus of 50–60 GPa (*12*).

Irradiation of Polyethylene

It was shown by Dole (*13*) and by Charlesby (*14,15*) that polyethylene crosslinked when irradiated by high energy ionising radiation. The most dramatic effect was an improvement in thermal properties but increased tensile strength was also observed. Under prolonged irradiation the material became brittle.

The process provided a facile alternative to the difficult and sometimes wasteful process of chemical crosslinking which is used to enhance the performance of the insulation of electrical cables in high temperature environments or under conditions of temporary electrical overload. It was also shown that a "memory" effect could be obtained in irradiated polyethylene. If after irradiation the material were heated, deformed and cooled it retained a memory of what shape it was when it was irradiated. On subsequent heating without a restraint it reverted to that shape. This has been applied in the manufacture of shrinkable film and tubing and a variety of fixing devices. Even above its melting point irradiated polyethylene has a consistency and integrity which can be utilised in the manufacture of fine textured foams. Subsequently processes have been developed for radiation crosslinking the poly(vinyl chloride) insulation of telephone cables and the "green" rubber used in automobile tyre lay-ups. These all represent sizeable industries and in addition there are other minor applications. Investigations of the extent of crosslinking in samples covering a range of crystallinities showed that crosslinking occurred predominantly in the amorphous regions. Radiation is absorbed randomly in the sample. Most of the "lesions" formed in the crystalline region are considered to migrate to the crystal surfaces. The distance apart of carbon chains in a perfect alkane crystal is 0.41 nm whereas the carbon–carbon distance in a crosslink is 0.154 nm. Crosslinking within a crystal is thus impossible except perhaps at an imperfection. Another possible outcome of energy absorption in the crystalline region is the formation of double bonds as has been demonstrated in alkane model compounds (*16*). No gel and thus no crosslinks are formed in an irradiated mat of single crystals of polyethylene unless the crystals are firmly pressed together. Thus the loops at the ends of cyrstals must be brought into contact before crosslinking can occur (*17*). The uneven distribution of crosslinks within the amorphous regions – some are concentrated at the crystal boundaries – leads to less improvement than would be expected from a statistical distribution of the same number of crosslinks because the deficient regions exhibit greater strain at a lower stress.

The concentration of crosslinks in the amorphous regions provides an explanation of the "memory" effect. On heating, the crystalline regions melt so that the amorphous regions take most of the strain when the material is deformed by an applied force. When cooled the deformed shape is held as the crystalline regions solidify locking in the strains. On subsequent warning they melt again and the strains in the amorphous region pull the material back into its original shape provided there is no constraint and the crystalline region again solidifies.

Main chain scission occurs in parallel with crosslinking in all polymers. Main chain scission produces a deterioration in physical properties. The extent of each of these two processes and which one predominates depends on both the chemical nature and on the structure of the material. Scission within a crystal is unlikely since the two ends remain in close proximity. On the other hand, rupture of a stressed tie molecule is less likely to heal and causes weakness at a vulnerable location.

Fibres melt–spun from higher molecular weight material do not give the expected increase in creep resistance upon irradiation. One factor may be that chain scission is more catastrophic than for lower molecular weights (*18*). As a first step in understanding this and developing techniques for usefully modifying the properties, Albert, Busfield and Pomery (*19*) have examined the decay of alkyl radicals and the resultant alkyl and polyenyl radicals in ultra–high modulus polyethylene as a function of draw ratio. A draw ratio of ~ 9 appears to be the optimum.

Additives

Oxygen is soluble in polyethylene and diffuses through it – certainly through the amorphous regions. Oxygen reacts readily with the free radicals formed by the absorption of radiation energy sometimes initiating a short chain reaction involving neighbouring molecules. Oxygen appears to interfere with the crosslinking process and with the healing of chain scission events. In both cases the net result is a deterioration in physical properties. Often a brown colour develops in irradiated material and the depth of this colour is greater in the more highly crystalline material (*20*). Because of the effect of oxygen and peroxy compounds in ageing and on long term performance there have been many investigations. Certainly oxygen must be removed before irradiation and trapped radicals in irradiated material should be annealed out to minimise post–irradiation reactions.

Crosslinking agents have been added in order to enhance the crosslinking and minimise the extent of main chain scission and thus use a lower dose to achieve the same strength. The hope has been that some scission sites could become crosslinking sites. The ideal would be a substance that was compatible with polyethylene in the amorphous regions and at the crystal surfaces but did not interfere with the perfection, morphology and extent of the crystalline regions.

Acetylene

At the 9th Japan Conference on Radioisotopes in 1969 Mitsui, Hosoi and Kagiya (*21*) reported on the use of acetylene (with and without the addition of fluorine compounds such as carbon tetrafluoride) as a crosslinking agent during the irradiation of polyethylene film. Further details were published in 1972–7 (*22–26*) and several patents were taken out (*27*). These and subsequent workers noted a small increase in weight during irradiation whereas material irradiated in vacuum lost weight due to the evolution of gaseous products. The small increase has been taken to be due to the incorporation of acetylene as a single molecule event in contrast to the grafting of a polymer chain. However, Appleby (*28*) has considered that acetylene may act as a catalyst and not be incorporated. In 1976 Grobbelaar *et al (29–32)* reported investigations on the use of acetylene in radiation crosslinking of blocks of polyethylene to be used as prostheses in knee joints. They found that the material was crosslinked at the surface to a depth of 0.3 mm. This conferred superior abrasion– and wear– resistant properties but left the interior with some resilience–desirable properties in such an application. They rejected the addition of fluorine containing compounds for material to be used within the body. Acetylene would not penetrate a highly crystalline high molecular weight material except where it was sufficiently deformed at the surface of the block to provide some amorphous material in which crosslinking could occur.

Busfield joined Woods and Ward and established conditions under which irradiation in the presence of acetylene could improve the mechanical behaviour of ultra high modulus polyethylene fibres (*33–35,18,36–37*). Although there was a marked reduction in ductility there were improvements in creep properties and in fatigue cycling. Appleby and Busfield (*38–40*) investigated the crosslinking of polypropylene tapes in the presence of acetylene and found a worthwhile improvement in the mechanical properties such as high temperature creep. The morphology of the material is a more important factor than is the tacticity. Exposure to acetylene after irradiation in a vacuum was almost as effective as having acetylene present during irradiation.

Large cross–section rods

When Sangster and Barry (*1*) started their work on solid polyethylene rods of diameters 10 mm and 25 mm there existed no reports on the successful acetylene – enhanced radiation crosslinking of massive polyethylene. Research at the BHP Melbourne Research Laboratories had determined that the optimum billet preparation conditions for medium molecular weight Hoechst G polyethylene incorporating an antioxidant were to extrude into a heated cylindrical mold held at 130°C for 2h followed by cooling at 10°C/h to room temperature. After machining to size the material was drawn at 100°C at 50 mm/min giving a draw ratio of 14. The product certainly did not look like normal translucent polyethylene and the crystallinity as calculated from the melting endotherm in a differential scanning calorimeter was 85%. This material was strong enough (initial modulus 22 GPa at a strain rate of 6.4 x $10^{-2}s^{-1}$) for the projected use of replacing steel rock bolts for wall or roof stabilisation in mines but the creep properties were unsuitable for long term use.

It was reasoned that such a high crystallinity would slow down but not prevent removal of dissolved oxygen from and access of acetylene to the amorphous regions buried deep in the interior of the rods. It was therefore decided that both evacuating and exposure to acetylene should be prolonged to seven days. No attempt was made to determine how long was really necessary. However, if a sample was evacuated for two hours and then closed off, there was a noticeable build up of gas, presumably oxygen, overnight in the vessel but not subsequently. The sample was flushed several times to remove traces of oxygen and finally left in acetylene for seven days. After irradiation at ambient temperature in a cobalt–60 gamma source the samples remained in acetylene for a further seven days before opening. No attempt was made to remove any remaining free radicals by annealing. The samples had a light brown appearance which could indicate some oxidation. Often there was a white deposit of "polyacetylene" on the walls of the glass irradiation vessel. Earlier it had been established that the initial tensile modulus measured at the moderately high strain rate of 6.4 x $10^{-1}s^{-1}$ gave a good rapid indication of the creep properties as determined over several months. Presumably the stretch mechanisms of early extension at this strain rate are similar to those which obtain at very low strain rates. A maximum initial tensile modulus of 47 GPa was found at a dose of 170–180 kGy (*1*). This was judged to be adequate for the intended use of the material. It should be noted that the presence of an antioxidant in the material to prevent deteriorating during the preparation of the samples required that a higher dose be given to produce the maximum effect.

The gel content which was taken as a measure of the extent of crosslinking increased to ~ 40% at ~ 60 kGy showing that a network had been formed. Thence it increased more slowly to ~ 55% at 200 kGy. The change at ~ 60 kGy indicates at least two different mechanisms dominating before and after 60 kGy. This is also reflected in the Charlesby–Pinner plot but since that treatment can be applied strictly only when the starting material is homogeneous and has a uniform distribution of molecular weights one can not draw any quantitative conclusions from that relationship. Both the melting temperature and crystallinity decreased and the initial tensile modulus increased slightly up to ~ 60 kGy and then did not change much after 100 kGy. The decrease in crystallinity from the initial 85% to 75% at 55 kGy indicates some crystal degradation and this presumably supplies material in which crosslinking can occur. Cross linking and degradation then appear to be in balance until beyond 120 kGy when the melting temperature and initial modulus increased to reach peaks at ~ 175 Gy at which dose the gel content is 50%. The melting point, crystallinity and tensile modulus then fall while the gel content keeps increasing. This has been taken as indicating that chain scission is now the dominant effect. The temperature at the onset of crystallisation may provide a better measure of the extent of radiation damage to the polymer chains (*41*).

It should be noted that beyond ~ 30 kGy the gel content exceeded the non-crystalline component. This could mean either that enthalpy-of-melting measurements are not a perfect indication of crystallinity or that some of the melted crystalline material is being trapped in the gel (crosslinking of tie molecules or crystal end loops is a possible mechanism) and gel content is not a true measure of the extent of crosslinking. There is no method of absolute measurement of the extent of crosslinking - C13 NMR techniques are not yet sensitive enough (*17,41-42*).

One can envisage there being a limited number of effective crosslinking sites but an explanation is required for the decrease in strength and other parameters at doses beyond the peak while the three dimensional network, as indicated by the gel content, still keeps on increasing. Tabata and his coworkers (*16*) have shown that, while no crosslinks are formed in the interior of irradiated crystals of model alkane compounds, double bonds are. This could lead to imperfections both at the long axis surfaces and within the interiors of the crystals and these might play a part in the decrease in the tensile modulus.

Future Work

These studies demonstrated the feasibility of producing large cross-section materials using a combination of high density highly oriented material treated by irradiation in the presence of acetylene. Further work would be required to develop a commercial process and greater enhancements should be possible.

- Polyethylene treatment. The grades of material and its subsequent treatment were arrived at to give the optimum for unirradiated material. The requirements for the optimum performance from the process as a whole could be different.
- Safety considerations. Acetylene is not considered to be a safe material to handle in large quantities. One alternative would be to irradiate the polyethylene in a vacuum and subsequently treat with acetylene. Another would be to use a mixture of acetylene and a halogen compound which would not only increase the effectiveness of the acetylene but might also stabilise it against hazardous decomposition.
- Other crosslinking agents. There is a range of other crosslinking agents which might be used. Careful consideration of the right degree of compatability could be important.
- Times required. Shorter times for exposure to vacuum and to gases should be equally effective.
- Temperatures. Evacuating and gas impregnating at higher temperatures could further reduce the times required and irradiation at 80–90°C might be more effective (*25,34*).
- Order of operations. Irradiation below the gel point before drawing may make drawing a more difficult process but could produce a better product.
- Specialised processes. Irradiation under high pressure has been claimed to increase the crosslinking yield (*44*).

Conclusion

The combination of drawing, additive cross linking agent and irradiation gives a product of high tensile modulus, good creep properties, lightness in weight, readily handled, high corrosion resistance and inertness and reasonable temperature properties. The product should find applications as reinforcing tendons, fabrics and ropes, load bearing members in tension and in prostheses. This remains a field in which fundamental research is still needed to elucidate many of the mechanisms which are operating.

Acknowledgements

It was a pleasure to collaborate with Dr D.B.Barry on the rock bolt project. Thanks are due to Dr W.K. Busfield for providing details of his work prior to publication.

Literature Cited

1. Sangster, D.F.; Barry, D.B. *J.Appl. Polym. Sci.* **1991,** *42*, 1385–93.
2. Capaccio, C.; Ward, I.M. *Polymer* **1974**, *15*, 233–238.
3. *Structure and Properties of Oriented Polymers;* Ward, I.M.; Ed.; Applied Science Publishers Ltd: London, 1975.
4. Wilding, M.A.; Ward, I.M. *Polymer* **1978**, *19*, 969–976.
5. Wilding, M.A.; Ward, I.M. *J. Mater. Sci.* **1984**, *19*, 629.
6. Ward, I.M.; Wilding, M.A.; *J. Poly. Sci. Polym. Phys. Ed.* **1984**, *22*, 561–575.
7. Ward, I.M. In *Advances in Polymer Science*; Springer–Verlag: Berlin, 1985; vol 70, pp. 1–70.
8. Richardson, A.; Parsons, B.; Ward, I.M. *Plast. Rubber Proc. Appl.* **1986,** *6*, 347.
9. Keller, A. *J. Polym. Sci., Symposia* **1977**, *58*, 395.
10. Pennings, A.J.; Smook, J.; de Boer, J.; Gogolewski, S.; van Hutten, P.F. *Pure and Applied Chem.* **1983**, *55*, 777–798.
11. Smith, P.; Lemstra, P.J. *J. Mater. Sci.* **1980**, *15*, 505–514.
12. Busfield, W.K. *Chemistry in Australia* **1985**, *52*, 66–70.
13. References in *The Radiation Chemistry of Macromolecules*; Dole M., Ed.; Academic Press: New York, **1972**; Vol. 1.
14. Charlesby, A. *Proc. Roy. Soc (London)* **1952,** *A215,* 187.
15. Charlesby, A. *Atomic Radiation and Polymers,* Pergamon: New York, **1960.**
16. Seguchi, T.; Arakawa, K.; Tamura, N.; Katsumura, Y.; Hayashi, N.; Tabata, Y. *Radiat. Phys. Chem.* **1990**, *36*, 259–266. Seguchi, T.; Katsumura, Y.; Hayashi, N.; Hayakawa, N.; Tamura, N.; Tabata, Y. *Radiat. Phys. Chem.* **1991,** *37*, 29–35.
17. Keller, A.; Ungar, G. *Radiat. Phys. Chem.* **1983**, *22*, 155–158.
18. Klein, P.G.; Woods, D.W.; Ward, I.M. *J Polym. Sci. Polym. Phys. Ed.* **1987,** *25,* 1359–1379.
19. Albert, C.F.; Busfield, W.K.; Pomery, P.J. *Polym. Int.* (accepted).
20. Hikmet, R.; Keller, A. *Radiat. Phys. Chem.* **1987**, *29*, 15–19.
21. Mitsui, H.; Hosoi, F.; Kagiya, T. *Abst. of 9th Japan Conference on Radioisotopes* **1969**, 203.
22. Mitsui, H.; Hosoi, F.; Kagiya, T. *Polym. J.* **1972**, *3,* 108.
23. Hagiwara, M.; Tagawa, T.; Tsuchida, E.; Shinohara, I.; Kagiya, T. *J. Polym. Sci. Polym. Lett.* **1973**, *11*, 613–617.
24. Hagiwara, M.; Tagawa, T.; Tsuchida, E.; Shinohara, I.; Kagiya, T. *J. Macromol. Sci., Chem.* **1973**, *7*, 1591–1609.
25. Mitsui, H.; Hosoi, F; Kagiya, T. *Polym. J.* **1974**, *6*, 20–26.
26. Kagiya, T.; Yokayama, N. *Bull. Inst. Chem. Res., Kyoto Univ.* **1977,** *55,* 11–19.
27. Kagiya, T.; Mitsui, H.; Hosoi, F. *Japan Pat. 48034954*, **23rd May 1973**; *50063046*, **29 May 1975** and others.
28. Appleby, R.W. PhD Thesis; Griffiths University: Nathan Queensland, **1990.**

29. Grobbelaar, C.J.; Marais, F.; Lustig, A. *Proc. Sym. Radiat. Process.*; At. En. Board, Pretoria **1976**, 10–13.
30. Du Plessis, T.A., Grobbelaar, C.J.; Marais, F. *Radiat. Phys. Chem.* **1977**, *9*, 647–652.
31. Grobbelaar, C.J.; Du Plessis, T.A.; Marais, F. *J. Bone Jt. Surg., Br.* **1978**, *60B*, 370–374.
32. Du Plessis, T.A. *S.-Afr Tydski. Natuurwet. Tegnol.* **1984**, *3*, 16–21.
33. Woods, D.W.; Busfield, W.K.; Ward, I.M. *Polym. Commun.* **1984**, *25*, 298–300.
34. Woods, D.W.; Busfield, W.K.; Ward, I.M. *Plast. Rubber Proc. Appl.* **1985**, *5*, 157–164.
35. Woods, D.W.; Busfield, W.K.; Ward, I.M. *Plast. Rubber Proc. Appl.* **1988**, *9*, 155–161.
36. Klein, P.G.; Brereton, M.G.; Rasburn, J.; Ward, I.M. *Makromol. Chem. Macromol. Symp.* **1989**, *30*, 45–56.
37. Klein, P.G.; Gonzalez-Orozco, J.A.; Ward, I.M. *Polymer* **1991**,*32*, 1732–1736.
38. Appleby, R.W.; Busfield, W.K. *Polym. Commum.* **1986**, *27*, 45–46.
39. Busfield, W.K.; Appleby, R.W. *Br. Polym. J.* **1986**, *18*, 340–344.
40. Appleby, R.W.; Busfield, W.K. *Materials Forum* **1987**, *13*, 21–25.
41. Kamal, I.; Finegold, L. *Radiat. Phys. Chem.* **1985**, *26*, 685–691.
42. Tabata, M.; Sohma, J.; Yokota, K.; Yamoaka, H.; Matsuyama, T. *Radiat. Phys. Chem.* **1990**, 36, 551–558.
43. Sohma, J.; Qun, C.; Yuanshen, W.; Xue-Wen, Q.; Shiotani, M. *Radiat. Phys. Chem.* **1991**, 37, 47–51.
44. Milinchuk, V.K.; Klinshpont, E.R.; Kiryukhin, V.P. *Radiat. Phys. Chem.* **1986**, *28,* 331–334.

RECEIVED August 14, 1992

Chapter 8

Role of Homopolymer Suppressors in UV and Radiation Grafting in the Presence of Novel Additives

Significance of Processes in Analogous Curing Reactions

P. A. Dworjanyn, J. L. Garnett, M. A. Long, Y. C. Nho, and M. A. Khan

Departments of Chemistry and Industrial Chemistry, University of New South Wales, Kensington, New South Wales 2033, Australia

Recently developed novel additives for accelerating grafting reactions initiated by ionising radiation and UV are summarised. These include mineral acids, inorganic salts like lithium perchlorate and polyfunctional monomers, also combinations of these. Using the grafting of styrene in methanol to cellulose/polyolefins as model systems, homopolymerisation is shown to be a competing detrimental reaction in the presence of these additives. Inclusion of salts such as copper and ferrous sulfates retard homopolymerisation and lead to unexpected enhancement in grafting under some experimental conditions. A novel mechanism involving partitioning of reagents between monomer solution and substrate is proposed to explain both the accelerative effect of additives in grafting and also the role of salts in the retarding of homopolymerisation. The importance of this work in the related fields of EB and UV curing including both free radical and cationic processes is evaluated particularly the mechanistic role of multifunctional acrylates as common components in both processes. The significance of concurrent grafting during curing is discussed, particularly its relevance to commercial processing involving recycling of radiation cured products.

The use of novel additives for enhancing yields in grafting reactions initiated by ionising radiation and UV is of interest both practically and fundamentally *(1-5)*. Mineral acids *(1)*, inorganic salts *(6)* and polyfunctional monomers (PFMs) have been used for this purpose *(7)*. Synergistic effects involving acids and salts with PFMs, particularly multifunctional acrylates (MFAs) and methacrylates (MFMAs) have also been reported *(7)*. A variety of backbone polymers and monomers have been investigated in these studies *(1,7-10,11)*. A novel mechanism to explain the behaviour of these acid and salt additives involving partitioning effects has been developed *(12)*. This work is of value in a preparative context since, in the presence of additives, lower doses of radiation and UV are required to achieve a particular percentage graft. Thus the additive technique is particularly useful for grafting reactions involving radiation sensitive backbone polymers typically cellulose, the polyolefins and PVC, and also monomers of low reactivity.

In all of these additive studies homopolymer formation remains a significant competing detrimental reaction. Methods for overcoming this problem involve the use of

0097–6156/93/0527–0103$06.00/0

copper, iron *(12-15)* and ceric salts *(16)*. Such compounds generally reduce both homopolymer and grafting yields, the former preferentially, thus overall there is an increase in grafting efficiency in the presence of these homopolymer suppressors. However under these conditions, higher doses of radiation are generally required to achieve a particular percentage graft and compensate for the lower yields in the presence of the suppressors. In the present paper the effect of these inorganic salts as homopolymerisation suppressors in the presence of acid, salt and PFM additives is reported using the grafting of styrene in methanol to both polypropylene and cellulose as model systems. UV and ionising radiation are used as initiating sources. Under some conditions, an actual enhancement in grafting yield with homopolymer reduction is observed. From the results, a refinement in the original partitioning mechanism to explain the additive effect is proposed. In addition the partitioning concept is extended to include an explanation of the role of homopolymerisation suppressors in these grafting processes. The possible significance of the work in analogous radiation curing processes including cationic systems is discussed.

Experimental

The procedures used for grafting were the following modifications of those previously reported *(1)*. For the irradiations, polypropylene film (isotactic, 0.10 cm thickness, 5.0 x 3.0 cm) or strips of cellulose (Whatman No 41 double acid washed chromatography filter paper) of comparable size were fully immersed in the monomer solutions and irradiated in either a 1200 Ci cobalt 60 source or with a 90 W medium pressure mercury vapour lamp. Homopolymerisation was determined by previously published methods *(7,12)*. Percent grafting efficiency was then estimated from the ratio of graft to graft plus homopolymer x 100. For the radiation curing experiments, appropriate resin mixtures containing oligomers, monomers, flow additives and photoinitiators (UV) were applied to the substrate as thin coatings, the material placed on the conveyor belt and exposed to EB and UV sources. Two UV systems were used namely a Primarc Minicure unit and a Fusion unit with lamps of 200 W/inch. Two EB facilities were utilised namely a 500 KeV Nissin machine and a 175 KeV ESI unit.

Results

Radiation Grafting Using Acid Additives with Homopolymer Suppressors. The results in Table I for the grafting of styrene in methanol to polypropylene show that when sulfuric acid is present in the monomer solution, the well known acid enhancement effect is observed at certain concentrations, particularly in solutions corresponding to the Trommsdorff peak *(1,12)*. When copper sulfate is used as additive the following observations can be made. (i) In the absence of acid, grafting yields are reduced at all styrene concentrations studied, particularly in the more dilute solutions. Homopolymer yields are also strongly reduced from both the visual observation of the monomer solutions after irradiation and also from the polystyrene precipitation studies from these same solutions. (ii) In the presence of acid a novel result is observed with Cu^{2+}. Not only is homopolymer reduced, grafting is enhanced at all monomer concentrations studied and there is a synergistic effect with acid and copper sulfate up to monomer concentrations of 40% (the gel peak). This synergistic effect is independent of the salt structure in terms of the salts studied (cupric sulfate, ferrous sulfate and Mohr's salt), although copper appears to be marginally the best for the enhancement process (Figure 1). For these three salts optimum grafting yields for 30% monomer solutions occur at salt concentrations of between 5×10^{-4} M and 1×10^{-2} M, with 5×10^{-3} M a compromise (Table II) in acid concentrations of 2×10^{-1} M (Table III).

Table I Effect of Copper Sulfate and Sulfuric Acid on Radiation Grafting of Styrene to Polypropylene[a]

Styrene (%v/v)	Graft (%)			
	Control	Cu^{2+}	H^+	$Cu^{2+} + H^+$
20	10.6	4.6	14	11.5
30	42	25	40	64
40	67	41	108	94
50	64	32	73	41
60	51	27	40	45

[a] Methanol used as solvent. Radiation dose of $2.5x10^3$ Gy at $5.2x10^2$ Gy/hr; $Cu^{2+} = CuSO_4.5H_2O$ (5 x 10^{-3}M); $H^+ = H_2SO_4$ (0.2M).

Table II. Effect of Cationic Salt Concentration on Radiation Grafting of Styrene to Polypropylene in Presence of Acid[a]

Cationic Salt (M)	Graft (%)		
	$Cu^{2+} + H^+$	$Fe^{2+} + H^+$	$M + H^+$
$5x10^{-4}$	53	59	47
$5x10^{-3}$	64	56	35
$1x10^{-2}$	70	36	3.3
$1x10^{-1}$	67	5.4	-

[a] Conditions as in Table I; styrene in methanol (30% v/v).

Table III. Effect of Acid Concentration on Radiation Grafting of Styrene to Polypropylene in Presence of Cationic Salts[a]

H_2SO_4 (M)	Graft (%)		
	Cu^{2+}	Fe^{2+}	M
0	34	41	-
$5x10^{-2}$	43	29	29
$1x10^{-1}$	52	40	50
$2x10^{-1}$	64	56	68
$3x10^{-1}$	63	58	49

[a] Conditions as in Table I; styrene in methanol (30% v/v).

UV Grafting with Acid Additives and Homopolymer Suppressors. Two categories of UV grafting are relevant to this discussion. The distinction involves the use of photoinitiator to accelerate the rate of grafting. Although the rates of grafting are slower in the absence of photoinitiator, under these conditions the possibility of photoinitiator contaminating the final product copolymer is eliminated. For some applications this is important.

Photoinitiator Absent. The data in Table IV show that photografting styrene in methanol to polypropylene without photoinitiator is low under the UV conditions used confirming previous work *(17)*. Inclusion of mineral acid in these grafting solutions leads to an increase in photografting at almost all monomer concentrations studied,

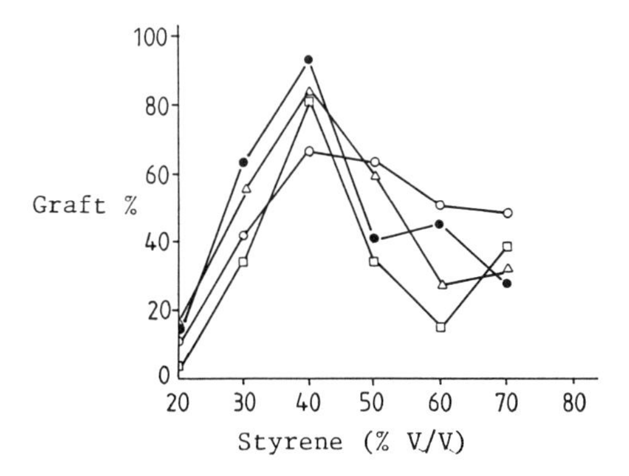

Figure 1 Comparison of Copper Sulfate with Ferrous Sulfate and Mohr's Salt in Radiation Grafting Styrene in Methanol to Polypropylene in Presence of Acid. (○) control; (●) Cu^{2+} + H^+; (△) Fe^{2+} + H^+; (□) Mohr's Salt + H^+ Radiation dose of $2.5x10^3$ Gy at $5.2x10^2$ Gy/hr; Cationic salts ($5x10^{-3}$ M); H^+ = H_2SO_4 (0.2 M).

particularly with the monomer solution corresponding to the Trommsdorff peak (50%). Addition of copper sulphate to these acidified monomer solutions lowers the grafting yield marginally at the lower concentrations up to 50%, except for the monomer solution corresponding to the Trommsdorff peak (30%) where a significant enhancement is observed. In the presence of acid the position of the Trommsdorff peak shifts from 50% to 30% monomer.

Table IV. Effect of Photoinitiator, Acid and Cationic Salt in Photografting Styrene in Methanol to Polypropylene[a]

Styrene (%v/v)	Graft (%)			
	Without B			With B
	N.A.	H^+	Cu^{2+} + H^+	
20	<2	4.0	3.1	7.2
30	<2	5.4	9.0	29
40	<2	6.1	5.0	39
50	<2	7.4	5.1	26
60	<2	4.0	4.3	25

[a] Irradiated for 18h at 30cm from 90W lamp;
B = benzoin ethyl ether, photoinitiator (1% w/v); N.A. = no additive;
H^+ = H_2SO_4 (0.2M); Cu^{2+} = $CuSO_4.5H_2O$ (5×10^{-3}M).

Photoinitator Present. When photoinitiator is included in the monomer solution without additives, the following changes in grafting patterns occur. (i) Significant increases in grafting yields are observed in all styrene solutions (Table IV). (ii) With the addition of ferrous sulfate, homopolymer is suppressed and grafting is almost eliminated especially in the low monomer concentrations (Figure 2). (iii) Inclusion of acid in the ferrous sulfate solution leads to continued suppression of homopolymer and

Table V. Comparison of Cationic Salts in Presence of Acid in Sensitised Photografting of Styrene in Methanol to Polypropylene[a]

Styrene (% v/v)	Graft (%)				
	B	$B+H^+$	$B+H^++Cu^{2+}$	$B+H^++Fe^{2+}$	$B+H^++M$
20	7.2	19	25	22	14
25	22	48	49	83	72
30	29	65	117	165	102
35	41	31	72	30	39
40	39	19	64	35	36
50	26	18	31	25	41

[a] Conditions as in Table IV; M = Mohr's salt ($FeSO_4(NH_4)_2SO_4.6H_2O$, 5 x 10^{-3}M).

Table VI. Effect of Copper Sulphate on Grafting Efficiency for the Sensitised Photografting of Styrene in Methanol to Polypropylene in Presence of Acid[a]

Styrene (% v/v)	B		$B+H^++Cu^{2+}$	
	Graft (%)	Homopolymer (%)	Graft (%)	Homopolymer (%)
20	7.2	12.8	25	31
30	29	17.5	117	42
40	39	35	64	12.2
50	26	14.4	31	7.7

[a] Conditions as in Table IV.

a large enhancement in grafting in virtually all solutions studied, an extraordinary increase occurring at the Trommsdorff peak (30%). A similar result is obtained when copper sulphate or Mohr's salt replaces ferrous sulfate (Table V). Not only is the grafting yield improved in the presence of copper sulfate but also the grafting efficiency (Table VI) which demonstrates that grafting is preferentially favoured over homopolymerisation in these systems i.e. in the presence of both salts and acids.

Acid and PFMs as Additives with Homopolymer Suppressors in Radiation and Photografting. When a typical MFA such as tripropylene glycol diacrylate (TPGDA) replaces acid in the monomer solution the increase in styrene grafting yield to polypropylene with ionising radiation is more pronounced (Figure 3) consistent with earlier work *(7)*. Inclusion of both acid and TPGDA in the same monomer solution leads to large synergistic effects in grafting enhancement particularly in the styrene solution (40%) corresponding to the Trommsdorff peak. When copper sulfate is included in these solutions, not only is homopolymer suppressed but the following additional observations can be made when compared with the results for the controls (no Cu^{2+}). (i) In the controls, grafting is marginally reduced in all solutions. (ii) With TPGDA present and in the absence of acid, grafting yields are increased for all monomer concentrations except 20%. (iii) With TPGDA and acid present, grafting is increased only at the 20% solution, being marginally reduced elsewhere, however the grafting yields are much higher than in the solutions containing no additives. Importantly, the graft at 20% with these two additives is the highest for any of the 20% solutions containing copper. This is valuable in a preparative context since there are advantages in grafting under low monomer concentration conditions.

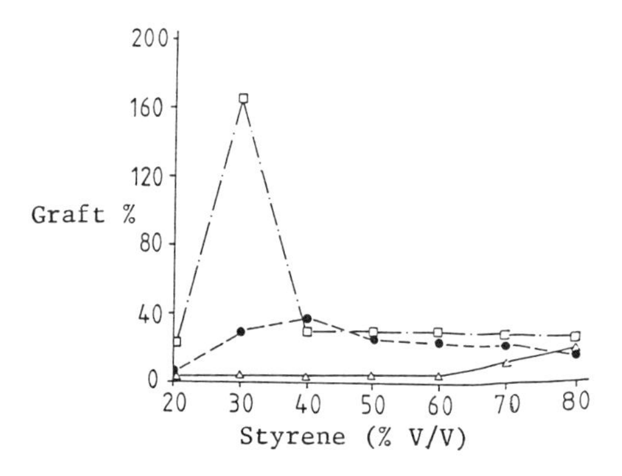

Figure 2 Effect of Ferrous Sulfate on Sensitised Photografting of Styrene in Methanol to Polypropylene in Presence of Acid.
(●) Benzoin ethyl ether (B); (▲) B + Fe^{2+}; (□) B + Fe^{2+} + H^+
Irradiated 18 hr at 30 cm from 90W lamp; B (1% w/v); $H^+ = H_2SO_4$ (0.2 M); $Fe^{2+} = FeSO_4.7H_2O$ ($5x10^{-3}$ M).

Table VII Effect of Multifunctional Acrylates (TPGDA) and Acid as Additives in Presence of Copper Ions on Radiation Grafting Styrene to Cellulose in Methanol

Styrene (% v/v)	Graft (%)						
	N.A.	Cu^{2+}	H^+	T	T + H^+	T + Cu^{2+}	T+H^++Cu^{2+}
20	37	36	68	149	91	32	45
40	62	57	87	205	218	82	93
60	55	48	69	145	87	69	67
80	60	40	61	164	90	62	72

[a] Total dose = $5.0x10^3$ Gy; dose rate = $5.0x10^2$ Gy/hr;
N.A. = no additive; $H^+ = H_2SO_4$ (0.2M); $Cu^{2+} = CuSO_4.5H_2O$ ($5x10^{-3}$);
T = tripropylene glycol diacrylate (1% v/v).

When cellulose is used as backbone polymer instead of polypropylene, similar trends in reactivity to polypropylene for the radiation grafting of styrene in methanol are observed (Table VII). Thus inclusion of acid in the control (no additives) increases the yields at all monomer concentrations studied but not to the same degree as for polypropylene, consistent with earlier work *(18)*. Addition of TPGDA to the control leads to a large enhancement in graft which is accentuated at the Trommsdorff peak in a synergistic effect with the acid. Inclusion of copper sulfate in these solutions leads to suppression of homopolymer and the following additional observations when compared

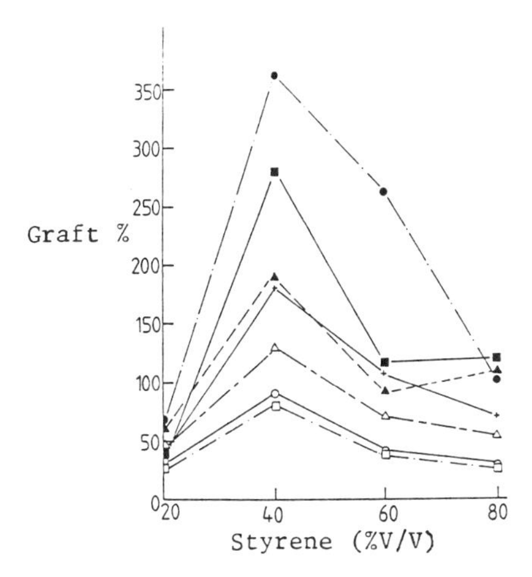

Figure 3 Effect of Tripropylene Glycol Diacrylate (TPGDA) on Radiation Grafting of Styrene in Methanol to Polypropylene in Presence of Acid and Copper Sulfate. (○) no additive; (□) Cu^{2+}; (△) H^+; (+) TPGDA; (●) TPGDA + H^+; (■) TPGDA + Cu^{2+}; (▲) TPGDA + Cu^{2+} + H^+ Radiation dose of 5.0×10^3 Gy at 5.0×10^2 Gy/hr; TPGDA (1% v/v); $H^+ = H_2SO_4$ (0.2 M); $Cu^{2+} = CuSO_4.5H_2O$ (5×10^{-3} M).

to the controls (no Cu^{2+}). (i) In the controls, grafting yields are slightly reduced. (ii) In the presence of TPGDA with no acid, grafting yields are strongly reduced more markedly than with polypropylene. (iii) With TPGDA and acid, the yields are higher for all concentrations studied, except 60%, when compared with TPGDA alone. The grafting yields are also higher than the control with and without Cu^{2+}. In all of these studies TPGDA is used as typical MFA. This monomer is predominantly used in radiation curing reactions as a reactive diluent and thus the data from these studies may be of value in the conditions required to observe concurrent grafting during cure.

In the corresponding photografting work when TMPTA is used as representative MFA, analogous results to the above ionising radiation studies for grafting styrene in methanol to polypropylene are observed. The effects are more dramatic with UV and the magnitude of the enhancement is extraordinarily high especially in the presence of copper ions (Tables VIII and IX). Under these conditions homopolymer formation remains suppressed. In the absence of acid and in the presence of Cu^{2+}, grafting yields are strongly reduced even with TMPTA at all monomer concentrations studied up to 60%

Discussion

Acid Enhancement in Radiation and Photografting. In earlier preliminary studies it has been shown that for a typical system such as the grafting of styrene in methanol to polyethylene initiated by either cobalt-60 or UV, inclusion of mineral acid in the monomer solution leads to an increase in the concentration of monomer in the solution absorbed within the backbone polymer when compared with the original bulk grafting monomer solution. This partitioning of monomer within the backbone polymer leads to the observed grafting enhancement. Analogous results have also been obtained when inorganic salts such as lithium perchlorate replace mineral acid *(12)*. Evidence to support this partitioning of monomer concept originated in experiments involving

Table VIII. Synergistic Effects in Sensitised Photografting of Styrene in Methanol to Polypropylene in the Presence of Acid and TMPTA at Different UV Intensities[a]

Styrene (% v/v)	Graft (%)						
	Series 1[b]				Series 2[c]		
	B	B+H$^+$	B+T	B+H$^+$+T	B	B+H$^+$	B+H$^+$+T
20	28	14	28	41	7.2	19	25
30	101	126	52	78	29	65	70
40	189	193	321	266	39	19	300
50	124	107	412	525	26	18	200
60	37	31	133	188	25	-	150

[a] B, H$^+$ as in Table IV; T = TMPTA = trimethylolpropane triacrylate (1% v/v);
[b] Irradiated 24h at 24cm from 90W lamp at 20°C.
[c] Irradiated 18h at 30cm from 90W lamp at 20°C.

Table IX. Synergistic Effect of Acid, TMPTA and Cationic Salts for Enhancing Yields in Photosensitised Grafting of Styrene in Methanol to Polypropylene[a]

Styrene (% v/v)	Graft (%)			
	B	B+Cu^{2+}+T[b]	B+T+H$^+$	B+T+H$^+$+Cu^{2+}
20	7.2	1.5	25	30
30	29	1.6	70	380
40	39	9	306	121
50	26	11	200	121
60	25	10	150	98
70	23	51	75	80
80	15	121	70	59

[a] Conditions as in Tables IV and VIII.
[b] Level of graft for B + Cu^{2+} low, similar to data for B + Fe^{2+} in Table VIII.

styrene labelled with tritium. More recent data (Figure 4) indicate that partitioning of nonpolar monomer into nonpolar media may be significantly improved by the presence of dissolved electrolyte. This partitioning may be interpreted as an example of the salting out technique employed in solvent extraction. In the present grafting systems the driving force for the increased partitioning of monomer into substrate is the reduced solubility of styrene in the bulk solution due to the presence of dissolved electrolyte. The net result of this driving force is higher rates of diffusion into, and higher rates of diffusion within, the substrate. This permits higher concentrations of monomer to be available for grafting at a particular backbone site in the presence of these additives. The extent of this improvement in monomer partitioning depends upon the polarities of monomer, substrate and solvent, as well as type of acid and the concentration of the acid. It is thus the effect of these ionic species on partitioning that is essentially responsible for the observed increase in radiation grafting yields in the presence of these acid additives. Radiolytically generated free radicals can be expected to make some contribution to this effect in a system in which initiation occurs by ionising radiation, however this radiation

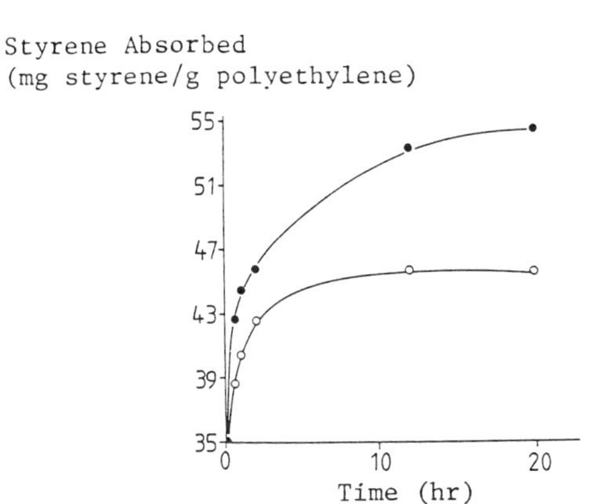

Figure 4 Time dependence for Swelling of Polyethylene Film by Styrene at 25°C. Technique involved using tritiated styrene with 30% styrene (v/v) in methanol solutions and polyethylene film (0.12 mm) at 25°C. The styrene was tritiated by platinum-catalysed exchange with T_2O. Scintillation counting was used to determine tritium. (○) no additive; (●) H_2SO_4 (0.1 M).

chemistry explanation does not appear to be the dominant pathway for the enhancement *(12,19)*.

In addition to partitioning of monomer in these systems, acid additives, themselves, can also be partitioned between the bulk grafting solution and the backbone polymer. If polypropylene and cellulose are again used as model backbone polymers for grafting styrene in methanol for this discussion, such partitioning of electrolyte would affect the grafting in the following manner. (i) With nonpolar polypropylene, acid will remain predominantly partitioned into the more polar bulk grafting monomer solution and thus becomes available for increasing the partitioning of styrene into the backbone polymer and thus enhancing grafting. (ii) With polar cellulose, the reverse occurs and the electrolyte is strongly partitioned into the backbone polymer. The magnitude of the enhancement in grafting will then be determined by the combined polarities of the components in the backbone polymer relative to those in the bulk solution. (iii) With cellulose, during the initial stages of grafting, styrene is grafted to pure cellulose which becomes increasingly enriched in styrene as grafting proceeds. Subsequent styrene grafting will thus be to styrenated cellulose. As this process occurs the bulk monomer solution will become progressively enriched in electrolyte due to the onset of the back reaction involving electrolyte partitioning from the styrenated cellulose and the relative magnitude of the grafting enhancement will increase. Data shown support this conclusion, a small finite enhancement in grafting being achieved during the initial stages of the reaction *(18)*, then as grafting proceeds the magnitude of the graft enhancement increases.

Effect of Cu^{2+}, Fe^{2+} and Mohr's Salt as Homopolymer Suppressors in Grafting. The effect of Cu^{2+}, Fe^{2+} and Mohr's salt in preferentially reducing homopolymer formation at the expense of grafting has been attributed to the scavenging of monomer radicals by the inorganic ion *[20]*. The efficiency of this process also appears to be controlled by partitioning phenomena which can be followed visually in the grafting system especially when copper salts are used. When grafting styrene in methanol to nonpolar polypropylene, inclusion of Cu^{2+} in the monomer solution leads to a blue colouration of the solution which virtually stops at the polypropylene interface. By contrast, with polar cellulose, the Cu^{2+} is preferentially absorbed out of the monomer solution into the cellulose, the backbone polymer being strongly coloured as evidence of this effect. The significance of this partitioning of Cu^{2+} in grafting with the two backbone polymers possessing extremes in polarities can be explained as follows.
(i) With polypropylene, the bulk of the salt suppressor remains in the grafting solution. This Cu^{2+} scavenges the monomer radicals in the solution by the mechanism proposed *[20]* hence minimising homopolymerisation and leaving more styrene available for grafting. The low concentration of Cu^{2+} in the backbone polymer results in little scavenging of monomer radicals absorbed within the backbone polymer. The unscavenged radicals would then be expected to graft rather than homopolymerise as a result of cage effects in the backbone polymer.
(ii) By contrast, with polar cellulose, a high proportion of the Cu^{2+} ions are partitioned out of the bulk solution into the backbone polymer. The relatively high concentration of Cu^{2+} in the backbone polymer reduces grafting because of scavenging of radicals from both the monomer and the backbone polymer (grafting sites). The depletion of Cu^{2+} in the bulk solution reduces scavenging of monomer radicals in the monomer solution and thus homopolymerisation is increased. If more Cu^{2+} ions are added to the bulk grafting solution to compensate for this loss, then both grafting and homopolymer yields are reduced. Evidence to support this proposal is available from the relative efficencies of grafting of polypropylene and cellulose, where, in the case of cellulose in the presence of Cu^{2+} and the additives, the enhancement is considerably less than with polypropylene under comparable radiation grafting conditions (Figure 3 versus Table VII).
(iii) With cellulose, during the initial stages of grafting, styrene is reacting with pure cellulose, whereas as grafting proceeds, the monomer is grafting to a cellulose which is becomes progressively enriched in styrene. Thus as grafting proceeds the cellulose system becomes more hydrophobic and resembles more the behaviour of polypropylene discussed in (i). This conclusion is substantiated by previous work with cellulose which has shown that as grafting proceeds the enhancement effect of acid becomes increasingly more predominant, i.e. as the cellulose becomes increasingly more styrenated. Acid enhancement effects with cellulose are thus accentuated after some finite grafting has occurred.
(iv) In the specific case of UV grafting, reduction in homopolymer using the salts leads to lower turbidity in the monomer solution due to minimising polystyrene precipitation in the methanol as the irradiation proceeds, thus UV transmittance in the grafting solution is improved and grafting increases.

Synergism of Acid and PFM Enhancement in Radiation and Photografting with Homopolymer Suppressors. In the simplest model, when both acid and homopolymer suppressors such as copper ions are present in the same grafting solution, the essential effect of the homopolymer suppressors will be to minimise homopolymer in the graft solution, thus maintaining in this solution a larger pool of styrene monomer which can then be partitioned into the backbone polymer by the acid additives leading to grafting enhancement. This model would explain why under certain experimental conditions there is a strong grafting enhancement when mineral acids are added to a monomer solution which already contains homopolymer suppressors. The mechanism of enhancement in radiation grafting due to the presence of acids and PFMs is different for each class of additive so that when both are included in the same monomer solution synergistic effects in yield are observed under certain experimental conditions. In earlier work, the grafting enhancement effect of PFMs was attributed to branching of the growing polystyrene chains *(12)*. This was a free radical process as was competing homopolymerisation. In the presence of Cu^{2+} and Fe^{2+}, homopolymerisation is retarded presumably by the radical scavenging mechanism already discussed *(20)* leaving a larger pool of PFM in the monomer solution which is available for accelerating the grafting reaction. Grafting is thus enhanced with the PFMs mainly through branching of the growing polystyrene chains. The MFAs would also be expected to be partitioned in the monomer solution in a manner similar to that proposed for the other additives including acid, however the concentration of MFA used is low, 1% v/v, and the effect of partitioning of this component on the mechanism of the enhancement at this stage would appear to be minimal. This aspect of partitioning is currently being investigated.

Significance of Grafting During Curing - Role of Additives. The results of the present grafting work are of value in radiation curing processes. Grafting and curing reactions initiated by either ionising radiation or UV are mechanistically related generally via free radical processes *(21-27)*. Grafting is the copolymerisation of a monomer/oligomer mixture (M/OL) to a backbone polymer (equation 1) whereas curing is the rapid polymerisation of M/OL onto the surface of a substrate (equation 2).In this context, curing can also cover reactions including crosslinking since the properties of the film that is cured will depend upon the oligomer/monomer components which frequently contain multifunctional monomers to achieve rapid polymerisation, and also lead to crosslinking.

$$\text{backbone} \xrightarrow{\text{M/OL}} \text{backbone}\sim\sim\text{M/OL} \qquad (1)$$

$$\sim\sim\sim \xrightarrow{\text{M/OL}} \frac{\text{M/OL}}{\sim\sim\sim} \qquad (2)$$

There is no time scale theoretically associated with grafting reactions which can take minutes, hours or even days, whereas curing reactions generally occur in a fraction of a second. One of the differences between grafting and curing is the nature of bonding occurring in each process. In grafting, covalent C-C bonds are formed from free radical reactions, whereas in curing bonding usually involves weaker van der Waals or London dipersion forces. Two types of curing systems are relevant to this discussion (a) free radical and (b) cationic processes.

Free Radical Curing Reactions. Curing reactions initiated by ionising radiation are considered to occur predominantly via free radical intermediates, however recent work *(21,22)* with Fourier Transform ion cyclotron resonance mass spectrometry (FTICRMS) indicates that under electron beam (EB) conditions at higher dose rates, ions may contribute to these processes. With UV, using conventional photoinitiators (aromatic and aliphatic ethers and ketones) free radical processes are also considered to be responsible for the polymerisation processes. In both grafting and curing a valuable common component in both systems is the MFA. In the preceeding grafting work, TPGDA and TMPTA have been used to represent such materials. In grafting, the MFA is used to improve yields and in curing its function is analogous since it is a reactive diluent needed to reduce the viscosity of the mixture and speed up cure.

When considering curing processes, the role of MFA is thus important, particularly its effect on concurrent grafting with cure. A similar problem exists for the commercial additives used in curing and their effect on grafting. These additives are used to improve viscosity, flow, slip, adhesion and gloss and include materials like the fluorinated alkyl ester (FC-430), the silane (Z-6020) and urea. The data in Table X for the photografting of styrene in methanol to polypropylene demonstates the effect of these commercial additives on a typical grafting system, per se, and, by implication, grafting during cure. Thus urea and the fluorinated alkyl ester are mild activators in photografting whereas silane is a retarder, presumably due to the presence of the silicon atom. When TMPTA is added to the monomer solution containing all of the preceding additives, a dramatic increase in grafting by almost two orders of magnitude is observed. These results indicate that the MFA will over-ride the presence of other additives particularly retarders such as the silane and thus enhance concurrent grafting with cure. Careful choice of MFA in curing is thus essential.

Table X Effect of Organic Additives (Urea, Silane, Fluorinated Alkylesters) in Presence of TMPTA on Grafting of Styrene to Polypropylene Initiated by UV[a]

Styrene	Graft (%)					
(% v/v)	N.I.	B	B+U	B+U+Si	B+U+FE	B+U+FE+T
20	< 5	< 5	< 5	< 5	< 5	260
30	< 5	35	30	18	23	588
40	< 5	39	46	31	53	711
50	< 5	17	19	13	16	368
60	< 5	14	13	9	19	283
70	< 5	14	11	7	10	131

a Irradiated 8 hr at 24 cm from 90W lamp at 20°C; N.I. = no initiator; B = benzoin ethyl ether (1% w/v); Si = silane (1% v/v), Z-6020 supplied by Dow; FE = fluorinated alkyl ester (1% w/v) FC-430 supplied by 3M; T = TMPTA.

Cationic Curing Reactions. In addition to free radical processes, more recent work using cationic photoinitiators indicates that curing can be achieved in UV/EB via ionic mechanisms *(28,29)*. Typical of the cationic photoinitiators used are triarylsulfonium salts, such as the hexafluorophosphate which yield both ions and free radicals (equations 3,4) and subsequent polymerisation can theoretically occur via ionic or free radical pathways.

$$Ar_3S^+ X \xrightarrow{h\nu} [Ar_3S^{+\bullet}\ X^-]^* \tag{3}$$

$$[Ar_3S^{+\bullet}\ X^-]^* + RH \longrightarrow Ar_2S + Ar^\bullet + R^\bullet + H^+ X^- \tag{4}$$

The essential advantages of the ionic process is that non-acrylate chemistry can be used as an alternative in curing to produce new polymers and thus new products not capable of being obtained by other methods *(15,21,22)*. Vinyl ethers and epoxides as typified by triethylene glycol divinyl ether (TEGDVE, GAF) and cycloaliphatic diepoxide (CADE, Cyracure UVR6110, Union Carbide) are currently the most important. The results show that, using the draw down bar technique for application, neat TEGDVE and CADE when cured also concurrently graft to cellulose with the cationic photoinitator (Table XI) but not, with Irgacure 184 free radical initiator.

Table XI Grafting and Curing of Vinyl Ether and Epoxied Monomers to Cellulose with Cationic and Free Radical Photoinitiators[a]

		Monomer	Graft (%)			
System [b]	Passes [c]	(% v/v)	I	I + T	C	C + T
TEGDVE	1	30	0	2.6	0	3
		70	0	0.9	4	5
		100	-	-	27	-
	3	100	<1 (M)	-	100 (M)	-
		100	<1 (F)	-	123 (F)	-
CADE	3	100	0 (M)	-	140 (M)	-
		100	0 (F)	-	150 (F)	-

[a] All photoinitiators (1% w/w); T=TMPTA(1% v/v);
I = Irgacure 184 (1-hydroxy-1-cyclohexylphenyl ketone, Ciba-Geigy);
C=cationic FX512 (triaryl sulfonium hexafluorophosphate,3M);
D = Darocur 1173 (2,2-dimethyl, 2-hydroxy acetophenone, Merck).
[b] TEGDVE = triethylene glycol divinyl ether; CADE = cycloaliphatic diepoxide (Cyracure UVR6110, Union Carbide); methanol used as solvent.
[c] Number of passes under Minicure (M) of Fusion (F) lamp systems of 200 W/inch at speed of 15 meters/min (one pass) and 30 meters/min (3 passes).

Importance of Grafting Results in Commercial Curing. The present work involving concurrent grafting during curing has implications in industrial processing. The data show that the concept is applicable to both conventional free radical and cationic systems. For many curing processes involving organic substrates, it is generally accepted that concurrent grafting is an advantage. However, with the rapid expansion of UV/EB technology, the necessity to recycle products processed by this technology is becoming increasingly important. In certain industries such as printing and packaging, there are problems with stripping UV/EB cured coatings and inks from substrates during recycling. In this respect concurrent grafting can be an impediment since the covalent bond, between film and substrate, is relatively difficult to break. There is thus a need to either develop coatings with good adhesion after curing without concurrent grafting or develop a balance in the properties of additives in the coatings to achieve this bonding e.g. for this purpose adjust the relative amounts of the silane which is seen to be a retarder and TMPTA which accelerates grafting. In paper recycling one method for overcoming the above problem is not only to use additives in the correct proportions but also to incorporate -COOH and $-SO_3H$ groups in the oligomers which tend to retard grafting and also solubilise the coating in dilute alkali, thus assisting recycling. In contrast to the printing and packaging industry, the requirements of the plastics field are generally different. The type of recycling problem mentioned previously is not so critical and the covalent bonding associated with concurrent grafting is an advantage in many cured products since this property minimises delamination of films after curing.

Partitioning of Reagents in Grafting and Curing. The similarities in mechanisms between grafting and curing have already been discussed. The role of partitioning of reagents, particularly additives, in grafting has been shown to be important. In like manner, analogous reactivity in curing systems is being observed which may be attributed to partitioning. In a typical example, in UV curing of a mixture containing two monomers, one trifunctional and the other monofunctional onto paper board, several hours after curing the monofunctional monomer could be detected at the back of the board where no coating had been applied. Presumably on application, the lower molecular weight monomer had been preferentially partitioned out of the film into the board, this monomer then migrated through the board and, because of penetration difficulties with UV, some of the monomer escaped cure and diffused through the board to atmosphere. Numerous other similar examples can be cited of radiation cured systems where the behaviour of the reagents in curing may be interpreted in terms of proportional partitioning of components from the oligomer into the substrate. More work is needed to clarify the possible role of partitioning in curing, since the application of the concept can lead to the design of films possessing specific properties, particularly if conditions can be designed either to accelerate or minimise the partitioning of specific additives or components in a curing process.

Conclusion

In the present work, further mechanistic studies of the role of mineral acids, salts and PFMs in accelerating radiation and UV grafting has confirmed the original proposal that partitioning of reagents between monomer solution and substrate is important. The grafting of styrene dissolved in methanol to cellulose/polyolefins is used as model system for these studies since the monomer is well behaved and contains no active functional groups. The data from the styrene system can be readily extrapolated to other monomers. The inclusion of copper and iron salts as retarders for homopolymerisation in these solutions leads to unexpected graft enhancement under certain experimental conditions. This is an observation of value in a preparative context for the synthesis of graft copolymers since maximum yields for minimum radiation doses are obtained under these conditions. The mechanism of the salt retardation process in homopolymerisation has been related to partitioning concepts similar to those already proposed for analogous grafting reactions. The significance of this grafting work in the related field of curing using both free radical and cationic processes has been evaluated. The influence of concurrent grafting during curing in determining the properties of the finished cured product in industrial radiation processing has been considered, particularly its relevance to ease of recycling.

Acknowledgments

The authors thank the Australian Institute of Nuclear Science and Engineering, the Australian Research Council and the Korean Science and Engineering Foundation (YCN) for support, also the International Atomic Energy Agency for the award of a Fellowship (MAK) and Firenews Pty Ltd for financial assistance.

Literature Cited.

1. Garnett,J.L.*Rad.Phys.Chem.* **1979**, *14*, 79.
2. Charlesby,A. *Atomic Radiation and Polymers*, Pergamon Press: Oxford, 1960.
3. Chapiro, A. *Radiation Chemistry of Polymeric Systems*, Wiley-Interscience: New York, 1962.
4. Hebeish,A.; Guthrie,H.T. *The Chemistry and Technology of Cellulosic Copolymers,* Springer-Verlag: Berlin, 1981.
5. Kabanov,V.Y. *Rad.Phys.Chem.* **1989**, *33*, 51.
6. Garnett,J.L.; Jankiewicz,S.V.; Long,M.A.; Sangster,D.F. *J.Polym.Sci., Polym.Lett.Ed.* **1985**, *23*, 563.

7. Dworjanyn,P.A.; Garnett,J.L. In *Progress in Polymer Processing*. Silverman,J.; Singh.A., Eds; Hanser: New York, 1992, Vol 3, 93.
8. Gupta,B.D.;Chapiro,A. *Eur.Polym.J.* **1989**, *25(11)*, 1137.
9. El-Assy,N.B. *J.Appl.Polym.Sci.* **1991**, *42*, 885.
10. Zaharan,A.H.; Zohdy,M.H. *J.Appl.Polym.Sci.* **1986**, *31*, 1925.
11. Misra,B.N.; Rawat,B.R. *J.Polym.Sci..Polym.Chem.Ed.* **1985**, *23*, 307.
12. Garnett,J.L.; Jankiewicz,S.V.; Sangster,D.F. *Rad.Phys.Chem.* **1990**, *36*, 571.
13. Chapiro,A.; Seidler,P. *Eur.Polym.J.* **1965**, *1*, 189.
14. Huglin,M.B.; Johnson,B.L. *J.Polym.Sci. A-1*, **1969**, *7*, 1379.
15. Dworjanyn,P.A.; Garnett,J.L., Bellobono,I.R.,Ed.; *Proceedings Int. Meeting on Grafting Processes onto Polymeric Fibres and Surfaces: Scientific and Technical Aspects,* Milan, Universita Degli Studi di Milano, 1990, Supplemento n.85, p.63.
16. Kiatkamjornwong,S.; Dworjanyn,P.A.; Garnett,J.L., *Proceedings of Radtech Asia '91*, RadTech Japan: Osaka, **1991**, p.384.
17. Dworjanyn,P.A.; Garnett,J.L. *Rad.Phys.Chem.* **1989**, *33*, 429.
18. Garnett,J.L.; Phuoc,D.H.; Airey,P.L.; Sangster,D.F. *Aust.J.Chem.* **1976**, *29*, 1459.
19. Chappas,W.J.; Silverman,J. *Rad.Phys.Chem.* **1979**, *14*, 847.
20. Collinson,E.; Dainton,F.S.; Smith,D.R.; Trudel,G.J.; Tazuke,S. *Disc.Far.Soc.* **1960**, *29*, 188.
21. Garnett,J.L.; Dworjanyn,P.A.; Nelson,D.J.; Bett,S.J., Tabata,Y., Ed.; *Proceedings of Radtech Asia*, Tokyo, **1988**, p.343.
22. Bett,S.J.; Dworjanyn,P.A.; Greenwood,P.F.; Garnett,J.L., *Proceedings RadTech '90 - North America*, RadTech International North America: Northbrook, Illinois, 1990, Vol 1, p.313.
23. Geacintov,N.; Stannett,V.; Abrahamson,E.W.; Hermans,J.J. *J.Appl.Polym.Sci.* **1960**,*3*, 54.
24. Reine,A.H.; Arthur,J.C. *Text. Res. J.* **1962**, *32*, 918.
25. Needles,H.L.; Wasley,W.L. *Text. Res. J.* **1969**, *39*, 97.
26. Kubota,H.; Murata,T.; Ogiwara,T. *J.Polym.Sci.* **1973**, *11*, 485.
27. Davis,N.P; Garnett,J.L.; Urquhart,R. *J.Polym.Sci. Polym.Lett.Ed.* **1976**, *14*, 537.
28. Crivello,J.V. *Adv.Polym.Sci.* **1984**, *62*, 1.
29. Lapin,S.C.; Snyder,J.R., *Proceedings RadTech '90 - North America*, RadTech International North America: Northbrook, Illinois, 1990, Vol 1 p.410.

RECEIVED August 14, 1992

Chapter 9

Photoinitiated Polymerization of Liquid Crystalline Monomers Bearing Nonaromatic Mesogenic Units

C. E. Hoyle[1], A. C. Griffin[2], D. Kang[1], and C. P. Chawla[1]

[1]**Department of Polymer Science, University of Southern Mississippi, Hattiesburg, MS 39406**
[2]**Melville Laboratory for Polymer Synthesis, University of Cambridge, Pembroke Street, Cambridge CB2 3RA, United Kingdom**

There have been a number of papers appearing in the literature over the past two or three decades which have focused on the effect of conducting polymerization in liquid crystalline and organized media. Two excellent reviews of the subject describe the state of the literature of the early 1980s (*1, 2*). The basic conclusions of the two review articles indicated that, although there was some evidence to support the supposition that polymerization of a liquid crystalline monomer or polymerization of an isotropic monomer in an inert liquid crystalline solvent could result in enhanced polymerization rates, no definitive investigations dealing with the consequences of polymerization of mesomorphic systems had been conducted. This was the case until the classic work by Broer and coworkers (*3-6*) in the late 1980's. The work of Broer leaves little question that polymerization of liquid crystalline monomers can result in quite different kinetics for chain formation. In addition, Broer provided results which indicated that orientation induced in monomeric liquid crystalline films can be retained in the polymers subsequently formed. Furthermore, in a recent report, Broer, Mol, and Challa (*7*) show that chain transfer processes which normally occur in diacrylates are minimized upon polymerization in organized media.

All of the compounds investigated by Broer have groups comprised of arylbenzoate mesogenic units separated by alkyl spacer groups from the reactive acrylate moiety. In this paper we report results for the kinetics of the photoinitiated polymerization of two liquid crystalline monomers which have non-aromatic mesogens. The first monomer has a cholesteryl mesogen separated by a flexible alkyl spacer from a reactive methacrylate. The second is a semifluorinated acrylate monomer.

0097–6156/93/0527–0118$06.00/0

Photoinitiated polymerization of the cholesteryl methacrylate monomer has been described by Shannon in a series of articles (*8, 9*) and patents (*10, 11*) which, along with the work of Broer, must be classified as truly exceptional collections describing the virtues and opportunities for polymerization of mesogenic monomers. In an initial paper, we have reported on some of the kinetic aspects of a cholesteryl methacrylate monomer with a long alkyl chain spacer group (*12*). In this paper, we report on the use of a pulsed laser to provide additional information. By using a laser fired at quite wide intervals, it is possible to clearly identify the rate acceleration which accompanies changes in the polymerization medium. In the case of the functionalized semifluorinated acrylate monomer, we find an unusually rapid polymerization rate which, if optimized, can give conversion efficiencies on the order of 70-80 percent from a single laser pulse.

Experimental

The synthesis of CMA-10 followed a literature procedure (*8,9*). F12H10A was synthesized by a method similar to that reported in reference *13*. The laser employed was an excimer laser from Questek or Lumonics. Single shot power-per-pulse values were obtained from power measurements recorded at higher repetition rates and are therefore subject to some error. The DSC was either a modified Perkin Elmer 1B or Perkin Elmer 2B. In all cases except for the data in Figure 5, samples were cast in indented DSC pans. For the results in Figure 5, samples were cast in DSC pans after a simple washing procedure. Results in Figure 5 are thus only qualitative since the sample was not uniformly distributed across the pan but agglomerated near the edges in a ring. α,α-Dimethoxyacetophenone (DMAP) was obtained from Ciba-Geigy. For quantum yield measurements a filtered (366 nm) medium pressure mercury lamp was employed with a nominal output of 2.16×10^{-2} mJ cm^{-2} sec^{-1} at the sample pan. The heat of reaction of CMA-10 was taken as 13.7 kcal mole^{-1}. The heat of reaction of F12H10A was estimated as 22.5 kcal mole^{-1} by assuming 99% conversion after 30 laser pulses and utilizing the resultant area (in kcal) under the exotherm curve. (We note that absolute quantum yield values may be subject to considerable error).

Results and Discussion

In order to illustrate the types of behaviour which can be observed during the photoinitiated polymerization of a liquid crystalline monomer to give a liquid crystalline polymer, two monomers which have liquid crystalline phases have been evaluated. The first is a methacrylate monomer with a pendant cholesteryl mesogenic group decoupled from the methacrylate functionality, and subsequently the polymer generated, by a sequence of ten methylene spacers. Some of the properties of this monomer have been presented in the past (*8, 9*). The cholesteric monomer presents an opportunity to investigate the polymerization kinetics in liquid crystalline media characterized by mobility that is somewhat limited. The second monomer investigated is a semifluorinated acrylate which exists in a highly ordered smectic B type phase. The kinetics for the semifluorinated methacrylate results are reflective of greater restrictions on the polymer kinetic chain mobility.

Cholesteryl Monomer. The cholesteric methacrylate monomer, which we designate as CMA-10, exists in a smectic phase below 52 °C and a cholesteric

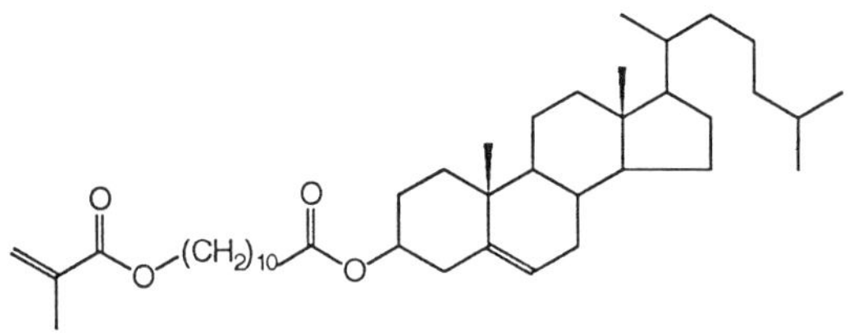

phase from 53 °C to 64 °C as determined by differential scanning calorimetry and polarized optical microscopy. (Transition temperatures are slightly lower when low concentrations of photoinitiator are added to the pure monomer). In the first set of experiments designed to elucidate the polymerization profiles of CMA-10 as a function of the percent conversion, single pulses from an excimer laser (XeF=351 nm) were focused onto a DSC pan containing CMA-10 with 1 wt percent α,α-dimethoxyacetophenone (DMAP). Figure 1 shows results for the cumulative percent conversion versus time plots at 45 °C obtained from intergration of the individual exotherms generated by laser pulses of approximately 10 nanoseconds duration and spaced 15 seconds apart. The percent conversion was calculated using a value of 13.7 kcal mole^{-1} as the enthalpy of polymerization of CMA-10. The most notable feature of the percent conversion versus time plot in Figure 1 is the continuous and smooth buildup in the primary curve followed by a limiting percent conversion of 88 percent. We have reported that during polymerization at 45 °C there is little or no change in the smectic texture (see references *8* and *9* for original reports on texture at this temperature) of the medium as viewed by cross-polarized optical microscopy, even at higher percentages of conversion. The absence of any significant abrupt change in the plot in Figure 1 is consistent with the observation that the medium texture does not change appreciably during the photoinitiated polymerization process.

When the polymerization of CMA-10 at 58 °C (Figure 2) is initiated with laser pulses spaced 15 seconds apart, at very short times the conversion is significantly lower than for subsequent pulses. This is a reasonable result if one keeps in mind that the polymerization of CMA-10 results in an abrupt change to a birefringent medium with different texture from the monomer at 58 °C (cholesteric) as identified by cross-polarized at conversions of about 1-2 percent. The birefringence after the medium change has been postulated to arise from a smectic structure.

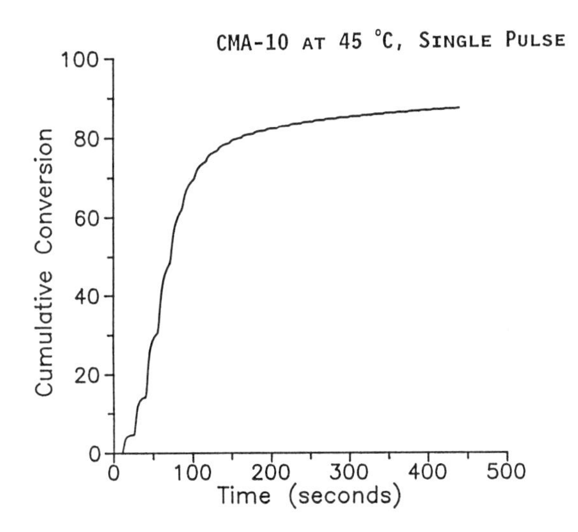

Figure 1. Cumulative conversion plot for CMA-10 film (1 wt percent DMAP) at 45 °C from exotherms initiated by XeF (351 nm) laser pulses (0.64 mJ cm^{-2} $pulse^{-1}$; 15 seconds apart).

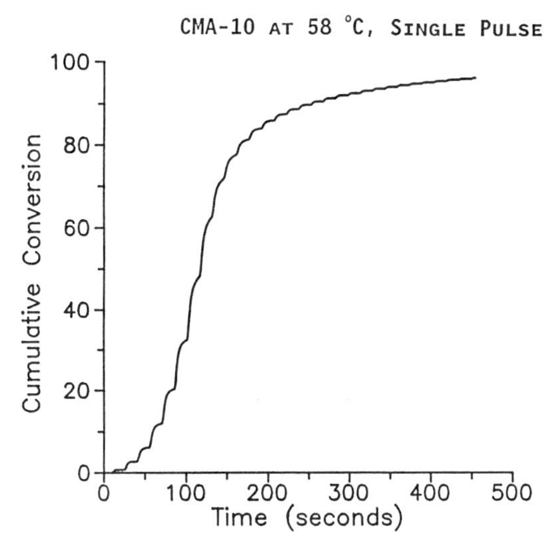

Figure 2. Cumulative conversion plot for CMA-10 film (1 wt percent DMAP) at 58 °C from exotherms initiated by XeF (351 nm) laser pulses (0.64 mJ cm^{-2} $pulse^{-1}$; 15 seconds apart).

If the laser-initiated polymerization is conducted with the CMA-10 monomer initially in the pure isotropic phase, i.e, at 75 °C and 88 °C, two observations are made from an examination of the percent conversion versus time plots in Figures 3 and 4. First, in each case the percent conversion for the first few laser pulses is markedly lower than for subsequent pulses. This is an equivalent way of stating that there is a discontinuous break in the percent conversion versus time plot for polymerizations conducted at temperatures in which the monomer is initially in the isotropic or cholesteric phase. As in the case of polymerization at 58 °C where the monomer is initially in the cholesteric phase, the medium converts to a low order liquid crystalline medium with grainy texture similar to that observed at 58 °C during the course of the polymerization process. This no doubt correlates with the increase in the polymerization efficiency after the first few laser pulses. If we project that the medium change begins at the intersection of tangent lines drawn through the percent conversion versus time plots before and after the increase in polymerization rate, it is apparent that the percent conversion required to obtain the texture change and accompanying rate increase is higher at greater temperatures.

In order to provide additional characterization for the polymerization of CMA-10 at 75 °C and 88 °C, significantly reduced (in intensity) pulses from the laser (λ_{ex} = 351 nm) were spaced 200 milliseconds apart, i.e., the laser was operated at a repetition rate of 5 Hz. The exotherm curves were then recorded for delivery of selected numbers of total pulses to the CMA-10 sample in the DSC. Figure 5 shows results for CMA-10 at 75 °C and 88 °C. If only 10 pulses are delivered to each CMA-10 sample at 75 °C and 88 °C, the exotherm curve at 88 °C is larger and has accordingly a higher area (extent of conversion) than the exotherm curve for the sample at 75 °C subject to 10 pulses. If 20 pulses are delivered to each sample, the exotherm areas are approximately identical, even though the peak maximum of the exotherm curve for CMA-10 at 88 °C is still somewhat higher. The results obtained for 10 pulses reflect polymerization of the monomer strictly in the isotropic phase of the monomer prior to any (or at least very little) change in the reaction medium: at both 75 °C and 88 °C the percent conversion is well below 3 percent. We conclude that polymerization in the isotropic phase is faster at 88 °C than 75 °C. However, for 20 pulses, the percent conversion is greater than for 10 pulses and the areas under the exotherms are approximately identical. This reflects the fact that part of the polymerization at 75 °C occurs after the abrupt change in the medium, and the polymerization rate is accelerated. In the case where a larger number of pulses (25 or 35) are injected into the sample, the conversion is higher at 88 °C than at 75 °C. At 75 °C a significant part of the polymerization actually takes place in the liquid crystalline medium since the total percent conversion is well above 5 percent in both cases and the change from an isotropic to a birefringent liquid crystalline medium occurs at about 2 percent conversion. Thus, the plots in Figure 5 illustrate, as do the result in Figures 3 and 4, that polymerization after the medium change is more efficient than in the isotropic phase: even if the sample system is at a lower

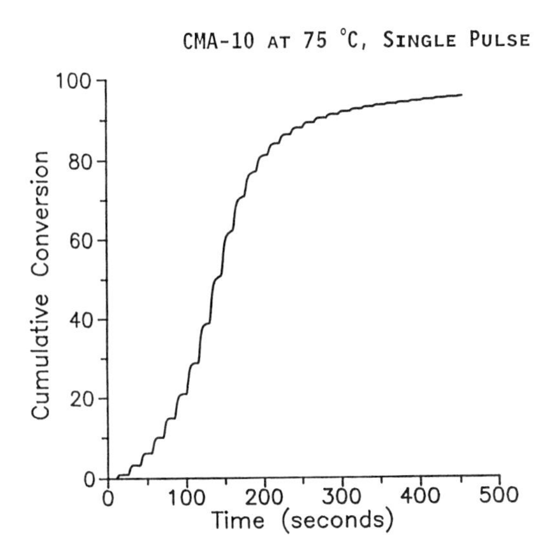

Figure 3. Cumulative conversion plot for CMA-10 film (1 wt percent DMAP) at 75 °C from exotherms initiated by XeF (351 nm) laser pulses (0.64 mJ cm^{-2} $pulse^{-1}$; 15 seconds apart).

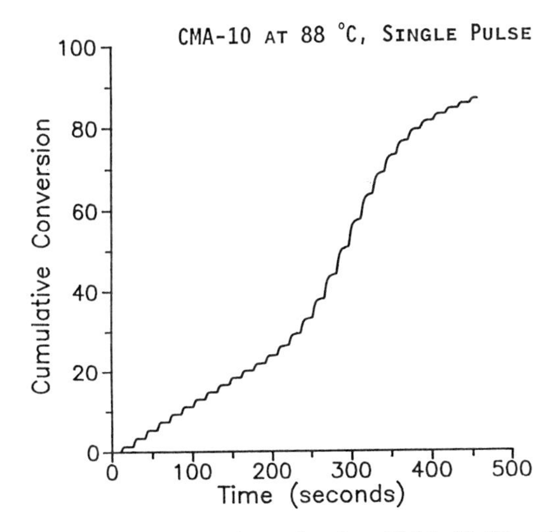

Figure 4. Cumulative conversion plot for CMA-10 film (1 wt percent DMAP) at 88 °C from exotherms initiated by XeF (351 nm) laser pulses (0.64 mJ cm^{-2} $pulse^{-1}$; 15 seconds apart).

temperature. This is interesting since the natural effect of temperature, as illustrated by the results in Figure 5 for only 10 pulses, is to increase the rate.

To provide additional, and somewhat more quantitative, information about the efficiency of the polymerization of CMA-10, we have calculated polymerization quantum yields for CMA-10 at 88 °C. Figure 6 provides a pictorial of our procedure for obtaining quantum yields. Initally, each sample is exposed to the output of a medium prerssure mercury lamp (filtered with a 366 nm band pass filter) via a rapid shutter until a given percent conversion is attained: the percent conversion is calculated from the integrated area under the exotherm curve. After allowing the exotherm to return to a baseline value, the sample is exposed by shuttering the mercury lamp for a short period of time. This generates a small exotherm from whence the percent, and thus molar quantity, of monomer converted by the second shuttered exposure can be calculated. By knowing the exact intensity of the lamp at 366 nm using a black body absorber in the DSC pan, the quantum yield of polymerization can be calculated. Approximate lower limit quantum yields for CMA-10 at 88 °C obtained via the method in Figure 6 are lower in the isotropic phase ($\Phi_{pol, iso} \approx 3700$) than in the medium ($\Phi_{pol, sm} \approx 6500$) after the texture change occurs: this is consistent with the results in Figure 4. (We note that the quantum yield of 6500 is a lower limit value due to some scattering in the medium which can reduce the effective irradiation fluence).

Semifluorinated Monomer. Figure 7 shows a pictorial description of the thermal behaviour of the semifluorinated acrylate monomer designated as F12H10A (11 CF_2 groups, 1 CF_3 group, 10 CH_2 groups, 1 acrylate functionality). Both optical microscopy and DSC confirm the transition temperatures labelled in Figure 7. The smectic phase designated with an S in the diagram has the texture of a smectic B phase (or closely related variant thereof) when examined by polarized optical microscopy. For each transition (isotropic to smectic and smectic to crystalline), there is supercooling, which seems quite reasonable in view of the difficulty in organizing into a highly ordered phase. Figure 8 shows DSC exotherms resulting from single pulse of the excimer laser spaced 15 seconds apart. For Figure 8, the scale is in mcal sec^{-1}, although it is not explicitly shown on the figure. Results shown for polymerization at two temperatures, 80 °C and 100 °C, are dramatically different. At 80 °C, the exotherm generated by the first laser pulse is extremely large with the total area corresponding to about 44% conversion while at 100 °C only about 1% of the monomer is converted by the first laser pulse. This difference between the results for F12H10A and CMA-10 are striking.

There are two differences between the polymerization of F12H10A and CMA-10 which may help to explain their respective responses. First CMA-10, a methacrylate monomer, would be expected to have a lower propagation rate constant than F12H10A, an acrylate monomer, since methacrylate monomers have an intrinsically lower propagation rate than acrylate monomers if the side chains are identical. This factor is minimized in the present case since we have noted in recent work that the methacrylate analog of F12H10A is as equally efficient as F12H10A when polymerization is initiated with the monomer in the

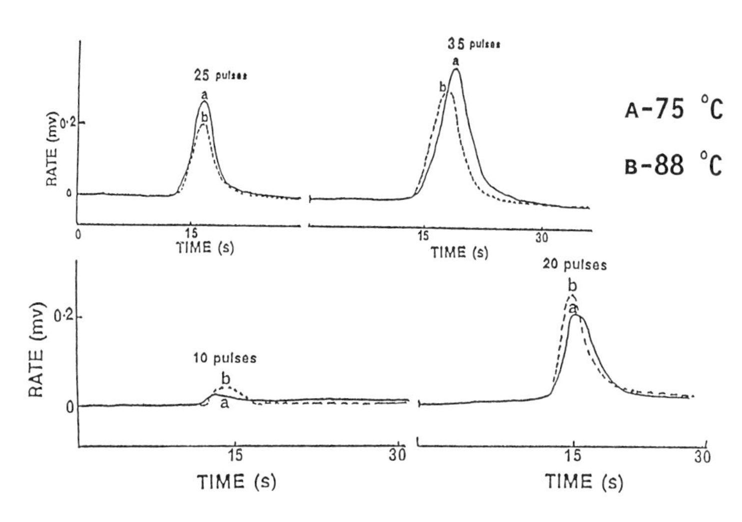

Figure 5. DSC polymerization exotherms for CMA-10 film (1 wt percent DMAP) at 75 °C (a curves) and 88 °C (b curves) initiated with a pulsed XeF (351 nm) laser operating at 5 Hz (exact fluence per pulse not determined but much lower than that used in Figures 2-4).

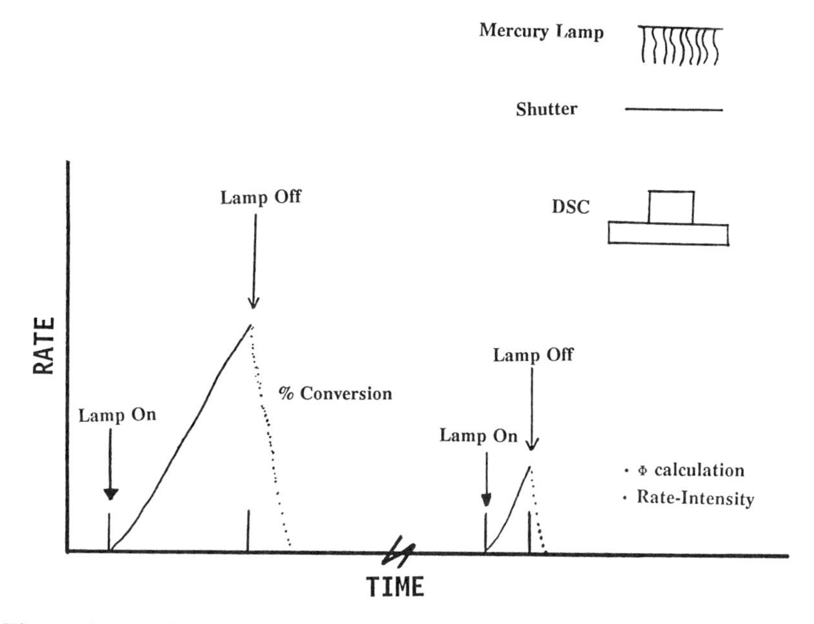

Figure 6. Pictorial of method employed to determine polymerization quantum yields for CMA-10.

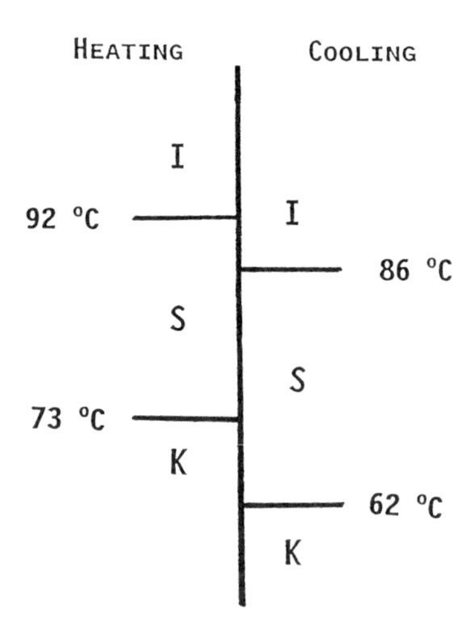

Figure 7. Transition temperatures for F12H10A.

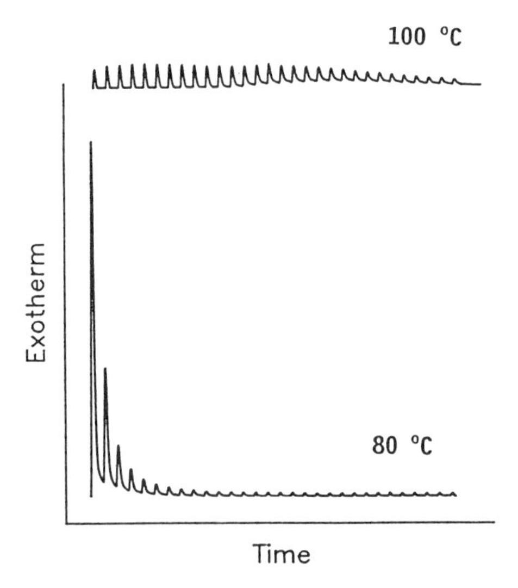

Figure 8. DSC polymerization exotherms for F12H10A film (1 wt percent DMAP) initiated by XeF (351 nm) laser pulses (3.89 mJ cm^{-2} $pulse^{-1}$) spaced 15 seconds apart at 80 °C and 100 °C.

mesophase: a description of this phenomena is relegated to a separate publication. The second, and perhaps most important, difference noted between polymerization of F12H10A and CMA-10 is the nature of the highly ordered F12H10A polymerization medium in which a smectic B type texture is maintained during the polymerization process: the CMA-10 polymerization medium maintains a smectic texture during polymerization when viewed by a cross polarized optical microscope. The very high order of the F12H1OA system may well result in a large reduction in the termination rate constant and a concomitant increase in the polymerization rate: such a supposition is in concert with literature suggestions (4,15).

Another aspect of the exotherm curves in Figure 8 deserves attention. The exotherm maxima for polymerization at 100 °C show an increase after the 12th pulse. The resultant plot in Figure 9 of percent conversion versus time shows a distinct discontinuity at about 25-30 percent (pulses 12-14). At this conversion there is a change in the phase of the medium from a purely isotropic to a mesomorphic medium: probably smectic. However an exact identification of the texture and determination of whether the phase is a pure monophase or a two-phase system, at least immediately after the rate increase occurs (25-30 percent conversion), has not been accomplished.

Although the percent conversion on the first pulse from the laser for F12H10A is indeed quite large in Figure 8, it is of interest to see if an even higher percent conversion can be attained. One way of increasing the conversion efficiency is to use a different photoinitiator which has a higher quantum efficiency for initiation. Schnabel and Sumiyoshi [14] have reported that the quantum efficiency for radical generation by 2,4,5-trimethyl benzoylphosphine oxide (TPO) is very high. Figure 10 shows a typical plot of exotherm rate versus time for F12H10A at 80 °C and 100 °C with 1 weight percent TPO. The exotherm obtained for exposure to the first laser pulse is extremely high at 80 °C corresponding to a percent conversion of about 74 percent. Plots in Figure 11 of cumulative conversion versus time, derived from Figure 10, confirm the rapid polymerization rate and high efficiency at 80 °C. The plot of 100 °C, while exhibiting a more efficient process than in Figure 9 where DMAP was employed, still shows a delay in the onset of the rate acceleration process which occurs at about 25-30 percent conversion. The effect of using TPO is thus seen to increase the rate of polymerization of F12H10A in both the liquid crystalline and isotropic systems.

Conclusions

In this paper, the kinetic rates of polymerization of two methacrylate (acrylate) monomers with nonaromatic mesogenic groups have been explored using a DSC to record polymerization exotherms. Overall, we observe that polymerization rates in the liquid crystalline phase are faster than in the isotropic phase. Specific conclusions are:

1. For the CMA-10 monomer, polymerization at temperatures in which the monomer is initially in a lower order cholesteric or

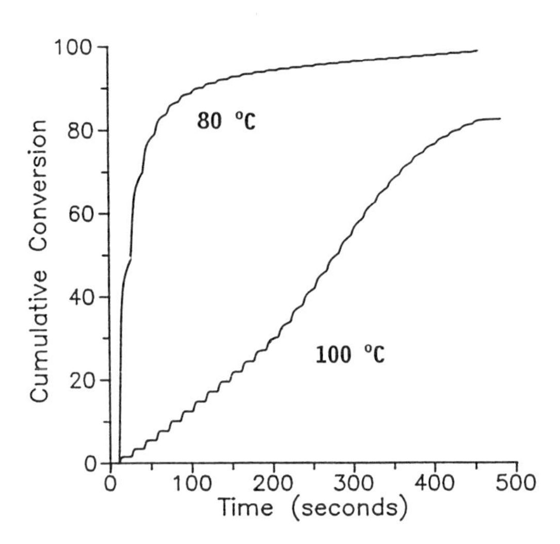

Figure 9. Cumulative conversion plot for F12H10A film (1 wt percent DMAP) at 80 °C and 100 °C.

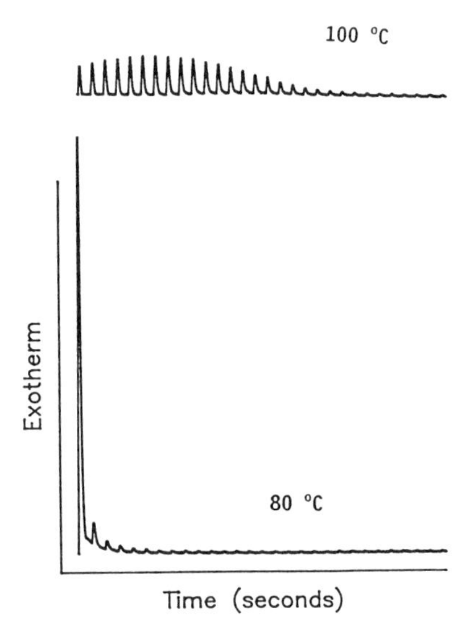

Figure 10. DSC polymerization exotherms for F12H10A film (1 wt percent TPO) initiated by XeF (351 nm) laser pulses (3.79 mJ cm^{-2} $pulse^{-1}$) spaced 15 seconds apart at 80 °C and 100 °C.

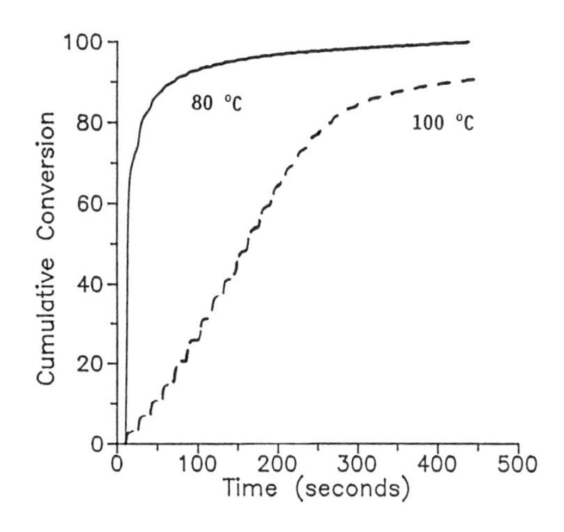

Figure 11. Cumulative conversion plot for F12H10A film (1 wt percent TPO) at 80 °C and 100 °C.

isotropic phase leads to an abrupt change in the medium texture during the polymerization process and an accompanying rate acceleration.

2. Polymerization of the highly ordered smectic B type semifluorinated F12H10A is extremely rapid and conversions of between 70 and 80 percent can be obtained by subjection to a single laser pulse.

Acknowledgements

This research is supported by NSF Grant DMR8917485 (Polymers Program). We also acknowledge the NSF EPSCoR program through grant No. R11-8902064, the State of Mississippi, and the University of Southern Mississippi.

Literature Cited

1. Paleos, C.M. *Chem. Soc. Rev.*, *14*, 45 (1985).
2. Barrall, E.M. and Johnson, J.F. *J. Macromol. Sci., Rev. Macromol. Chem.*, *17*, 137 (1979).
3. Broer, D.J.; Finkelman, H.; Kondo, and K. *Makromol. Chem.*, *189*, 185 (1988).
4. Broer, D.J.; Mol; G.N. and Challa, G. *Makromol. Chem*, *190*, 19 (1989).
5. Broer, D.J.; Boven, J.; Mol, G.N. and Challa, G. *Makromol. Chem.*, *190*, 2255 (1989).
6. Broer, D.J.; Hikmet; A.M. and Challa, G. *Makromol. Chem.*, *190*, 3201 (1989).
7. Broer, D.J.; Mol, G.N. and Challa, G. *Makromol. Chem.*, *192*, 59 (1991).
8. Shannon, P.J. *Macromolecules, 16*, 1677 (1983).
9. Shannon, P.J. *Macromolecules*, *17*, 1873 (1984).
10. Shannon, P.J. U.S. Patent 4,614,619 (1986).
11. Shannon, P.J. U.S. Patent 4,637,896 (1987).
12. Hoyle, C.E.; Chawla, C.P. and Griffin, A.C. *Mol Cryst. Liq. Cryst. Inc. Nonlin. Opt.*, *157*, 639 (1988).
13. Höpken, J.; Faulstich; S. and Möller, M. in *Intergration of Fundamental Polymer Science and Technology-5*, Lemstra, P.J. and Kleinjjes, L.A., Eds., Elsevier Appl. Sci.,
14. Schnabel, W. and Sumiyoshi, T. in *New Trends in the Photochemistry of Polymers*, Allen, N.S. and Rabek, J.F., Eds., Elsevier Applied Science Publishers, p. 73, New York (1985).
15. Blumstein, A. *Midland Macromol. Monographs*, 3, 133 (1977).

RECEIVED November 11, 1992

Chapter 10

Electronic Structure, Photophysics, and Photochemistry of Polysilanes

Josef Michl and Ya-Ping Sun[1]

Department of Chemistry and Biochemistry, University of Colorado, Boulder, CO 80309–0215

The current state of understanding of the electronic structure, photophysics, and photochemistry of poly(di-*n*-alkylsilane)s is surveyed with emphasis on behavior in room-temperature solution.

Polysilanes, $(RR'Si)_n$, are the silicon analogs of polyalkenes. They have recently elicited considerable attention as potential new photoresists, charge conductors, non-linear optical materials, polymerization initiators, and ceramic precursors (*1,2*). The sigma-sigma* electronic absorption of these formally fully saturated structures occurs at unusually long wavelengths, 300 - 400 nm depending on substitution, and frequently shows striking thermochromism. In solution, polysilanes are quite sensitive to ultraviolet light and their molecular weight decreases rapidly under irradiation.

In the following, we first discuss the electronic structure, then the spectroscopy and photophysics, and finally, the photochemistry of poly(di-*n*-alkylsilane)s. We deal primarily with the properties of room-temperature solutions, with emphasis on our own contributions. Most of the results pertain to high molecular weight polymers, but some to low molecular weight models, with chains as short as six silicon atoms.

Electronic Structure

While it is not our intention to provide an exhaustive review of the theoretical work that has been done on polysilanes, as such reviews are available elsewhere (*2,3*), it will be useful to summarize briefly the way in which molecular orbital theory accounts for the remarkable excited state properties of these compounds.

[1]Current address: Department of Chemistry, Clemson University, Clemson, SC 29634

0097–6156/93/0527–0131$07.00/0

What is Special about Silicon? The first issue to address is the striking difference in the excitation energies of saturated hydrocarbons, which absorb only below 200 nm, and of their saturated silicon analogues, whose solutions have an absorption peak at 300 - 350 nm. The contrast is stark even in the case of a single C-C versus Si-Si bond. It cannot be attributed to a significant difference in bond strengths, since these are nearly equal. Simple arguments (*2,4*) show, however, that the bond energy, i.e. the energy required for homolytic bond dissociation, is related to the triplet and not the singlet sigma-sigma* excitation energy. The latter, which is germane to the issue at hand, is instead related to the heterolytic dissociation energy. Since silicon is considerably easier to ionize, and actually also has a slightly higher electron affinity, the energy required for the formation of the Si^+ Si^- ion pair is much lower than that required to form the C^+ C^- pair. The low sigma-sigma* excitation energy of the polysilanes is thus ultimately attributable to the electropositive nature of silicon, and similar properties can be anticipated also for other metalloid and metallic elements.

Backbone Orbitals and States. Next, we need to consider the effect of linking a large number of Si-Si bonds into a chain. For most chemists, the natural way to develop an understanding of the electronic structure of a polymer is to extrapolate from the properties of the first dozen or so members of the homologous series, while the use the band theory for an infinite regular chain comes naturally to a physicist. We shall see in the following that the behavior of the polysilane chain in room-temperature solution is most readily interpreted as due to a sequence of linearly concatenated oligosilane segments that behave as nearly independent chromophores and have lengths ranging from a few to a few dozen SiR_2 units. As a result, an understanding of the electronic structure of relatively short polysilane chains actually may be more pertinent than the understanding of a perfectly periodic infinite chain that is offered by band theory.

Recent theoretical studies that have taken the chemist's approach used ab initio (*2,3,5*) or semiempirical (*6*) methods of calculation, and their findings support a simple model based on the use of sp^3 hybridized silicon orbitals as the basis set. The origin of this model goes back to Sandorfy's C and H models for the structure of alkanes (*7*), but electron repulsions are now included explicitely.

The simplest approximation (model C) considers only the two hybrid orbitals on each silicon that are directed towards neighboring silicon atoms in the chain. They are orthogonal and interact with each other through a geminal resonance integral. The interaction between hybrids aimed at each other and located on neighboring silicon atoms occurs through a more strongly negative vicinal resonance integral. Such a linear chain of orbitals interacting by means of resonance integrals of alternating magnitude is isoconjugate with the chain of orbitals participating in the pi electron system of a polyene with alternating bond lengths, and the same two groups of molecular orbitals result, one bonding and one antibonding. In the case of the silicon chain, they are of sigma and sigma* type, respectively, and are said to result from sigma conjugation or delocalization.

It should be noted that the analogy to polyenes is incomplete in that in the sigma conjugated chain the alternation of the magnitudes of the resonance integrals is an inevitable consequence of having two hybrid orbitals per silicon atom, and its sense cannot be altered by a simple geometrical distortion. In contrast, in the pi conjugated polyene chain, with a single orbital per carbon atom, such inversion of the sense of alternation of the resonance integrals requires merely an adjustment of bond lengths, and permits the formation of solitons, polarons, etc., after electronic excitation.

One- and Two-Photon Excited States. A property that polysilanes and polyenes share at this level of approximation is alternant symmetry (*8*), also sometimes referred to as charge conjugation symmetry. A consequence of this symmetry is the pairing of orbital energies, the degeneracy of the energies of certain configurations, and the separation of electronic states into two categories, "plus" and "minus". One-photon transitions from the ground state, which is of the "minus" type, are only allowed into the "plus" states, and two-photon transitions are only allowed into the "minus" states.

In both series, polysilanes and polyenes, the excitation from the highest occupied to the lowest unoccupied molecular orbital (HOMO-LUMO) is almost unaffected by configuration interaction and is responsible for the appearance of an intense sigma-sigma* transition in the electronic spectrum (note that the LUMO has one more node than HOMO, thus the transition moment for a one-photon absorption is large). The excitation energy of this state decreases with increasing chain length. Since the resonance orbitals alternate, it converges to a finite value as the chain length grows beyond limits, similarly as in polyenes. Experimentally, in the case of a silicon chain carrying alkyl substituents, this value is about 30 000 cm^{-1} in room-temperature solution.

Again in both series, polysilanes and polyenes, the excitations from the second HOMO to the LUMO (SHOMO-LUMO) and from the HOMO to the second LUMO (HOMO-SLUMO) are degenerate at this level of theory, because of the presence of alternant symmetry. The configurations mix to give rise to a "minus" and a "plus" combination, and the former mixes strongly with the doubly excited $HOMO^2$-$LUMO^2$ configuration, which also is of the "minus" type. Three totally symmetric states result. They are a "minus" state at low energy, a "plus" state above it, and another "minus" state at the highest energy. In long polyenes, the lower "minus" state ($2A_g$) actually lies about 4 000 cm^{-1} below the HOMO-LUMO ($1B_u$) state. In long sigma-conjugated polysilanes, where the alternation of the resonance integrals is more pronounced, it lies about 8 000 cm^{-1} above the HOMO-LUMO state, and the energies of the two states can be fitted quantitatively by adjusting the assumed degree of alternation (*6*). What cannot be adjusted within this model is the exact one-to-one mixing of configurations that are degenerate because of alternant symmetry, such as SHOMO-LUMO and HOMO-SLUMO.

Note that all three configurations whose mixing results in the three states just discussed carry two-photon absorption amplitude. Two of them (SHOMO-LUMO and HOMO-SLUMO) are obtained from the ground configuration by promoting an electron into an orbital which has two more nodes, and one ($HOMO^2$-$LUMO^2$) by promoting two electrons, each into an orbital that has one more node. However, the resulting three states do not share the two-photon absorption intensity equally. As noted above, the "plus" state can have none, and it so happens that the contributions of the three configurations largely cancel in the lower "minus" state and add in the upper one. Indeed, in polyenes, the lower "minus" state has been found to be two-photon allowed, but only rather weakly.

In contrast, the lower "minus" state of polysilanes has been observed to carry one of the largest two-photon cross-sections known (*8*). It might be possible to account for this within the simple model used so far since the cancellation of the contributions from the three configurations may be less strong than in the polyenes, but it is probably preferable to admit the inadequacy of a procedure that ignores two of the four hybrid orbitals on each silicon atom.

The Effect of the Presence of Bonds to Substituents. When the remaining two hybrids on each silicon are introduced, and also the two substituent orbitals that are used to make the Si-C (or Si-H) bonds (Sandorfy model H), alternant symmetry is lost, and this level of description agrees well with the results of ab initio calculations (*3*). The loss of orbital energy pairing leads to a loss of configuration degeneracy and of the separation of states into two categories. Now, the three configurations that carry two-photon absorption amplitude are widely separated in energy and hardly mix at all when configuration interaction is introduced. Three states still result, but all are two-photon alowed.

The details of the orbital mixing that destroys the alternant symmetry will not be discussed here (*2,3*). Briefly, for a planar zig-zag chain the newly introduced orbitals combine into four groups: sigma, sigma*, pi, and pi*. The interaction of the newly introduced sigma orbitals with backbone sigma orbitals already present in the Sandorfy model C compresses the latter closer in energy and reduces the energy difference between the HOMO and the SHOMO. The interaction of the newly introduced sigma* orbitals with the backbone sigma* orbitals already present spreads the latter farther apart in energy and increases the energy difference between the LUMO and the SLUMO. The interactions are such that the HOMO, with a node at every Si atom and nearly pure Si 3p character (pure in the case of an infinite chain), cannot participate efficiently and largely remains restricted to the Si-Si bonds as it was in the Sandorfy C model. The LUMO, on the other hand, with an antinode at every Si atom, and significant Si 3s character, participates heavily and acquires significant contributions from the orbitals of the Si-C bonds. As a result, the SHOMO-LUMO excitation takes much less energy than the HOMO-SLUMO excitation, the two configurations no longer are degenerate, and alternant symmetry is lost.

Pi Symmetry Orbitals and States. When the chain is planar, the pi and pi* orbitals newly introduced in the Sandorfy H model are precluded by symmetry from mixing with those already present. The pi orbitals have a large contribution from the carbon orbitals and are low in energy. Excitations out of them require high energy and are not of interest in the present context. In contrast, the pi* orbitals have a large contribution from the Si 3p orbitals perpendicular to the plane of the chain. They are relatively low in energy, and are interleaved with the sigma* orbitals. As a result, excitations out of them require only moderate energy, and sigma-pi* configurations are energetically close to sigma-sigma* configurations. The ab initio results (*3*) suggest that for the shortest chains a sigma-pi* state actually is the lowest of all the valence excited states, but that its energy does not drop as rapidly as that of the sigma-sigma* states when the polysilane chain increases in length. Therefore, for chain segments of most interest in polysilane spectroscopy, the lowest excited state is of the sigma-sigma* (HOMO-LUMO) type. The calculations are not reliable enough to predict exactly at which chain length the change in the state order occurs, but we shall see below that there are experimental indications that it is somewhere between a 6-membered and a 10-membered chain.

Effects of Chain Conformation. Both models considered so far, Sandorfy C and H, ignore interactions between hybrid orbitals that are not nearest neighbors. Therefore, they are incapable of describing properly the effects of changes in the SiSiSiSi dihedral angles in the polysilane chain. These are included automatically in the descriptions provided by ab initio (*3,5*) or standard all-valence electron semiempirical (*9*) calculations of the excited states. The geminal and vicinal resonance integrals between neighbors clearly do not change their values upon rotation along a Si-Si bond, nor do the resonance integrals between next nearest neighbors. However, those between a hybrid orbital and another separated from it by two other hybrids do, in a way well known from studies of syn and anti periplanar interactions. In the trans geometry, this resonance integral is positive (overlap is negative), while in the eclipsed cis geometry it would be negative (overlap is positive). Somewhere near the gauche geometry, the value of the resonance integral goes through zero.

The introduction of the relatively small 1,4 resonance integrals represents a minor perturbation that does not change the overall bonding in polysilanes dramatically. However, it does change the energies of the frontier orbitals in a way that is easily understood in qualitative terms (*9*). When the 1,4 resonance integral is positive, as in the ordinarily favorable trans geometry, its introduction destabilizes the HOMO and stabilizes the LUMO. When the integral is negative, as in the cis geometry, the opposite is true. Of course, a cis geometry is normally not sterically favorable because of interactions between the alkyl substituents. When the 1,4 resonance integral vanishes, as in the neighborhood of the ordinarily favorable gauche geometry, the consideration of the 1,4 interactions makes no difference in orbital energies.

Note that it is only the 3p part of each hybrid orbital that is sensitive to changes in the dihedral angle, and not the 3s part. The conformationally driven changes in the energy of the LUMO, which has significant 3s character as noted above, are therefore smaller than the changes in the energy of the HOMO.

The changes in the orbital energies that occur as a result of 1,4 interactions are reflected in corresponding changes in state energies, which now also become conformationally dependent. In agreement with the simple arguments, both ab initio and semiempirical calculations predict higher excitation energies for the gauche than for the trans arrangement of the Si-Si bonds.

Qualitative arguments suggest that this gauche - trans difference will cause a localization of orbital densities and transition densities on trans chain segments when these are separated by one or more gauche links, and this is indeed supported by numerical calculations at the INDO/S level (*9*). The localization observed in the results of these calculations is not complete, and its magnitude may well depend on the approximations in the method of calculation (*10*), but there is little doubt that the tendency of the polysilane chain to act as a collection of nearly independent chromophores, deduced from experimental results, is related to the conformational dependence of the 1,4 resonance integrals, although it might depend critically on other factors as well.

Spectroscopy and Photophysics

As already noted, room-temperature solution spectra of high molecular weight poly(di-*n*-alkylsilane)s are characterized by a strongly allowed one-photon transition that dominates both the absorption and the emission (*11*). The absorption peak is located at 31 000 - 32 000 cm^{-1}, the full width at half height is about 3 000 - 3 500 cm^{-1} and the extinction coefficient is about 10^4 per silicon unit. Its position depends only slightly on the nature of the alkyl substituent. When excited at energies above the absorption maximum, the characteristics of the emission are independent of the excitation energy. Its peak is located at 29 000 cm^{-1}, the half-width is about 1 900 cm^{-1}, the lifetime is about 165 ps, the quantum yield is 0.3 - 0.4. The properties of the emission are virtually totally independent of the nature of the alkyl substituent, except that the quantum yield increases from 0.3 to 0.4 as the alkyl is extended from *n*-butyl to *n*-tetradecyl. In low-temperature glasses the absorption and emission bands are usually found near 28 000 cm^{-1} and are much sharper.

The photophysical properties are not affected significantly by isotopic substitution in the alpha or beta methylene groups of the alkyl chain (*11*). As the molecular weight is reduced, the absorption and emission peaks shift to higher energies, but even chains as short as 16 or only 10 Si atoms still resemble the high polymer very strongly, as is discussed in more detail below.

The direction of the transition moment responsible for the transition is parallel to the chain direction (*9,12-14*), in line with the sigma-sigma* assignment.

As noted above, in addition to the one-photon allowed HOMO-LUMO transition just discussed, a two-photon allowed transition located about 8 000 cm^{-1} higher has also been characterized experimentally by simultaneous two-photon absorption (*8,15*). This clearly is the lowest of the three transitions expected to result from the three two-photon allowed configurations discussed above. Measurements of the successive absorption of two photons confirmed the results and revealed an additional two-photon allowed state at nearly twice the energy of the one-photon allowed state (*16*). It has been proposed on the basis of ab initio calculations (*3*) that this higher state corresponds to the doubly excited $HOMO^2$-$LUMO^2$ configuration. The third expected two-photon state remains to be observed. Of course, if the alternant symmetry preserving Sandorfy C description (*6*) is correct, the upper observed state would have to correspond to the upper "minus" state and the third expected state ("plus") would have zero two-photon intensity.

Spectral Inhomogeneity. There is much experimental evidence that relatively low-energy electronic excitations are localized over a fairly small number of silicon atoms in the polymer backbone at any one time, so that in effect the polysilane molecule behaves as an inhomogeneous collection of related but not quite identical chromophores, communicating by energy transfer. All of them contribute to the observed absorption, with weights dictated by their oscillator strengths and relative occurence, but only the lowest-energy ones dominate the observed fluorescence. The experimental manifestations of such spectral inhomogeneity are numerous.

Spectral Shape of Fluorescence. The spectral shape of the fluorescence, which is narrower than absorption to start with, is invariant with the excitation energy as long as the latter lies above the absorption maximum. However, it narrows down strikingly at lower excitation energies, as its blue edge recedes while the red edge hardly changes at all (*11,13*). Apparently, at the lowest excitation energies only the lowest-energy chromophores are excited and emit. At higher excitation energies, these chromophores still contribute heavily to the emission since they are populated by energy transfer, but now, at least some of the fluorescence originates from higher-energy chromophores as well.

Fluorescence Lifetime. As might be expected from this description, the fluorescence decay time is a function of the emission energy monitored [e.g., for poly(di-n-hexylsilane) excited at 34 800 cm^{-1} it varies from 130 ps for observation at 30 800 cm^{-1} to 170 ps at 27 800 cm^{-1}] (*11*). The 165 ps emission decay observed with a broadband filter, and thus averaged over all emission energies, is close to exponential and does not vary significantly with the excitation energy, presumably because very rapid energy transfer establishes a very similar "standard" distribution of emitting chromophores and the averaged decay curve is not affected much by the presence or absence of emission from the shorter-lived higher energy chromophores (*11*).

Fluorescence Quantum Yield. The lack of dependence of the average emission lifetime on the excitation energy contrasts with the observation that the total fluorescence quantum yield depends on it strongly (*11-13*). For excitation at the energy of the absorption maximum or above, the quantum yield is on the order of 0.3 - 0.4, and as the excitation energy is lowered to the red edge of the absorption band, it increases by about a factor of two. Apparently, radiationless deactivation competes more efficiently with the fluorescence from higher-energy chromophores and with energy transfer from these to the lower-energy chromophores. This cannot be due merely to a greatly increased rate of energy transfer, which would not affect the total fluorescence quantum yield. Possibly, it is due to an increase in the rate constant of a non-radiative process such as internal conversion, intersystem crossing, or photochemical reaction. It has been proposed as more likely (*11*) that the high-energy chromophores responsible for the bulk of absorption at and above the absorption maximum have a much reduced fluorescence rate constant due to a change in state order, so that they are essentially non-fluorescent and both energy transfer and the other non-radiative processes can compete better. The quantum yield of 0.3 - 0.4, observed at high excitation energies, reflects the efficiency of energy transfer to the lower-energy chromophores. This fits the observation that the shape of the fluorescence no longer depends of the excitation wavelength when the latter lies above the absorption maximum. Unfortunately, the quantum yields of internal conversion, intersystem crossing, and photochemical reactions have not been measured.

We shall see below that in line with this postulate, linear permethylhexasilane (*17*) is nearly non-fluorescent in room-temperature solution, its weak emission is out of line with expectations based on its very normal absorption band, and it is very different from that of the polymer and those of linear permethyldecasilane (*18*) and linear permethylhexadecasilane (*19*).

However, at room temperature, even linear permethyldecasilane (*18*) has a significantly smaller fluorescence quantum yield than permethylhexadecasilane (*19*) or polysilanes. As we shall see below, this is attributable to a thermally activated radiationless deactivation channel, presumably internal conversion, that is available only to the shorter oligosilanes. This, too, probably contributes to the absence of emission from the high-energy chromophoric segments in the polymers. The observed lifetime and quantum yield of emission from the lowest-energy chromophores can be combined to yield a radiative lifetime of about 250 ps, corresponding to an oscillator strength of about 1.8. Since the contribution of each SiR_2 unit to the oscillator strength can be estimated at about 0.09 from the absorption spectra, it follows that the average emitting chromophore stretches over roughly 20 silicon atoms in poly(di-*n*-hexylsilane) (*13*). Other authors arrived similarly at somewhat larger estimates of 30 - 60 Si atoms (*20*), depending on substituents. From measurements of excited state absorption saturation, the average value of 25 Si atoms per chromophore in poly(di-*n*-hexylsilane) was deduced (*16*).

Fluorescence Polarization. Further evidence for spectral inhomogeneity and energy transfer is provided by steady-state (*9,11,13,21*) and time-resolved (*8,22*) measurements of fluorescence polarization. The steady-state degree of polarization is near the theoretical upper limit when the red edge of the absorption band is excited and it drops to nearly the isotropic value of zero for excitation energies above the absorption maximum. This is as expected if energy transfer from the high-energy to the low-energy chromophores competes well with fluorescence, and if there is little correlation between the orientation of the polymer chain at various points in space, as expected for a random-coil (*23*) chain polymer. The time-resolved measurements demonstrated that the energy transfer occurs on a ps scale.

Fluorescence Yield and Lifetime Quenching. Finally, fluorescence quenching studies (*11,24*) also offer evidence for spectral inhomogeneity in room-temperature solutions of polysilanes. The quenching of the fluorescence of the high molecular weight polymers and of the model compounds, permethylhexadecasilane and permethyldecasilane, with CCl_4 is predominantly of the static kind. The Stern-Volmer plots are essentially identical in hexane and hexadecane, in spite of the ninefold difference in their shear viscosities. The plot for the lifetime is linear, that for the quantum yield is strongly curved upwards. The usual analysis yields large static quenching volumes and shows that a CCl_4 quencher molecule located anywhere within touching distance of either model oligosilane molecule quenches the excitation instantaneously. As is usual, the Stern-Volmer plots for the low-molecular weight model compounds do not depend on the excitation energy significantly.

The static quenching volumes measured for poly(di-*n*-hexylsilane) (*11*) are an order of magnitude smaller than those of poly(methyl-*n*-propylsilane) (*24*), presumably due to a blocking of access to the silicon backbone by the bulky n-hexyl chains. For the permethylated model compounds, the static quenching volumes are particularly large although in this case long-range energy transfer along the backbone is impossible. It appears that contact of the quencher molecule with an alpha CH_2 or CH_3 group is sufficient for instantaneous quenching, while contact with more distant parts of the alkyl chains is not.

In contrast to the short-chain model compounds, for both polymers, the static quenching volumes depend strongly on the excitation energy used. The values measured with excitation into the red edge of the absorption band are much smaller than those measured for higher-energy excitation. This could mean that the lower-energy chromophores have smaller spatial extent than the higher-energy ones, and thus have a smaller probability for a quencher molecule to be located in close vicinity. However, it could also simply reflect the fact that a higher-energy excitation results in energy transfer over longer distances, so that the total volume visited by the excitation before fluorescence occurs is larger, even if the higher-energy chromophores are actually smaller.The latter interpretation is strongly supported by the fact that the quenching volumes for the short-chain model compounds are independent of the excitation energy.

The two cases, low-energy chromophores being smaller or larger than high-energy chromophores, can be distinguished by considering the changes in the spectral shape of the fluorescence induced by the addition of increasing amounts of the quencher. The chromophores that extend the most in space should have the least chance to emit in the presence of a large quencher concentration, since they are nearly guaranteed to border on a quencher molecule. When compared with fluorescence shape in the absence of quencher, the emission shape in the presence of quencher should therefore be distorted by more heavily weighting the spatially smaller chromophores. This should be a more sensitive measure of the quenching volumes than the fitting of curvature in Stern-Volmer plots.

Measurement showed a quite clear relative enhancement of the higher-energy components of the fluorescence by the quencher when the excitation energy was at or above the absorption maximum, and it has been concluded that in room-temperature solution the higher-energy chromophores have the smaller spatial extent. The higher-energy chromophores in the polymers presumably contain fewer than about 20 (*13*) silicon atoms. When the excitation energy is as low as possible, the addition of the quencher has no effect on the emission shape. This is understandable, since only the lowest-energy chromophores are then excited.

Note that in the case of the polymer it could be argued that the larger effective quenching volume of the low-energy chromophores is solely a reflection of the fact that they are populated predominantly by energy transfer, so that the volume located along the transfer path gets to be at least partially included. After all, this presumably is the reason why the higher-energy chromophores have a larger measured static quenching volume.

However, the same fluorescence shape dependence on quencher concentration is exhibited by dilute solutions of permethylhexadecasilane, in whose emission energy transfer clearly does not play a major role, and whose static quenching volume determined from the Stern-Volmer plot indeed is independent of the excitation energy as a result, unlike those of the polymers. Thus, we conclude that it is the intrinsic physical size of the low-energy conformers of this molecule that is actually larger.

The Chain Segment and the Random Disorder Models. Two models for the spectral inhomogeneity are under active consideration presently, the segment model (*9,11-13,20,21,25-27*) and the random disorder model (*28,29*). There are indications that the former provides a good description of the dominant source of inhomogeneity in room-temperature solution, and that the latter properly describes the dominant source of the much smaller degree of inhomogeneous broadening present in the low-temperature form.

An identification of the dominant mechanism responsible for the spectral inhomogeneity is not facile. Perhaps the most striking difference between the two models is the spatial extent attributed by them to the chromophores as a function of their energy. According to the segment model, the lowest-energy chromophores are associated with the longest segments and thus have the largest extent in space. According to the random disorder model, the lowest-energy excitations are distributed over the smallest number of silicon atoms and have the smallest extent in space. We have just seen that there is some evidence that the higher-energy absorbing conformations of the silicon backbone are those with the smaller spatial extent.

According to the segment model, the dominating influence that determines the nature of the chromophores at any one time is the sequence of the trans and gauche dihedral angles. The gauche twists separate the chain into a series of short and long all-trans sequences, and these stand for the individual chromophores. Their excitation energies fall off with increasing length of the segments and are believed to reach a limiting value of roughly 30 000 cm^{-1} for about a dozen Si atoms in the segment. Beyond this point, the excitation energy decreases very little as the segment is extended further.

The distribution of segment lengths is a function of the temperature and in room temperature solution even quite short segments appear to be abundant. The relatively large differences between the excitation energies of the individual segments then represent the main source of inhomogeneity. In the low-temperature form, only quite long all-trans segments are present, and differences in their lengths have hardly any effect on the excitation energies, so that they cause a very small degree of observable inhomogeneity. Now, a second source of a dispersion of excitation energies, which was of only secondary importance at room temperature, may become dominant. This is an inhomogeneous distribution of the dihedral angles around their ideal values of 60 and 180 degrees, which undoubtedly also affects the excitation energy.

A random distribution of variations of the dihedral angles around their most probable values is the basic tenet of the random disorder model, which in itself need not attach any particular significance to a difference in sigma conjugation through a trans and a gauche link, as the segment model does. However, we have listed above some good theoretical reasons why this difference should be quite significant and it seems that a combination of the two models outlined presently represents the currently most satisfactory description of the situation.

Photophysics - a Summary. A schematic summary of the singlet state photophysics of poly(di-*n*-alkylsilane)s in room-temperature solution is provided in Figure 1, where n stands for the length of a chromophoric segment, and its excitation energy is plotted vertically (*11*). The dispersion of the excitation energies is the dominant source of spectral inhomogeneity. The distribution of the dihedral angles about their most probable values, and less important sources of inhomogeneity, such as distribution of conformations in the alkyl chains and randomness in the solvent surroundings, are not shown individually, but their integrated effect is indicated by the width of the absorption line associated with each segment as shown on the right-hand side. These lines have been weighted in a symbolic fashion to take into account two factors: the intrinsic oscillator strength, which increases with segment length, and the frequency with which a segment of a given length occurs in the sample. The observed absorption spectrum corresponds to the sum of contributions from all the segments.

The shortest few segments do not contribute significantly to emission, but the others do. The increase of the fluorescence rate constant with the increasing segment length is symbolized by the increasing thickness of the vertical line that indicates the emission process. The contributions of the individual segment lengths to the total fluorescence are shown on the right similarly as was done for absorption, with an estimated Stokes shift. The weights of the contributions take into account in a symbolic fashion the intrinsic fluorescence rate constants and the probability for excitation, dictated largely by energy transfer. The latter is shown in dotted lines as proceeding from a segment of any length to a nearby segment of greater length. The total fluorescence shape is obtained by summing the individual contributions and is clearly narrower than the absorption shape, both because the shortest segments do not emit and because of energy transfer from the shorter to the longer segments.

In addition to the processes shown in Figure 1, the singlet excited state of each segment is believed to be depopulated by intersystem crossing to the triplet

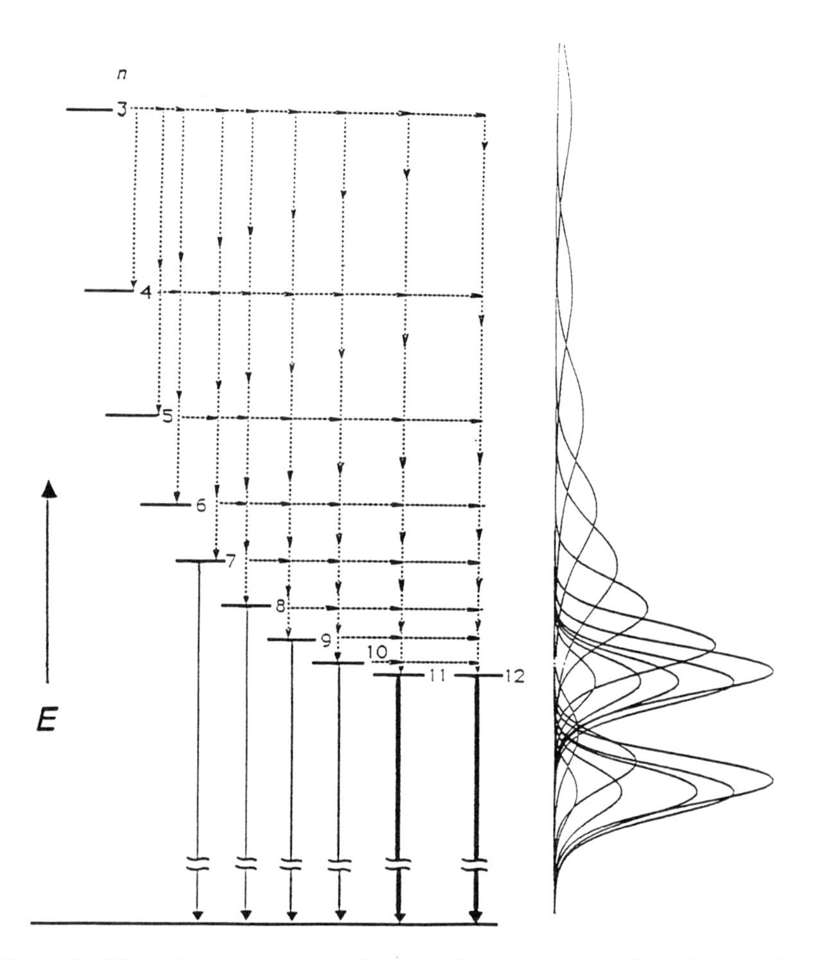

Figure 1. The primary processes in the solution photophysics of poly(di-*n*-alkylsilanes). See text.

state, by internal conversion to the ground state, and by photochemical reactions. These processes are believed to be jointly responsible for the deviation of the fluorescence quantum yield from unity. As we shall see below in the section dealing with photochemistry, it is considered likely that a good fraction of the triplet molecules undergo a subsequent photochemical reaction, too. The relative importance of these additional photophysical processes could well vary widely as a function of the length of the excited segment, particularly if the fluorescence rate constants for the short segments are indeed much smaller than those of the long ones. Moreover, it is known that short excitation wavelengths cause chain abridgement reactions in addition to chain cleavage reactions, while long excitation wavelengths do not (see below). This may well reflect different photochemical properties of segments of different lengths.

Model Compounds: $Si_{16}Me_{34}$, $Si_{10}Me_{22}$, and Si_6Me_{14}

It is of great interest to find out how long a polysilane chain needs to be before it acquires the spectroscopic, photophysical, and photochemical characteristics of the polymer. Also, if the segment model is correct, an investigation of the properties of short model chains should be very instructive for an understanding of the high polymer. These issues have been addressed recently in investigations of room-temperature solutions of permethylated linear silicon chains with 16, 10, and 6 silicon atoms (*17-19*), and we have already alluded to some of the results above.

Permethylated Hexadecasilane. This 16-silicon chain was found to be nearly identical to the high molecular weight polymer in almost all respects (of course, rotational Brownian motion will depolarize its solution fluorescence at room temperature). In particular, as already mentioned, the fluorescence shape depends on the excitation wavelength and on the presence of a quencher. Its quantum yield of 0.07 is constant at excitation energies above the absorption maximum and increases to 0.63 at the lowest excitation energies, while the emission lifetime remains constant at about 180 ps throughout. The radiative lifetime of 285 ps for the lowest-energy emitters is only slightly longer than those observed in the polymers, supporting the notion that the average emitting segments in the latter have close to 20 Si atoms. The fact that the fluorescence bandwidth is still much smaller than the absorption bandwidth, as in the polymers, might be in part due to intramolecular energy transfer from one segment to another, but is probably primarily due to changes in molecular conformation after excitation, which would be much easier than in the polymer, since both ends of the molecule are now free to move. Driving force for the extension of the segment length is absent in the ground state, but in the excited state it is provided by the concommitant decrease in the excitation energy.

There are two salient differences between $Si_{16}Me_{34}$ and the polymers (*19*):

(i) The absorption and emission peaks of $Si_{16}Me_{34}$ are shifted to about 34 100 and 30 500 cm^{-1}, compared with about 32 500 and 29 500 cm^{-1} for poly(methyl-*n*-propylsilane), poly(dimethylsilane) being too insoluble for measurement and comparison. This difference is attributed to a different segment length distribution: obviously, in $Si_{16}Me_{34}$, the longest linear segment can only contain 16 silicon atoms, and is present in only one out of many conformeric structures.

(ii) The fluorescence quantum yield of $Si_{16}Me_{34}$ measured at excitation energies above the absorption peak is only 0.07, whereas for poly(methyl-*n*-propylsilane), it is 0.4 (*19*), and its static CCl_4 quenching volume of 3 500 A^3 as determined from the Stern-Volmer plot does not depend detectably on the excitation energy, whereas that of the polysilanes that have been investigated does (*11,24*). The quenching volume corresponds roughly to van der Waals contact between the quencher and any part of the permethylated oligosilane.

These differences are rationalized by the absence of long-range energy transfer along the silicon backbone in the 16-silicon atom chain. The spectral inhomogeneity is still due to the presence of molecules with different straight segment lengths, i.e., of different conformers of $Si_{16}Me_{34}$. Now, however, excitation energy can only be transferred from one segment to another within an excited conformer and cannot be transfered very efficiently from those of higher energy to the more emissive ones of lower energy, as it can in the polymer. Thus, the intrinsicaly lower fluorescence quantum yield of the higher-energy chromophores, in which internal conversion, intersystem crossing, or photochemical reactions are apparently better able to compete, is clearly revealed and not masked as much by energy transfer. Also, their static quenching volume is not as much artificially increased by the partial addition of the volume located along the energy transfer path as it is in the polymers, and reflects the intrinsic size of the conformer itself. The more sensitive measure of conformer size, the changes of fluorescence shape with quencher concentration, are now undistorted by energy transfer and show that the higher-energy chromophores are less susceptible to static quenching, and therefore presumably are smaller, as already discussed above.

The photodegradation of $Si_{16}Me_{34}$ has not been investigated in detail, but appears to be significantly slower than that of the high polymers.

Measurements in a glass at 77 K had to be performed at concentrations as low as 2 x 10^{-7} M in order to avoid solute aggregation. They revealed the presence of many spectrally distinct non-interconverting conformers, two of which have been characterized fairly well by separate excitation and emission spectra. Another one, which is non-fluorescent, was characterized by absorption only. The exact conformations of the three species however remain unknown.

At 77 K, the observed bandwidths of the absorption and emission bands are quite comparable. The emission has a positive polarization degree of about one-third, somewhat lower than expected in the absence of energy transfer.

In summary, it appears that a 16-silicon chain is capable of existing in a variety of conformations with quite distinct photophysical properties dominated by the sigma-sigma* state. Since these conformers have quite widely varying excitation energies, it is probably useful to think at least of some of them as containing nearly independent chromophoric segments shorter than 16 silicon atoms.

Permethylated Decasilane. The spectroscopic behavior of this 10-silicon chain in room-temperature solution is still very much like that of the high polymers (*18*). The absorption and emission bands are shifted to even higher energies, 35 000 and 32 200 cm^{-1}, respectively, and the width of the former is twice that of the latter. In addition, the absorption spectrum contains a shoulder at 39 000 cm^{-1}. Upon cooling, the shoulder disappears, the absorption band red-shifts to 34 100 cm^{-1}, the emission band blue-shifts to 33 600 cm^{-1}, and both become much narrower.

Upon closer inspection, however, it is seen that the room-temperature solution photophysical behavior of this model compound deviates significantly from that of the 16-silicon chain and of the high polymers. Although the shape of the fluorescence depends very slightly on the excitation energy, it no longer follows the trend characteristic of the higher homologues, i.e., no change at excitation energies above the absorption maximum, the blue edge receding and the red edge constant as the excitation energy is reduced further (*18*). Moreover, the fluorescence shape is independent of the concentration of CCl_4 quencher (*24*).

When $Si_{10}Me_{22}$ is excited above its absorption maximum in room-temperature solution, the fluorescence quantum yield is only 5 x 10^{-4} and the lifetime is less than 20 ps. The quantum yield then increases up to 0.05 at the lowest excitation energies. Upon lowering the temperature, the quantum yield and lifetime of the emission increase continually and in parallel until about 200 K, and then remain constant. In the low-temperature limit, the quantum yield is 0.35 when the excitation is at energies above the absorption maximum, increasing up to about 0.8 at the lowest excitation energies, and the lifetime is 430 ps. In this limit, the fluorescence polarization degree is strongly positive, about one-third. The radiative lifetime of $Si_{10}Me_{22}$ is now roughly twice that observed for the polymers, in keeping with the estimate of 20 Si atoms per average emitting chromophore in the latter.

Compared to high molecular weight polysilanes, $Si_{10}Me_{22}$ is relatively stable to photodegradation in solution.

It has been concluded (*18*) that in room-temperature solution, and to a smaller degree perhaps at low temperatures as well, the 10-silicon chain is present as a mixture of conformers. After room-temperature excitation, they undergo radiationless processes or convert to one or a small number of conformers that do not differ much in their emission energy. The chief radiationless deactivation process is thermally activated; a preliminary estimate of the activation energy is 1.7 kcal/mol, with a frequency factor of 10^{11} s^{-1}, somewhat on the low side for a unimolecular process. It is likely that the thermal activation is due to the presence of a small barrier near the minimum in the lowest excited singlet potential energy surface, permitting escape along a geometrical coordinate that leads to a region of facile return to the ground state surface and regeneration of the starting material. It is presently not clear what the geometrical nature of this distortion might be.

If the barrier results from an avoided crossing of two first-order states, it is likely that it is present in longer and shorter polysilanes as well, since the same states are present in principle, but has a different size there. The absence of such a deactivation channel in the 16-silicon chain suggests that in this case the barrier is much higher, and this could be due to the fact that the local minimum in the sigma-sigma* state from which emission occurs is lower in energy relative to the other state involved in the suspected crossing. Its relative stabilization is presumably due to increased sigma delocalization. At this point, we can only speculate about the nature of the second state, but sigma-pi* is the logical candidate; this would indeed be higher in energy relative to the sigma-sigma* state in the longer chain. A tentative proposal is shown in Figure 2. Since we shall see below that the sigma-pi* state is most likely weakly emissive, and since dual fluorescence is not observed, the Figure provides a mechanism for an even faster depopulation of the sigma-pi* state to yet another state that communicates effectively with the ground state. At this point, this is purely a conjecture.

At low temperatures, when the radiationless decay channel is shut off, the fluorescence behavior of the 10-silicon chain is closer to that of the 16-silicon chain, but we have found no evidence for spectrally distinct conformers in 3-methylpentane glass. Apparently, all emissive conformers present at 77 K, if there is more than one, have similar emission spectra.

Permethylated Hexasilane. This 6-silicon chain compound has an absorption spectrum that corresponds quite closely to an extrapolation from the longer polysilanes and is still clearly dominated by an intense sigma-sigma* transition, but its fluorescence spectrum is strikingly different (*17*).

In room-temperature solution, the sigma-sigma* absorption band is now located at 38 500 cm^{-1}. Upon cooling to 77 K, it blue-shifts slightly and narrows down significantly.

The spectral shape of the fluorescence is independent of excitation energy and is not very different at room and low temperatures. The fluorescence appears quite ordinary and provides no indication of the spectral inhomogeneity characteristic of $Si_{16}Me_{34}$ and the high polymers. It is very likely that a mixture of conformers is present, but these apparently all adiabatically convert to the same conformer before emitting, or perhaps do not differ much in their emission spectra.

However, the spectral shape of the fluorescence is strikingly different from the fluorescence from all the higher homologs that have been investigated, in that it is much broader (at room temperature, 7 600 cm^{-1}, and at 77 K, 5 400 cm^{-1} full width at half height), and that it is very strongly Stokes shifted (about 10 000 cm^{-1}). The fluorescence is not easy to measure at room temperature, since the quantum yield is less than 10^{-4}. Upon cooling, the yield increases continuously, and at 77 K, it reaches 0.45. It is independent of the excitation energy, and the fluorescence excitation spectrum is virtually coincident with the absorption spectrum. The emission lifetime changes in parallel, and reaches 1.3 ns at 77 K. The temperature dependence of the quantum yield and of the lifetime indicate a thermal activation barrier of 2.1 kcal/mol and an unexceptional frequency factor of 2×10^{12} s^{-1} for the radiationless deactivation channel.

The radiative lifetime corresponds to a fluorescence rate constant of 3.5×10^{8} s^{-1}. In contrast, integration of the first absorption peak corresponds to a

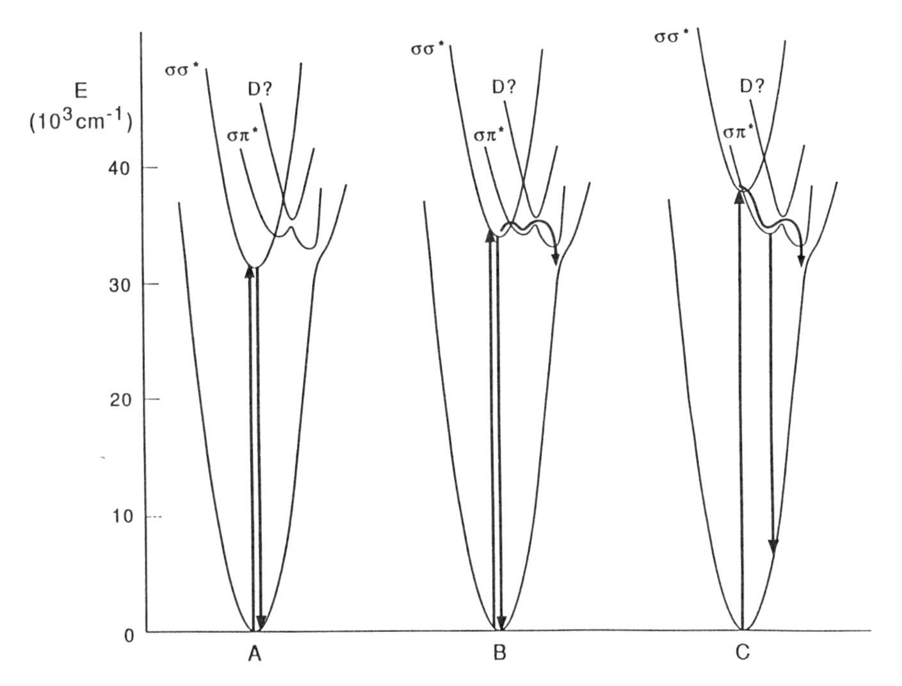

Figure 2. Potential energy surfaces. A tentative proposal accounting for the photophysics of A: $Si_{16}Me_{34}$, B: $Si_{10}Me_{22}$, and C: Si_6Me_{14}. See text.

value larger by a factor of three, $1 \times 10^9 s^{-1}$. This discrepancy is much larger than was found for the 10- and 16-silicon chains, where the two derivations agreed within 10 - 20 %. Taken together with the striking difference in the spectral shape, it suggests quite clearly that the emitting state in the 6-silicon chain compound is not the vertical sigma-sigma* state that is responsible for its intense absorption band, and also responsible for both the absorption and the emission bands in all the compounds with longer chains. The absorption counterpart of the emission is apparently a weaker band buried under the intense sigma-sigma* band. In such a case, the polarization of the emission may be expected to be dominated by intensity borrowed from the intense transition, and to be strongly positive regardless of the nature of the weaker transition. The observed degree of polarization indeed is 0.35, independent of the excitation and monitored energy.

We have presently no proof for the identity of the emitting state. The most likely candidate is a geometrically distorted sigma-sigma* state, perhaps with one long Si-Si bond. Another possibility is the lowest sigma-pi* state, calculated to lie below the sigma-sigma* state once the oligosilane chain is short enough. Perhaps the crossing of the sigma-sigma* and the sigma-pi* states suggested as responsible for the barrier located in the way of the radiationless decay process in the 10-silicon chain has now been changed further by an additional relative increase in the energy of the sigma-sigma* state in the even shorter 6-silicon chain, so that the barrier has disappeared altogether and there is no local minimum of sigma-sigma* character in the lowest excited singlet potential energy surface, but only one of sigma-pi* character. This picture is presented schematically in Figure 2.

The origin of the barrier responsible for the thermal activation of the radiationless decay of excited Si_6Me_{14} would then have to be different from that in $Si_{10}Me_{22}$, and this would account for the rather different frequency factors observed in the two cases. Figure 2 indicates a tentative proposal for the physical origin of both barriers, invoking a doubly excited state, but much additional work is needed before the nature of the states involved is clarified. The difficulty with the attribution of the emission to a sigma-pi* state is that the emissive state has an oscillator strength weaker by only a factor of three compared with the sigma-sigma* state, and yet is not detectable in absorption. The degree of vibronic intensity borrowing would have to be very different in absorption and in emission to account for this.

Regardless of the detailed interpretation of the origin of the extreme weakness of the emission from permethylated hexasilane, the near absence of room-temperature emission from this chromophore certainly fits well with the segment model for high molecular weight polysilanes, in which absence of fluorescent emission from short segments has been postulated (*11*). The photochemical degradation of Si_6Me_{14} has not yet been investigated quantitatively, but it does not appear to be particularly rapid. This suggests that the very efficient non-radiative channel may be internal conversion, as suggested in Figure 2.

In answer to the questions posed in the beginning of this section, it can now be stated that the properties of the short model chains support the segment model for the polymer, and that a chain as short as 16 silicon atoms is sufficient to emulate the photophysical behavior of the high polymer, while a 10-atom chain shows significant deviations, and a 6-silicon chain is totally different.

Photochemistry

At room temperature, photochemical degradation of polysilanes proceeds rapidly, and this has been one of the properties that originally directed the attention of chemists to this class of substances. Possible applications, say in the field of photoresists (*30*), mostly require the use of neat polymers, and attracted a fair amount of mechanistic investigation (for a leading reference to studies of direct laser ablation, which however appears to involve a thermal mechanism, see ref. *31*). Some applications, such as the use of polysilanes as photoinitiators (*32*), require an understanding of solution photochemistry. In the following, we shall concentrate on photochemistry in room-temperature solutions, as we did in the case of photophysics. In low-temperature glasses, the rate of photochemical degradation is greatly reduced.

Three primary photochemical processes have been postulated for poly(di-*n*-alkylsilanes) (*2,33-37*): (i) chain scission by homolytic cleavage, (ii) chain scission by reductive elimination, and (iii) chain abridgement by reductive elimination (Figure 3). Little is known about them quantitatively, and even less is known about the detailed course of the subsequent steps proposed in the Figure. Still, the three primary processes rest on reasonable experimental evidence, on precedent in the photochemistry of small oligosilanes, on and solid theoretical principles. They permit us to account for the final isolated products without invoking unprecedented reactions. When the process (ii) was originally proposed (*35*), there was not much precedent for it in the chemistry of low molecular weight oligosilanes, but this has since been obtained (*38*).

Experimental Evidence for Photoprocesses (i) - (iii). There is currently a dearth of interpreted direct observations of primary products, and all evidence for the occurence of the primary processes (i) - (iii) is indirect.

(i) Chain scission by homolytic cleavage is the only among the postulated primary processes that is expected to produce radicals. Evidence for the formation of some radical species is provided by the observation that the irradiation of polysilanes in the presence of olefins susceptible to radical-induced polymerization indeed causes such polymerization (*32*), and by the observation of persistent EPR signals from irradiated polysilane solutions (*35,36*). However, the former observation says nothing about the structure of the radicals, and the structure deduced from the EPR spectra of the persistent radicals corresponds to the cleavage of a C-Si and not a Si-Si bond. They are however clearly not formed by homolytic C-Si bond scission into a pair of radicals, since the alkyl radicals would yield an alkene and an alkane. The former would not survive the reaction conditions, as already mentioned above, but the latter would, and yet, it was not detected. The proposed mechanism of formation of the persistent radicals (*35,36*) is complex and is discussed below; it is consistent with a primary homolytic Si-Si bond cleavage. Their yield is only a few per cent.

Perhaps the least equivocal indirect evidence for homolytic Si-Si bond cleavage is provided by initial qualitative experiments in which polysilanes were subjected to exhaustive degradation by 254 nm irradiation (*33*). The final products were two- and three-silicon chains with two alkyls on each silicon atom and one hydrogen on each terminal silicon atom. These are just the products that would be expected if the primary radicals formed by Si-Si bond cleavage disproportionate. Deuterium labeling experiments demonstrated that radical disproportionation rather than indiscriminate hydrogen abstraction from the solvent or the alkyl side chains on the polysilane provides the source of the terminal hydrogen atoms in the disilane and trisilane products. When the irradiation is performed with light of wavelengths of 300 nm and longer, the polymer molecular weight is still degraded rapidly, but the resulting oligosilane fragments are longer, and disilanes and trisilanes are not formed. This can be understood readily, as light of these wavelengths is not absorbed efficiently by chains containing fewer than about seven silicon atoms, and these are therefore not degraded further.

Irradiation of a solution containing both a di-*n*-hexyl and a di-*n*-butyl polymer in the presence of a large excess of triethylsilane, which scavenges all silylenes, yielded evidence that short oligosilane chains containing both di-*n*-hexylsilane and di-*n*-butylsilane units are formed (*37*), and this is most readily interpreted as due to silyl radical recombination.

On theoretical grounds, it is likely that the homolytic cleavage process occurs in the triplet state, but there is almost no experimental evidence on this point. Piperylene, a known triplet quencher, reduces the degradation rate of poly(methyl-*n*-propylsilane) without affecting its fluorescence quantum yield (*37*), but other polysilanes, such as poly(di-*n*-hexylsilane), did not show this effect. It is conceivable that the rate of triplet quenching, like that of quenching by CCl_4, is reduced when the silicon backbone is shielded by bulky alkyl groups.

In summary, although the homolytic photocleavage of Si-Si bonds in polysilanes has not been observed directly, there is overwhelming indirect evidence in its favor.

(ii) Chain scission by reductive elimination has been first postulated in an attempt to rationalize the formation of the persistent radicals (*35,36*) and the scheme shown in Figure 4 was proposed. The proposal was supported by the observation of chain fragments carrying three alkyl groups on a terminal silicon atom. This would be difficult to account for in any other plausible way. Still, it was a relief to discover a completely analogous reaction for low-molecular weight oligosilanes (*38*). In both cases, this process plays only a relatively minor role relative to reactions (i) and (iii).

(iii) Chain abridgement by reductive elimination was postulated in the initial study of polysilane photochemistry (*33*), based the results obtained upon irradiation of the polysilanes in the presence of triethylsilane, a well-known silylene trapping agent. This photochemical process can be induced only with light of wavelengths shorter than about 300 nm (*34*), and it has been proposed

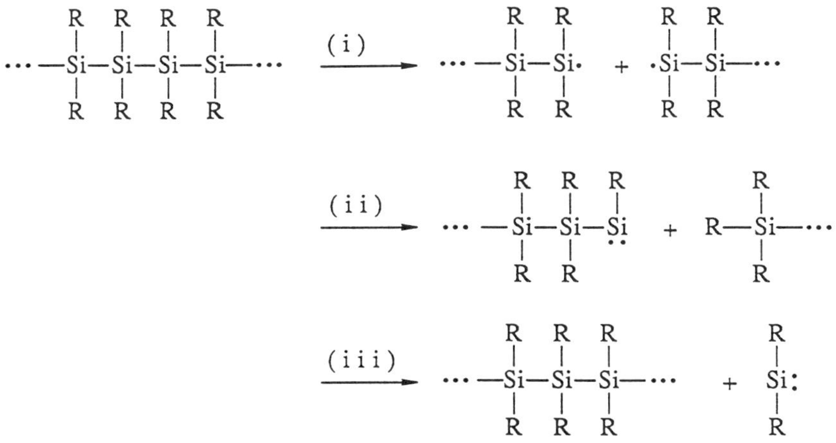

$$\cdots\!-\!SiR_2\!-\!SiR_2\!-\!SiR_2\!-\!SiR_2\!-\!\cdots \xrightarrow{(i)} \cdots\!-\!SiR_2\!-\!SiR_2\cdot \;+\; \cdot SiR_2\!-\!SiR_2\!-\!\cdots$$

$$\xrightarrow{(ii)} \cdots\!-\!SiR_2\!-\!SiR_2\!-\!\ddot{Si}R \;+\; R\!-\!SiR_2\!-\!\cdots$$

$$\xrightarrow{(iii)} \cdots\!-\!SiR_2\!-\!SiR_2\!-\!SiR_2\!-\!\cdots \;+\; SiR_2{:}$$

Figure 3. The three primary processes in the photochemistry of poly(di-*n*-alkylsilane)s.

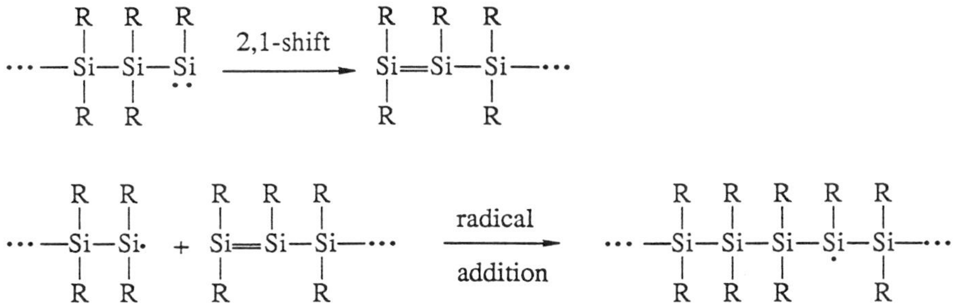

$$\cdots\!-\!SiR_2\!-\!SiR_2\!-\!\ddot{Si}R \xrightarrow{\text{2,1-shift}} RSi\!=\!SiR\!-\!SiR_2\!-\!\cdots$$

$$\cdots\!-\!SiR_2\!-\!SiR_2\cdot \;+\; RSi\!=\!SiR\!-\!SiR_2\!-\!\cdots \xrightarrow[\text{addition}]{\text{radical}} \cdots\!-\!SiR_2\!-\!SiR_2\!-\!SiR_2\!-\!\dot{Si}R\!-\!SiR_2\!-\!\cdots$$

Figure 4. The reaction sequence proposed to account for the formation of the persistent radicals characterized by EPR and ENDOR.

that it only occurs upon excitation of one of the shorter chromophoric segments. In addition to intermolecular trapping, there is some evidence indicating that the silylene can udergo intramolecular trapping by inserting into its own alkyl chain. This type of cyclization process is believed to be important in the laser-induced thermal ablation of neat polysilanes (*31*).

Theoretical Aspects of the Primary Photochemical Steps. The three primary photochemical steps are believed to follow paths that can be readily understood in terms of potential energy diagrams (*5*).

The triplet state of a polysilane chain, as well as the ground state, correlate with a silyl radical pair in the limit of an infinitely stretched Si-Si bond, while all other states correlate with higher energy limits. It is for this reason that the triplet state has always been under suspicion as being responsible for chain cleavage by homolytic scission.

Calculation of the potential energy surfaces along the paths corresponding to the two 1,1-elimination paths, (ii) and (iii), support the simple view that these are photochemically allowed singlet pericyclic reactions, with a pericyclic minimum half-way along the reaction path that provides an efficient funnel for return to the ground state.

Summary

Although much is now understood concerning the electronic structure, photophysics, and to a lesser degree, also the photochemistry of poly(di-*n*-alkylsilane)s, unresolved problems abound. This area of research is likely to provide a fertile ground for further experimental and theoretical work and for further development of theoretical understanding for a long time to come.

Acknowledgements

We are grateful to the Air Force Office of Scientific Research (Grant 91-0032), the National Science Foundation (Grant CHE-9020896), and IBM, Inc. (Grant 707312) for support of our work on polysilanes, and to two highly stimulating collaborators, R. D. Miller (IBM, Almaden) and R. West (University of Wisconsin, Madison).

Literature cited

1. West, R. *J. Organomet. Chem.* **1986**, *300*, 327.
2. Miller, R. D.; Michl, J. *Chem. Rev.* **1989**, *89*, 1359.
3. Balaji, V.; Michl, J. *Polyhedron* **1991**, *10*, 1265.
4. Michl, J. *Accounts Chem. Res.* **1990**, *23*, 127.
5. Michl, J.; Balaji, V. In *Computational Advances in Organic Chemistry*; Ogretir, C.; Csizmadia, I. C., Eds.; Kluwer: Dordrecht, The Netherlands, 1991, p 323.

6. Soos, Z. G.; Hayden, G. W. *Chem. Phys.* **1990**, *143*, 199.
7. Sandorfy, C. *Can. J. Chem.* **1955**, *33*, 1337.
8. Thorne, J. R. G.; Ohsako, Y.; Zeigler, J. M.; Hochstrasser, R. M. *Chem. Phys. Lett.* **1989**, *162*, 455.
9. Klingensmith, K. A.; Downing, J. W.; Miller, R. D.; Michl, J. *J. Am. Chem. Soc.* **1986**, *108*, 7438.
10. At the HF STO-3G level of approximation, the localization is less pronounced: Matsumoto, N.; Teramae, H. *J. Am. Chem. Soc.* **1991**, *113*, 4481.
11. Sun, Y.-P.; Sooriyakumaran, R.; Miller, R. D.; Michl, J. *J. Inorg. Organomet. Polym.* **1991**, *1*, 3.
12. Harrah, L. A.; Zeigler, J. M. *Macromolecules* **1987**, *20*, 601.
13. Michl, J.; Downing, J. W.; Karatsu, T.; Klingensmith, K. A.; Wallraff, G. M.; Miller, R. D. In *Inorganic and Organometallic Polymers;* Zeldin, M.; Wynne, K.; Allcock, J., Eds.; ACS Symposium Series 360; American Chemical Society: Washington, D.C., 1988, Chapter 5.
14. McCrary, V. R.; Sette, F.; Chen, C. T.; Lovinger, A. J.; Robin, M. B.; Stohr, J.; Zeigler, J. M. *J. Chem. Phys.* **1988**, *88*, 5925.
15. Schellenberg F. M.; Byer R. L.; Miller R. D. *Chem. Phys. Lett.* **1990**, *166*, 331.
16. Thorne, J. R. G.; Repinec, S. T.; Abrash, S. A.; Zeigler, J. M.; Hochstrasser, R. M. *Chem. Phys.* **1990**, *146*, 315.
17. Sun, Y.-P.; Michl, J. *J. Am. Chem. Soc.* in press.
18. Sun, Y.-P.; Hamada, Y.; Huang, L.-M.; West, R.; Michl, J. unpublished results.
19. Sun, Y.-P.; Hamada, Y.; Huang, L. H.; Maxka, J.; Hsiao, J.-S.; West, R.; Michl, J. *J. Am. Chem. Soc.* in press.
20. Kim, Y. R.; Lee, M.; Thorne, J. R. G.; Hochstrasser, R. M.; Zeigler, J. M. *Chem. Phys. Lett.* **1988**, *145*, 75.
21. Johnson, G. E.; McGrane, K. M. In *Photophysics of Polymers;* Hoyle, C. E.; Torkelson, J. M., Eds.; ACS Symposium Series 358; American Chemical Society: Washington, D.C., 1987, Chapter 36.
22. Ohsako, Y.; Thorne, J. R. G.; Phillips, C. M.; Zeigler, J. M.; Hochstrasser, R. M. *J. Phys. Chem.* **1989**, *93*, 4408.
23. Cotts, P. M.; Miller, R. D.; Trefonas, P. T., III; West, R.; Fickes, G. N. *Macromolecules* **1987**, *20*, 1046.
24. Sun, Y.-P.; Wallraff, G. M.; Miller, R. D.; Michl, J. *J. Photochem. Photobiol.* in press.
25. Harrah, L. A.; Zeigler, J. M. In *Photophysics of Polymers;* Hoyle, C. E.; Torkelson, J. M., Eds.; ACS Symposium Series 358; American Chemical Society: Washington, D. C., 1987, p 482.
26. Rauscher, U.; Bässler, H.; Taylor, R. *Chem. Phys. Lett.* **1989**, *162*, 127.
27. Walsh, C.; Burland, D. M.; Miller, R. D. *Chem. Phys. Lett.* **1990**, *175*, 197.
28. Tilgner, A.; Pique, J. P.; Trommsdorff, H. P.; Zeigler, J. M.; Hochstrasser, R. M. *Polymer Preprints* **1990**, *31*, 244.

29. Thorne, J. R. G.; Ohsako, Y.; Repinec, S. T.; Abrash, S. A.; Hochstrasser, R. M.; Zeigler, J. M. *J. Luminesc.* **1990**, *45*, 295.
30. Hofer, D. C.; Jain, K.; Miller, R. D. *IBM Tech. Discl. Bull.* **1984**, *26*, 5683.
31. Magnera, T. F.; Balaji, V.; Michl, J.; Miller, R. D.; Sooriyakumaran, R. *Macromolecules* **1989**, *22*, 1624.
32. West, R.; Wolff, A. R.; Peterson, D. J. *J. Radiat. Curing* **1986**, *13*, 35.
33. Trefonas, P., III; West, R.; Miller, R. D. *J. Am. Chem. Soc.* **1985**, *107*, 2737.
34. Karatsu, T.; Miller, R. D.; Sooriyakumaran, R.; Michl, J. *J. Am. Chem. Soc.* **1989**, *111*, 1140.
35. McKinley, A. J.; Karatsu, T.; Wallraff, G. M.; Miller, R. D.; Sooriyakumaran, R.; Michl, J. *Organometallics* **1988**, *7*, 2567.
36. McKinley, A. J.; Karatsu, T.; Wallraff, G. M.; Thompson, D. P.; Miller, R. D.; Michl, J. *J. Am. Chem. Soc.* **1991**, *113*, 2003.
37. Michl, J.; Downing, J. W.; Karatsu, T.; McKinley, A. J.; Poggi, G.; Wallraff, G. M.; Sooriyakumaran, R.; Miller, R. D. *Pure Appl. Chem.* **1988**, *60*, 959. This paper contains numerous errors introduced by editorial retyping.
38. Davidson, I. M. T.; Michl, J.; Simpson, T. *Organometallics* **1991**, *10*, 842.

RECEIVED December 7, 1992

Chapter 11

Excitation-Energy Migration and Trapping in Photoirradiated Polymers

K. P. Ghiggino, C. G. Barraclough, and R. J. Harris

Department of Chemistry, University of Melbourne, Parkville, Victoria 3052, Australia

A theoretical and experimental analysis of energy migration and trapping in vinyl aromatic polymers has been undertaken. A simplified model for energy migration in a finite polymer chain of chromophores based on a one dimensional random walk has been developed and used to examine the dynamics of excitation energy redistribution. For polymers containing an energy trap, the trapping efficiency and dynamics depend on the migration probability and chain length with marked deviations from single exponential fluorescence decay behaviour expected. Evidence for energy migration in poly(acenaphthylene) and poly(2-naphthyl methacrylate) containing energy trapping chromophores has been obtained from steady-state and time-resolved fluorescence measurements.

The fate of excitation energy residing in vinyl aromatic polymers following irradiation with light has been an active area of research interest over a number of years *(1-3)*. Due to the close proximity of the chromophores along the polymer backbone, there are possibilities for energy migration and final trapping of the excitation at sites far removed from the initial absorbing chromophore. This redistribution of energy within the polymer chain can have important consequences in determining the final site of photodegradation or in activating specific chromophores for further photochemical processes. The relaxation of the electronically excited chromophores in a polymer chain can be conveniently studied using fluorescence spectroscopy *(1-4)*. Figure 1 illustrates the possible intramolecular excited state relaxation pathways in a polymer containing pendant aromatic groups.

Once the energy is absorbed by a chromophore on the polymer backbone, it may be emitted as fluorescence from single chromophores (monomer fluorescence) or may migrate along the chain until it is trapped at a site of lower energy which may be a chromophore of a different type (an intrinsic acceptor) or an excited-state dimer (excimer) formed by the close association of an excited state and ground state moiety. The extent of energy migration in polymers and the role of excimers and other intrinsic trap sites remain matters of some controversy in polymer photophysics. In particular the dynamics of energy transfer processes within the polymer chain are not well characterized from either theoretical or experimental viewpoints. This article reviews recent theoretical modelling and experimental studies of energy migration and trapping dynamics in aromatic polymers carried out in our laboratories.

0097–6156/93/0527–0155$06.00/0

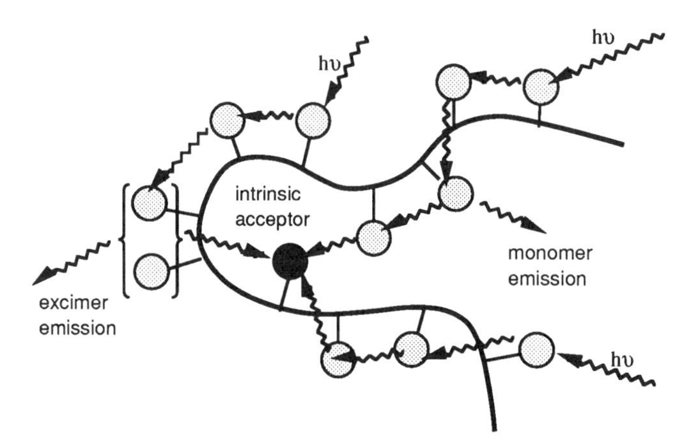

Figure 1. Pathways for excitation energy relaxation in synthetic polymers.

Modelling Energy Migration and Trapping

In order to obtain a simplified description of energy migration and trapping dynamics, energy migration can be assumed to occur via a one-dimensional random walk with only nearest-neighbour type transfer being possible. This allows a Monte Carlo simulation of the process to be employed as adopted in other work *(5-7)*. A schematic diagram of an idealized finite polymer chain of chromophores with one reflective barrier and incorporating an energy trap at the other end is depicted in Figure 2.

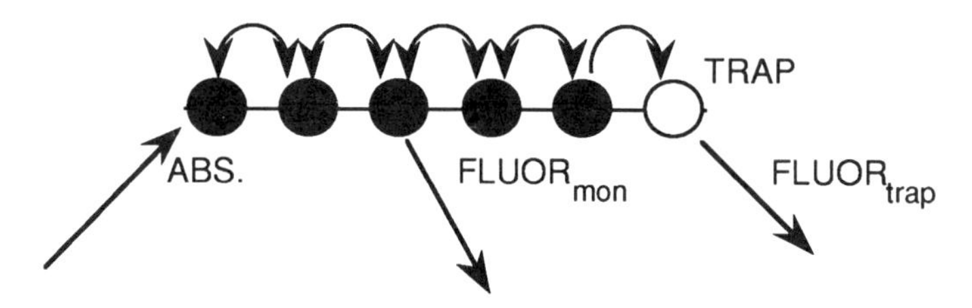

Figure 2: Energy migration model for a polymer containing an energy trap at one end of the polymer chain, using a one-dimensional random walk.

In this model energy migration, trapping or fluorescence are the only possibilities considered for excitation energy residing in the chromophores and transfer of energy into the end trap site is irreversible. This situation can be described by the transition matrix:

$$Q = \begin{pmatrix} 0 & p & 0 & 0 & \cdot & \cdot & \cdot & \cdot & \cdot & & & \\ p & 0 & p & 0 & \cdot & \cdot & \cdot & \cdot & \cdot & & & \\ 0 & p & 0 & p & \cdot & \cdot & \cdot & \cdot & \cdot & & & \\ & & & & \cdot & \cdot & \cdot & \cdot & \cdot & & & \\ & & & & \cdot & \cdot & \cdot & 0 & p & 0 \\ & & & & \cdot & \cdot & \cdot & p & 0 & p \\ & & & & \cdot & \cdot & \cdot & 0 & 0 & p \end{pmatrix}$$

where p = probability of migration to the chromophore on the left = probability of migration to the chromophore on the right (i.e. $p < 0.5$), and the i,j elements of the matrix give the starting and finishing positions of the excitation energy respectively.

The elements of the probability matrix after n hops, Q^n, can be derived using the procedure given by Feller *(8)* and are given by the algebraic expression:

$$q_{j,k}^{(n)} = (2p)^n \left(\sum_{r=0}^{a-1} \cos^n \frac{\pi(2r+1)}{(2a+1)} \frac{\left(\sin\frac{\pi(2r+1)j}{(2a+1)}\right)\left(\sin\frac{\pi(2r+1)k}{(2a+1)}\right)}{\sum_{v=1}^{a} \sin^2 \frac{\pi(2r+1)v}{(2a+1)}} \right)$$

where a = the number of chromophores in the polymer chain excluding the trap site and n = number of hops.

The corresponding expression for a finite chain where both ends are reflective boundaries (i.e. no energy trap) is:

$$q_{j,k}^{(n)} = (2p)^n \left(\frac{1}{a} + \frac{1}{a} \sum_{r=1}^{a-1} Sr \right)$$

$$Sr = \frac{\cos^n\left(\frac{\pi r}{a}\right) * \left(\sin\left(\frac{\pi rj}{a}\right) - \sin\left(\frac{\pi r(j-1)}{a}\right)\right) * \left(\sin\left(\frac{\pi rk}{a}\right) - \sin\left(\frac{\pi r(k-1)}{a}\right)\right)}{1-\cos\left(\frac{\pi r}{a}\right)}$$

The algorithm for the modelling requires as input variables the number of chromophores in the chain, the probability of a hop between like chromophores and to the trap, the time for one hop and the number of hops required. The program developed provides the total probability of emission, the probability of emission from the initially excited chromophore and the probability of populating the trap, as a function of the number of hops (hence time) after excitation.

Probabilities for energy hopping can be determined assuming the mechanism of energy migration/transfer between chromophores is a Förster-type induced dipole-dipole interaction *(9)*. The rate constant for this nonradiative energy transfer process is given below.

$$k_{ET} = \frac{9000 \ln(10) \kappa^2 \phi_d J}{128 \pi^5 n^4 \tau_D R^6 N}$$

In this equation ϕ_d is the fluorescence quantum yield of the donor in the absence of an acceptor, κ is the orientation factor ($\kappa^2 = 2/3$ for a random distribution of dipoles), J is the spectral overlap integral describing the overlap between the fluorescence spectrum of the donor and the absorption spectrum of the acceptor, n is the refractive index of the medium, τ_D is the lifetime of the donor in the absence of an acceptor and N is Avogadro's number. R is the mean separation between the donor and acceptor species.

The Förster equation can also be expressed as:

$$k_{ET} = \frac{3\kappa^2}{2\tau_D}\left(\frac{R_0}{R}\right)^6$$

where R_0 is known as the Förster critical transfer distance and is the distance between the donor and acceptor species at which excitation transfer and spontaneous deactivation of the donor are of equal probability. The rate of energy transfer depends on the orientation factor, κ, which can be calculated using:

$$\kappa = \cos\theta - 3\cos\theta_D\cos\theta_A$$

where θ_D is the angle the donor transition dipole makes with a line joining the midpoints of the donor and acceptor transition dipoles, θ_A is the corresponding angle for the acceptor transition dipole and θ is the angle between the donor and acceptor transition dipoles.

As an example, for the case of poly(acenaphthylene), R_0, determined from spectroscopic measurements, is 12 Å, the average interchromophore separation and orientation factor calculated from energy minimized diad structures (as would exist in a low temperature glass) are 5.4 Å and 1.27 Å respectively, and τ_D = 34.7 ns giving a value for k_{ET} of 8.4 ns^{-1}. Thus energy migration might be expected to be quite facile in this polymer although the migration distance will depend on the distribution of trap sites in the polymer.

Examples of results obtained from simulations using the model are given in Figures 3 and 4. In these cases the time interval for one migration step is taken to be 120 picoseconds. Figures 3 and 4 describe some general features of energy migration and trapping in such an idealized polymer. It can be readily shown that in the presence of migration along the chain of chromophores without a trap, that the expected fluorescence decay from the polymer will be single exponential with a decay constant identical to that of an individual chromophore. Figure 3 illustrates that incorporation of the trap site at one end of the chain increases the rate of fluorescence decay and this effect is more marked for shorter chain lengths where trapping becomes more efficient.

Of further interest is that marked deviations from single exponential fluorescence decay behaviour become apparent upon incorporating the trap. An indication of this is given in Figure 3 where the χ^2 goodness of fit parameter determined from single exponential fitting of the simulated decays is greater than unity. There are many reports in the literature *(1,2,10)* of non-single exponential decays found for vinyl aromatic polymers and the involvement of energy migration and trapping provides one explanation for these observations.

The cumulative probability curves for populating the trap site shown in Figure 4 indicate the time scale over which excitation can accumulate in the trap. As the migration hop probability decreases the probability of populating the trap site also declines as there will be more chance for radiative processes to deplete the excitation energy before trapping can occur. Additional information on the rate of energy dissipation away from the initial absorption sites can also be obtained. The population

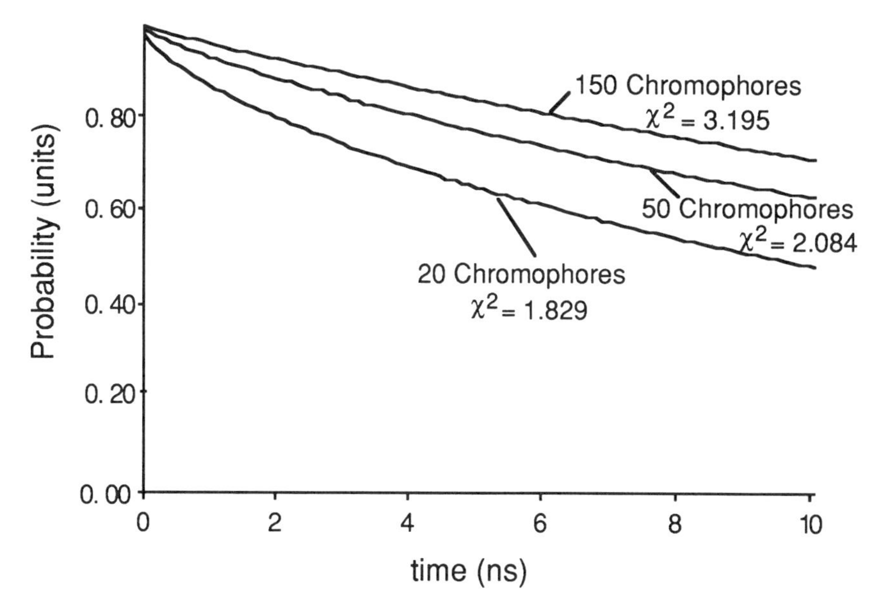

Figure 3. Simulated total monomer fluorescence decays for polymers containing an end-trap. p = 0.4983, trapping probability = 0.49994. τ_D = 34.7 ns for the polymer without a trap. The χ^2 values indicate the goodness of fit to a single exponential decay (χ^2 = 1 for a perfect fit with correct statistical weighting).

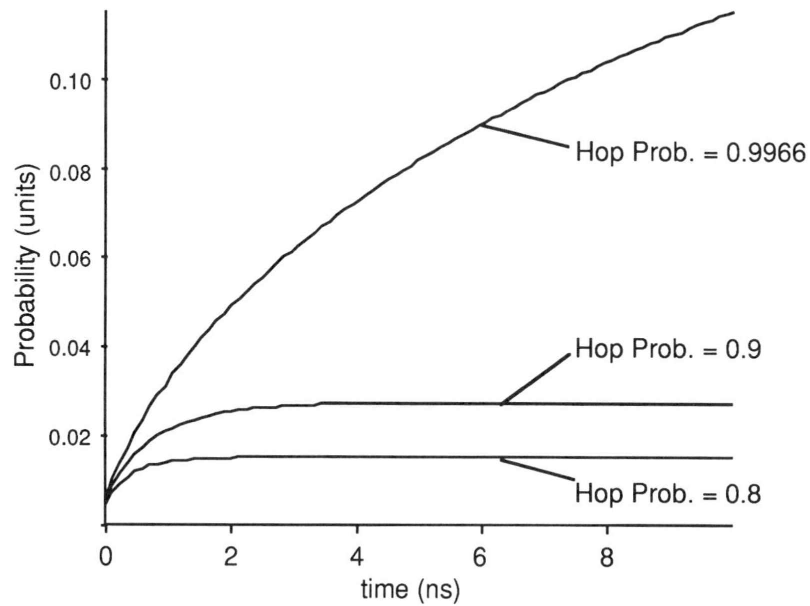

Figure 4. Cumulative probability of populating the trap site. Number of chromophores = 50. Hop probability indicated = 2p.

of initially excited chromophores with time are shown in Figure 5. For the set of parameters chosen there is an initial rapid decay followed by slower relaxation behaviour. The presence of a trap site increases the decay rate since there is now the opportunity for migrating excitation to be trapped and not return to the initial absorptioı. site. Such relaxation profiles would correspond closely to experimentally observable fluorescence polarization anisotropy decay measurements *(11)*. In the absence of rotational motion fluorescence from chromophores will be polarized following excitation by polarized light due to the fixed directions of transition dipoles in the molecules. However energy migration to chromophores having different relative orientations will result in their fluorescence being more depolarized. Thus the greatest contribution to polarized fluorescence will arise from the initially excited species. The depletion of the population of these chromophores with time depicted in Figure 5 could thus be followed from time-resolved fluorescence polarization measurements.

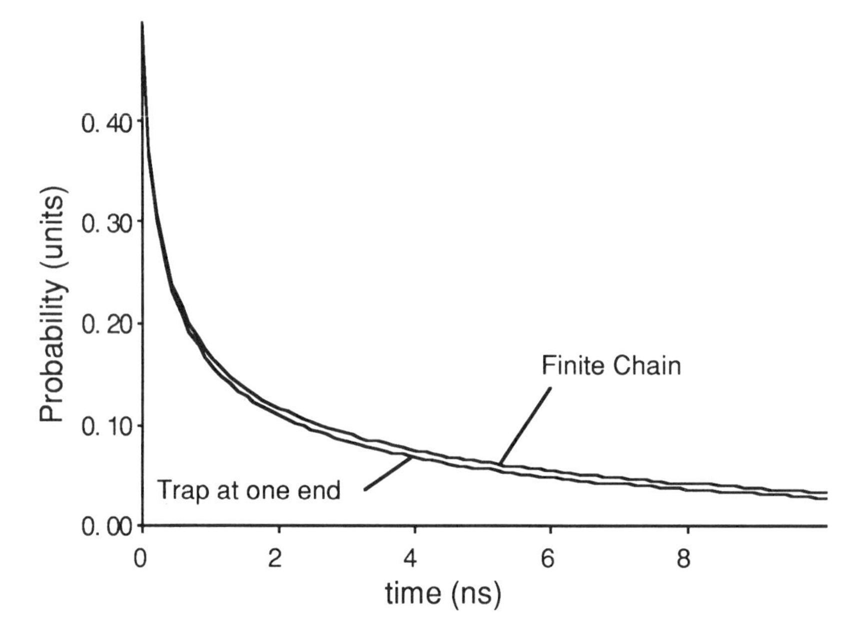

Figure 5. Decay of excitation on the initially excited chromophores for polymer chains with and without a trap. Hop and trapping probabilities as for Figure 3. Number of chromophores = 50.

At the present stage of development, the model discussed above can only give an approximate description of energy migration and trapping dynamics. For a real vinyl aromatic polymer chain non-nearest neighbour migration hops should be considered, the conformation of the polymer will be important and other possible fates, apart from emission or migration, of the excitation residing in a chromophore within the time interval chosen will influence the behaviour.

Experimental Studies of Polymer Energy Relaxation Processes

Poly(acenaphthylene) (PACE) In dilute solution at room temperature PACE, like many other naphthalene containing polymers, exhibits fluorescence from both monomer and excimer species with emission maxima at approximately 340nm and 400nm respectively. In low temperature solvent glasses and in solid poly(methyl methacrylate) host films at room temperature, excimer emission is almost absent indicating molecular motion is required to achieve the necessary excited state dimer geometry. An anthracene chromophore can be incorporated as an energy trap at one end of the polymer chain by the free radical polymerization of acenaphthylene in the presence of the chain transfer agent 9-chloromethyl anthracene. A polymer of average molecular weight 24,100 containing a 9-methyl anthryl (9MA) end group can be prepared in this way *(12)*. The fluorescence relaxation pathways in a dilute benzene solution of this polymer can be followed by recording time-resolved fluorescence spectral surfaces after excitation of the polymer by picosecond laser pulses at 290nm where only the acenaphthyl chromophores have significant absorbance (see Figure 6). The instrumentation and procedures used for recording and analysing these surfaces and fluorescence decay data have been described previously *(12)*.

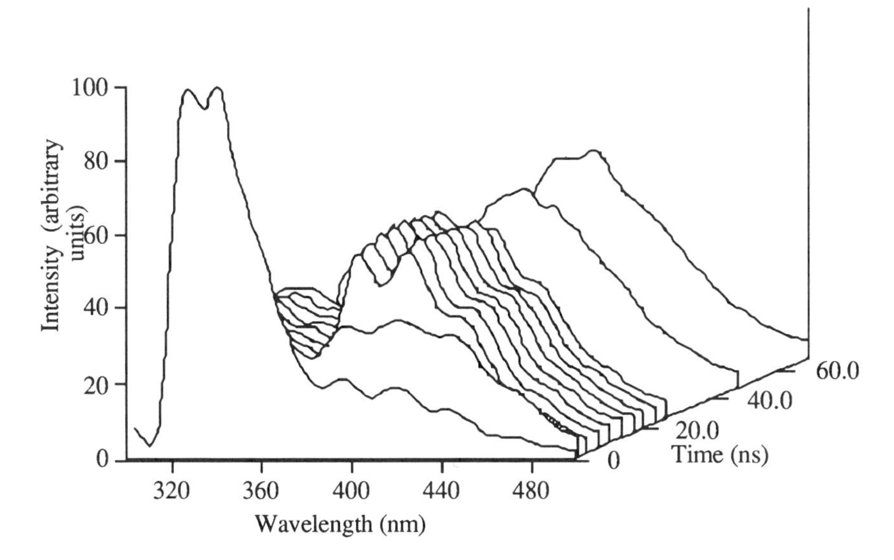

Figure 6. Area normalized time-resolved fluorescence spectra for PACE with a 9-methyl anthryl end group in benzene at 298K.

The area normalized time-resolved spectra in Figure 6 show the decay of the acenaphthylene monomer fluorescence and the rapid grow-in (within a nanosecond) of the characteristic structured 9MA emission with maxima near 420nm. Since direct excitation of the 9MA is not occurring, the end trap fluorescence arises due to energy transfer from the acenaphthyl groups. The efficiency of this energy transfer from the acenaphthyl moieties to the 9MA end group can be determined using the method described by Guillet and co-workers (*13*) and is found to be 16% for the polymer in benzene at 298K and 6% in a 2-methyl tetrahydrofuran (MTHF) glass at 90K (*14*).

These values for polymer to end-group energy transfer can be compared to the value calculated assuming transfer occurs by a single step Förster process (i.e. energy migration is not involved). The radius of gyration of a single PACE coil is 3.6 nm with a hydrodynamic volume of 9.7 x 10^{-20} cm^3 in benzene leading to an effective concentration of anthracene in the coil of 1.7 x 10^{-2} $mol.dm^{-3}$. An erroneous value of 1.7 x 10^{-4} was reported previously*(15)*. The theoretical transfer efficiency, E_{th}, is given by *(9)*:

$$E_{th} = \sqrt{\pi}\ g\ \exp(g^2)[1 - \mathrm{erf}(g)]$$

where $g = C/C_o$, C is the acceptor concentration and C_o is the Förster critical transfer concentration (corresponding to the acceptor concentration at which the probability of an energy transfer via the Förster mechanism is 0.5) given by $C_o = 3000/2\pi^{3/2}NR_o^3$. The value for C_o is calculated to be 0.056 M using the value for R_o of 20 Å determined for acenaphthyl to 9-methyl anthracene energy transfer.

Substitution of the effective anthryl chromophore concentration in this equation leads to an E_{th} value of 39%. This value may be compared with the experimentally determined transfer efficiency of 16%. Trapping of energy by excimer sites which can only undergo inefficient transfer to the anthryl group could reduce the overall experimental transfer efficiency.

Evidence for energy migration in this polymer comes from fluorescence polarization measurements carried out in a MTHF glass at 78K. As noted earlier, energy migration should lead to depolarization of the fluorescence from the polymer. With excitation by vertically polarized light, the ratio of vertically polarized emission to horizontally polarized emission, I_V/I_H, should be unity if complete depolarization occurs. For PACE-9MA the value of I_V/I_H measured at 330nm with 290nm excitation is 1.02 ± 0.03 indicative of energy migration in the polymer. In addition we have undertaken time-resolved fluorescence polarization measurements on the polymer and find that the depolarization of fluorescence occurs within the 80 picosecond time resolution of our instrument.

The fluorescence decay of both PACE and PACE-9MA are both highly non-single exponential when recorded in low temperature MTHF glasses at 79K. A fit to the sum of a minimum of 3 exponential terms was required to adequately describe the decay data at 330nm as shown in Table 1.

Table 1. Monomer fluorescence decay data for PACE and PACE-9MA in MTHF at 79K. λ_{em} = 330 nm. Functions of the form $I(t) = \Sigma A_i \exp(-t/\tau_i)$ have been used to fit the data.

POLYMER	τ_1 (ns)/A_1	τ_2 (ns)/A_2	τ_3 (ns)/A_3
PACE	0.54 / 8139	26.1 / 6546	43.2 / 6329
PACE-9MA	0.53 / 2828	5.17 / 4422	23.9 / 5209

The departure from single exponential behaviour for the monomer decay in these polymers would be expected if energy migration and trap sites are present as demonstrated in the random walk model presented earlier. Even in the homopolymer PACE there may be a low concentration of excimer or other adventitious trap species which could lead to the observed deviations from exponential kinetics. With further refinement of the random walk model to include non-nearest neighbour interactions and detailed energy relaxation pathways, fitting of such experimental decay functions should yield average hopping and trap probabilities.

Poly(2-naphthyl methacrylate) (P2NMA) P2NMA undergoes both photophysical and photochemical relaxation following light absorption. The polymer exhibits monomer and excimer fluorescence bands but is also susceptible to photoinduced discoloration due to the formation of acylnaphthol-based photoproducts as a result of photo-Fries rearrangements *(16)*.

For P2NMA and P2NMA containing a 9MA end trap, energy migration can again be demonstrated from steady state fluorescence polarization measurements *(14)* as shown in Figure 7.

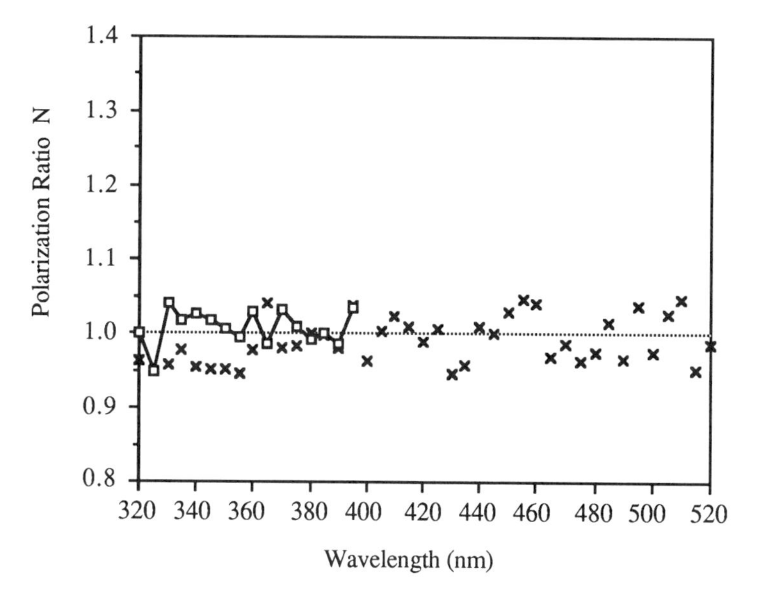

Figure 7. Fluorescence polarization ratio, $N = I_V/I_H$, as function of emission wavelength for P2NMA (squares) and P2NMA-9MA (crosses) at 90K in MTHF. Excitation wavelength 295nm. Total depolarization indicated by dotted line (N = 1).

The high degree of depolarization observed throughout the fluorescence spectrum for the polymer is indicative that energy migration and trapping has occurred. At room temperature the P2NMA to 9MA end group trapping efficiency is 9% *(14)*.

The results of incorporating a photostable energy trap within the polymer chain has also been studied *(16)*. In this case copolymers of 2-naphthyl methacrylate (2NMA) and 2-(2'-hydroxy-4'-methacryloxy-phenyl)-2*H*-benzotriazole (BDHM) were prepared *(16,17)* to produce polymers with the structure below. The BDHM chromophore has the structural characteristics of an ultraviolet absorber. The intramolecular hydrogen bond formed between the phenolic proton and the benzotriazole nitrogen facilitates excited state intramolecular proton transfer (ESIPT) *(16, 18)*. Through this pathway the molecule is able to dissipate excitation energy very rapidly (< 10 picoseconds) resulting in a very photochemically stable species. At the excitation wavelength of 285nm employed the majority of light is absorbed by the 2NMA species but the BDHM can act as a photostable trap for excitation energy in the polymer. This is illustrated in Figure 8 where the quenching of both monomer and excimer fluorescence by BDHM is shown.

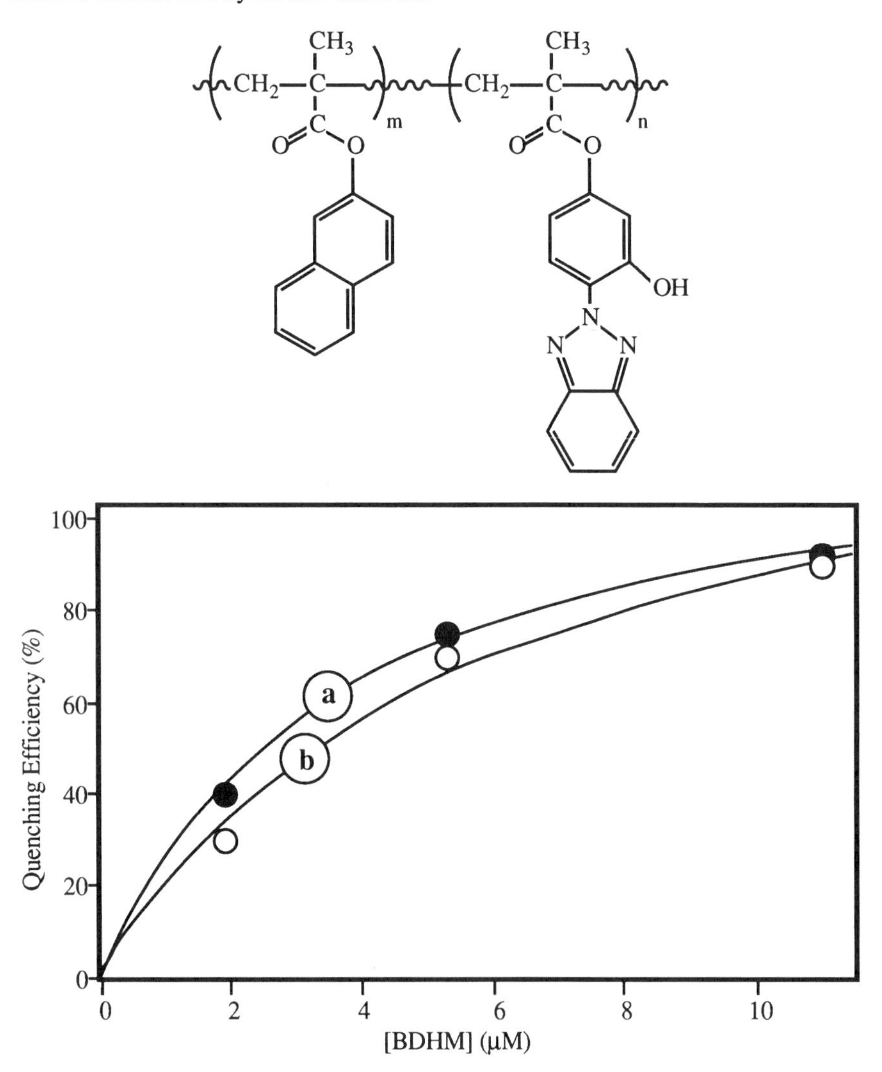

Figure 8. Plot of the quenching of (a) monomer and (b) excimer fluorescence in 2NMA-BDHM copolymers as function of the bulk BDHM concentration in benzene solution. Lines only indicate trend in the data. The concentrations correspond to 1.1, 3.2 and 6.2 mol% of BDHM in the polymer.

The Förster energy transfer R_0 and C_0 parameters, defined previously, are 20Å and 0.058M, and 17Å and 0.085M, for transfer to BDHM from the P2NMA monomer and excimer species respectively. For a polymer of molecular weight 8,330 and calculated radius of gyration of 23Å containing 6.2 mol% BDHM (i.e. approximately 2 BDHM chromophores/chain) the expected quenching efficiency, assuming Förster single step energy transfer is occurring, is about 90% for both the monomer and excimer *(16)*. These values are in reasonably close agreement with the measured values for this polymer given in Figure 8 indicating that the single step process competes effectively with energy migration in this low molecular weight polymer.

P2NMA can undergo photochemical degradation in solution by a photo-Fries rearrangement *(16)* as depicted in Figure 9. The acyl-naphthol photoproducts absorb strongly at about 320nm allowing the polymer degradation to be readily followed by absorption spectroscopy.

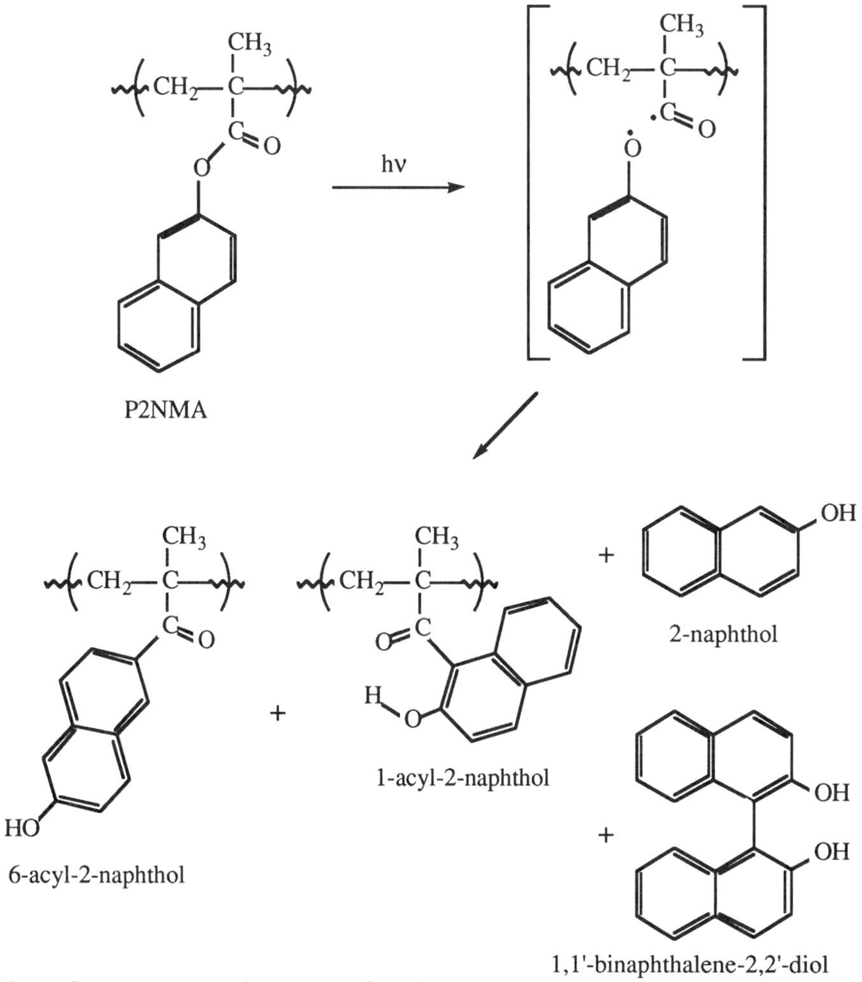

Figure 9. Photodegradation scheme for P2NMA.

The incorporation of the BDHM species into the polymer leads to a dramatic decrease in the degradation quantum yield. For the polymer containing 6.2 mol% BDHM in a degassed benzene solution the photodegradation yield decreases by a factor of 30 compared to the P2NMA homopolymer. It should be noted that at the irradiation wavelength employed (285nm) the BDHM in the polymer has a very low absorbance and thus the protection observed can be attributed to energy trapping and subsequent efficient energy disssipation by the BDHM chromophores rather than any UV screening effect. It is the high local concentration of the stabilizer within the polymer chain in solution that leads to efficient photostabilization by an energy trapping mechanism. The examples above show that in the P2NMA homopolymer energy migration can be demonstrated and that energy trapping by intrinsic acceptors can be used as an efficient means of photostabilizing such polymer systems.

Both the theoretical modelling and experimental observations discussed in this work indicate that energy migration and trapping can play an important role in redistributing excitation energy within vinyl aromatic polymer systems following absorption of radiation. The consequences of these processes should always be considered when discussing the overall impact of radiation effects on polymer systems.

Acknowledgments

KPG gratefully acknowledges the Australian Research Council and the Department of Industry, Technology and Commerce for financial support of this work. Assistance from Dr. T.A. Smith in the preparation of this manuscript is acknowledged.

Literature Cited

(1) *Polymer Photophysics;* Phillips, D., Ed.; Chapman and Hall: London, 1985.
(2) *Photophysics of Polymers;* Hoyle, C.E.; Torkelson, J.M., Eds.; ACS Symposium Series No.358; American Chemical Society: Washington, NY, 1987.
(3) Webber, S.E. *Chem. Review* **1990**, *90*, 1469.
(4) Guillet, J.E. *Photophysics and Photochemistry of Polymers;* Cambridge University Press: London, 1985.
(5) Fitzgibbon, P.D.; Frank, C.W. *Macromolecules* **1982**, *15*, 733.
(6) Blonski, S.; Sienicki, K.*Macromolecules* **1986**, *19*, 2936.
(7) Wang, Z.; Holden, D.A.; McCourt, F.R.W. *Macromolecules* **1991**, *24*, 893.
(8) Feller, W. *An Introduction to Probability Theory and its Application (2nd ed.)*; Wiley: NY, 1957; pp 388-396.
(9) (a) Förster, Th. *Z Nauturforsch* **1949**, *4A*, 321; (b) Förster, Th. *Disc. Faraday Soc.* **1959**, *27*, 7.
(10) Ghiggino, K.P.; Smith, T.A.; Wilson, G.J. *J. Modern Optics* **1990**, *37*, 1789.
(11) Genest, D.; Wahl, Ph. In *Time-Resolved Fluorescence Spectroscopy in Biochemistry and Biololgy,* Cundall, R.B.; Dale, R.E., Eds.; NATO ASI Series Vol. 69; Plenum Press: NY, 1983; pp 523-539.
(12) Ghiggino,K.P.; Bigger, S.W.; Smith, T.A.; Skilton, P.F.; Tan, K.L. In *Photophysics of Polymers,* Hoyle, C.E; Torkelson, J.M., Eds.; ACS Symposium Series No.358; American Chemical Society: Washington, NY, 1987; pp 368-383.
(13) Holden, D.A.; Guillet, J.E. *Macromolecules* **1980**, *13*, 289.
(14) Skilton, P.F., Ph.D. Thesis, The University of Melbourne, Australia, 1990.
(15) Ghiggino, K.P., *Macromol. Chem., Macromol. Symp.* **1992**, *53*, 355.
(16) Scully, A.D., Ph.D. Thesis, The University of Melbourne, Australia, 1990.
(17) Ghiggino, K.P.; Scully, A.D.; Vogl, O.; Bigger, S.W. In *Progress in Pacific Polymer Science*; Anderson, B.C.; Imanishi, Y., Eds.; Springer-Verlag: Berlin Heidelberg, 1991; pp 295-305.
(18) Ghiggino, K.P.; Scully, A.D.; Bigger, S.W. In *The Effects of Radiation on High Technology Polymers*, Reichmanis, E.; O'Donnell, J.H., Eds.; ACS Symposium Series No. 381; American Chemical Society: Washington, NY, 1988; pp 57-79.

RECEIVED August 14, 1992

Chapter 12

Structure of Polymer Films Studied with Picosecond Total Internal Reflection Fluorescence Spectroscopy

Minoru Toriumi[1] and Hiroshi Masuhara[2]

Microphotoconversion Project, Research Development Corporation of Japan, 15 Morimoto-cho, Shimogamo, Sakyo-ku, Kyoto 606, Japan

Pyrene-doped poly(p-hydroxystyrene) (PHST) thin films prepared by a spin-coating method are studied by time-resolved total-internal-reflection fluorescence spectroscopy. Inhomogeneities such as a concentration gradient of doped pyrene molecules, a gradient of polarity (hydrophobicity), and the existence of isolated pyrene molecules are observed. Stable ground-state dimers of pyrene are found in surface, bulk and interface layers of a PHST film. These results also depend upon the preparation process such as the baking condition.

The lithographic process of the semiconductor industry is one of the important applications of thin polymer films irradiated by electron beams, x-rays and ultraviolet light (*1*). In photolithography the resist material is applied as a thin coating over some substrate and subsequently exposed in an image-wise fashion through a mask such that ultraviolet light strikes selected areas of the resist material which is sensitive to radiation. The exposed area of resist materials is rendered less soluble, producing a negative tone image of the mask because of chemical reactions such as crosslinking. The material resists the etchant and prevents it from attacking the underlying substrate in those areas where it remains in place after development. Following the etching process, the resist is removed by stripping to produce a negative tone relief image in the underlying substrate. One of the most important steps is the formation of resist fine patterns in this process. The image formed by the light projection is degraded mainly by the diffraction. This light distribution is irradiated to the resist films, but much more improved fine patterns of resist materials are obtained after development. What is the reason why resist patterns recover the degraded projection image? Note the importance of chemistry in the micrometer region such as the inhomogeneity of irradiation intensity, and the inhomogeneous structure, distribution, and reactivity of polymer and incorporated photosensitive compounds. The chemistry in micrometer size has become of great important in many fields such as microelectronics, optical communication and molecular devices.

[1]Current address: Hitachi Central Research Laboratory, Hitachi, Ltd., Kokubunji, Tokyo 185, Japan
[2]Current address: Department of Applied Physics, Osaka University, Suita, Osaka 565, Japan

0097–6156/93/0527–0167$06.00/0

Time-resolved total-internal-reflection fluorescence (TIRF) spectroscopy is well suited to studies of dynamic chemical interactions and structures in the surface layers of the sub-micrometer region (*2-4*). We have developed variable-angle time-resolved TIRF spectroscopy to study thin film structures (*5-7*). We apply the TIRF technique to a thin film of pyrene-incorporated poly(p-hydroxystyrene) (PHST) which is a very important polymer for resist materials(*8,9*). There are many phenomena related to polymer structures that can be studied such as the efficiency of photodecomposition (*8*), the effect of the incorporation of photo-active compounds (PACs) (*10*) and casting solvents, the spatial distribution of PACs (*11,12*), dissolution behavior in the development process (*13,14*) and surface treatment (*15-17*) and the thin film itself (*18-20*). Understanding thin film structure is essential to elucidate these phenomena.

Experiment

Materials and film preparation. PHST (Polyscience, Inc.) was purified by a repeated petroleum ether/ethyl acetate precipitation sequence. Pyrene (Aldrich) was purified by column chromatography on silica gel followed by recrystalization from ethanol. Ethylene glycol monoethyl ether acetate (ECA) (Nacalai tesque, research grade) was used as received. Total-internal-reflection (TIR) substrate was constructed from a hemicylindrical sapphire (Shinkosha).

Here, we define the "interface" layer to mean the layer of a sample facing to the TIR substrate, and the "surface" layer to be the one facing the air.

The sample film was prepared to investigate the surface, interface, and bulk layers as follows. First, pyrene was incorporated into the polymer by dilution of stock solutions in ECA. Then, PHST films doped with pyrene were prepared on the TIR substrate by spin-coating and heating at 80 °C for ten min. to evaporate the casting solvent. The film thickness was determined to be about 2 μm by an optical interference method. Before fluorescence measurements, the films were coated with poly(vinyl alcohol) (Kuraray Co., Poval 224) to prevent quenching of the excited state of pyrene by oxygen in the air. The properties of the surface layer were measured as follows. The sample was spin-cast onto a quartz plate and baked at 80 °C for ten min. The air-side of the sample film was then pressed to the TIR substrate mechanically.

Total-internal-reflection fluorescence spectroscopy. In order to measure the chemistry of thin films in the micrometer region, we have developed a fluorescence spectroscopy using total-internal-reflection phenomena.

Principle. A sample of polymer film is coated on the flat surface of a TIR substrate of a cylindrical prism (Figure 1). The excitation laser beam is focused by a cylinder lens onto the focal plane of the TIR substrate. This arrangement allows a parallel beam of light with a specified angle of incidence, θ_i, to impinge on the interface between the TIR prism and the sample film. The excitation volume of sample depends upon the incident angle. When the incident angle is smaller than the critical angle, θ_c, which is determined by the refractive index of the system, the excitation light transmits through the sample. The information of the entire film is obtained by observing the fluorescence from the excited volume under this "ordinary" condition. On the contrary, for an incident angle larger than the critical angle, the incident light is totally reflected, and there is some penetration of the excitation light by an evanescent wave into the interface region of the sample film. In the case of total-internal-reflection, the information from interface layers of the sample is obtained by observing the fluorescence from the excited volume.

The penetration depth of the evanescent wave is one of the important physical parameters of TIR spectroscopy, and is used to determine the excitation volume of a sample. An ultraviolet light is used as an excitation source in TIRF spectroscopy. In many materials the absorption in the UV region is very large. This condition of optically absorbing samples is very different from the conventional ATR(Attenuated

Total Refelction)-IR spectroscopy, in which samples show a negligibly low absorbance in the IR wavelength. In the strict sense, total-internal reflection does not hold because of the large absorption effects, but the effective penetration depth, d_p, can be defined and given by

$$d_p = \frac{\lambda_0}{2\pi}\sqrt{\frac{2}{-n_2^2(1-\kappa^2) + n_1^2\sin^2\theta_i + \sqrt{[n_2^2(1-\kappa^2) - n_1^2\sin^2\theta_i]^2 + 4n_2^4\kappa^2}}} \quad (1)$$

where λ_0 is the wavelength of the incident wave in vacuum, n is the refractive index, the suffices 1 and 2 indicate the TIR substrate and the sample respectively and κ is the absorption index of the sample and relates to the normal Beer's law (*24*). Figure 2 shows the depth and incident angle dependences of the intensity of an excitation light with a wavelength of 300 nm. Zero depth corresponds to the interface between the sample and the TIR substrate. When the ultraviolet light impinges the polymer films with the incident angle near the critical angle (68°) under the condition of total internal reflection, the light intensity decreases exponentially with the increase of depth into the sample film. The evanescent wave decreases more rapidly with the increase of the incident angle. It should be noted that the penetration depth and the excitation volume are easily controlled by the incident angle and are limited to an interface region that is smaller than the wavelength of the excitation light. This high depth resolution is the great advantage of TIR spectroscopy.

The fluorescence intensity, I_f, is given for external reflection, by

$$I_f = E_0^2 \frac{4n_1^2\cos^2\theta_i}{(n_1\cos\theta_i + \mathrm{Re})^2 + \mathrm{Im}^2} \cdot \frac{4n_1^2\cos^2\theta_o}{(n_1\cos\theta_o + \sqrt{n_2^2 - n_1^2\sin^2\theta_o})^2} \cdot \int_0^{th} \varepsilon C \exp\left(-\frac{2z}{d_p}\right) dz \quad (2)$$

where E_0 is the field amplitude of an excitation light, θ_o is the observation angle, th is the film thickness, ε is molar absorption coefficient of pyrene and C is the pyrene concentration (Toriumi, M.; Yanagimachi, M.; Masuhara, H. Appl. Opt. in press.). Re and Im are the real and imaginary parts of $\mathbf{n_2}\cos\theta_t$ defines as,

$$\mathrm{Re} = \sqrt{\frac{n_2^2(1-\kappa^2) - n_1^2\sin^2\theta_i + \sqrt{[n_2^2(1-\kappa^2) - n_1^2\sin^2\theta_i]^2 + 4n_2^4\kappa^2}}{2}} \quad (3)$$

and

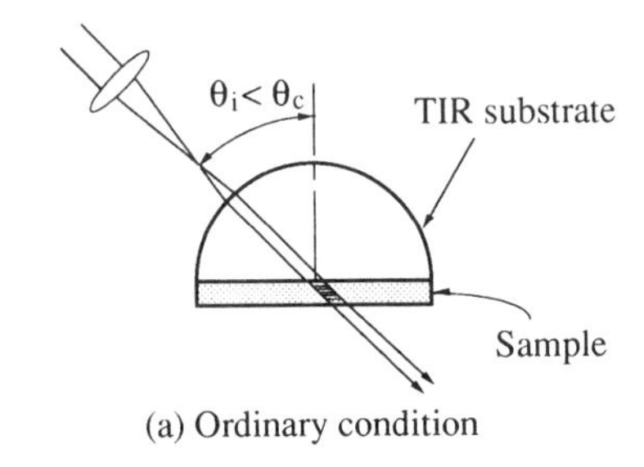

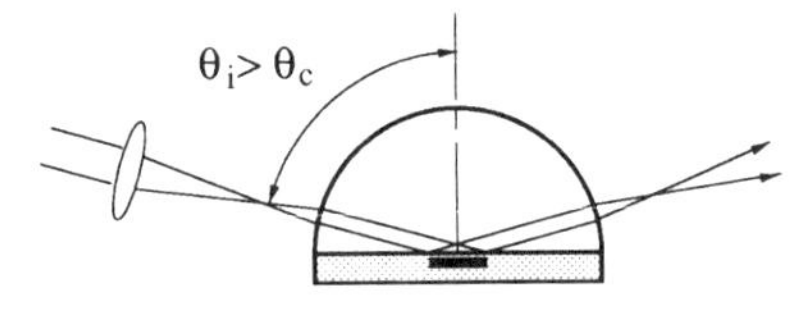

Figure 1. Excitation conditions. (a) Ordinary condition, (b) total-internal-reflection condition.

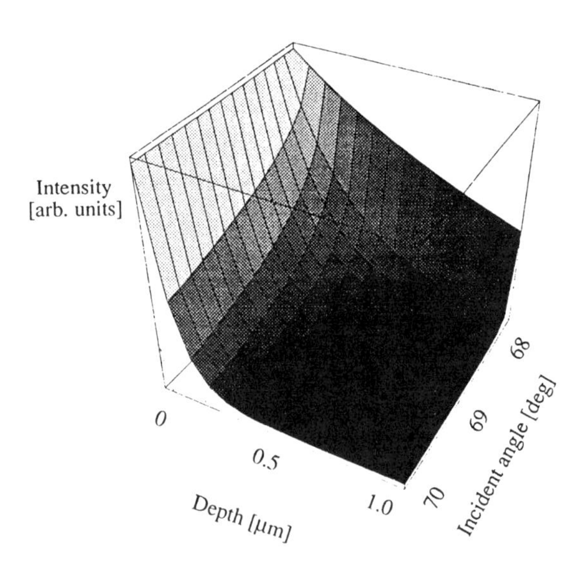

Figure 2. Depth and incident angle dependences of excitation light intensity.

$$\mathrm{Im} = \sqrt{\frac{-n_2^2(1-\kappa^2) + n_1^2 \sin^2\theta_i + \sqrt{[n_2^2(1-\kappa^2) - n_1^2\sin^2\theta_i]^2 + 4n_2^4\kappa^2}}{2}} \qquad (4)$$

Equation 2 indicates the angular distribution of fluorescence intensity is a Laplace transform of the a parameter, $4\pi\mathrm{Im}\ z/\lambda_0$. The depth profile of fluorescent molecules are experimentally determined using equation 2. This is another advantage of variable-angle TIRF spectroscopy.

Experimental equipment. Fluorescence was measured by a time-correlated single-photon counting technique. The frequency-doubled output of a CW mode-locked Nd^{3+}:YLF laser was used to synchronously pump a dye laser. The dye laser was cavity-dumped to produce pulses at a proper repetition rate, and the output was then frequency-doubled to give 2-ps excitation pulses at 300 nm with an average power of 100 mW. The laser beam was focused by a cylindrical lens onto the focal plane of the hemicylindrical TIR prism on which a sample film was spin-coated.

The hemicylinder was firmly mounted on a rotating table. Characteristic fluorescence from the interface and bulk layers of the film was observed under the TIR (θ_i=70°, θ_o=45°) and ordinary (θ_i=θ_o=45°) conditions. For instance, θ_i=70° corresponds to a penetration depth of 79 nm calculated from the equation 1 and θ_i=45° to 670 nm, with λ_0=300 nm, n_1=1.81, n_2=1.68 and a pyrene concentration of 100 mM / M HST unit of PHST films. The TIR condition, therefore, represents fluorescence from the 79-nm-thick interface layer adjacent to the TIR substrate, or from the 79-nm-thick interface layer in the PHST-pyrene system. In the ordinary condition, fluorescence is from the bulk layer 670 nm thick.

Fluorescence observed at angle θ_o was introduced by a cylindrical collection lens and a focusing-lens system into a scanning double monochromator. Light was detected using a multichannel-plate photomultiplier . The resolution of wavelength was around 0.5 nm.

Fluorescence rise and decay profiles were collected using a picosecond time-correlated single-photon-counting system, consisting of a constant-fraction discriminator, a time-to-amplitude converter, and a multichannel analyzer. The response time of the whole system was less than 10 ps.

Fluorescence spectra were also recorded by the single-photon-counting system.

Results and discussion

Micropolarity of the interface and surface layers. Pyrene was used as a fluorescence probe to investigate the micropolarity of a thin film structure. The solvent dependence of vibrational structure of pyrene fluorescence has been well studied (*21*). In the presence of the polar solvents there is an enhancement in the intensity of the 0-0, I_0, band whereas there are little effects on the 0-2, I_2, band. The relative intensity, I_0/I_2, of pyrene fluorescence is used as a good indicator of the micropolarity around a pyrene molecule (*22,23*).

Figure 3 shows the relative intensity, I_0/I_2, of the vibronic bands of pyrene fluorescence as a function of pyrene concentration both in the surface and interface layers and a bulk layer. The values of I_0/I_2 in the interface layer agree with those in the bulk layer within the experimental errors At high pyrene concentrations, above around 80 mM / M HST unit, the intensity ratio decreases with the pyrene concentration. This is due to the superposition of the structureless broad excimer emission with a peak wavelength of around 500 nm. The accurate ratio should be estimated by correcting for these contributions to the excimer fluorescence. At pyrene concentrations lower than 50 mM / M HST unit, the intensity ratio is almost independent of the pyrene concentration. In this range the value of I_0/I_2 is about 1.2. This value is reasonable, because aromatic compounds such as benzene and benzyl alcohol show the value of about 1.2 as shown in Table I. The value of I_0/I_2 determined for the surface layer is a bit smaller than that for the interface and bulk layers. This indicates that the environmental polarity around pyrene molecules in the surface layer is less polar than in the bulk and interface layers. It means that pyrene exists in the hydrophobic sites in the PHST films. What is the reason for the low polarity in the surface? The secondary structure of p-cresol tetramer, which has the same chemical structure as hydroxystyrene, has been reported to bring the hydroxy-groups together to some extent, so that they are linked by a series of intramolecular hydrogen bonds (*13,14*). The same explanation can be applied to PHST films. The polar sites of hydroxy-groups are packed by intramolecular interaction and surrounded by non-polar sites of aromatic parts. The lower polarity in the surface layer indicates that the effects of such hydrogen-bonding interaction are larger in the surface layer than in the bulk and interface layers. This discussion agrees with the dissolution behavior of PHST films in an alkaline solution. Several polymer films are reported to dissolve slowly in the surface layer than in the bulk layers during alkaline development (*For example,25*).

Inhomogeneous concentration distribution. The pyrene fluorescence spectrum has two components and at longer wavelengths a broad structureless emission band is assigned to excimers as mentioned above. The excimer-to-monomer fluorescence intensity ratio is almost the same in the bulk and surface layers at lower pyrene concentrations such as 90 mM / HST unit M, but the excimer contribution is larger in the interface layer than in the bulk and surface layers. At higher concentration the contribution of excimer fluorescence is smaller in the surface layer and medium in bulk layer and the largest in the interface layer as shown in Figure 4. Excimer fluorescence is produced by the interaction of excited and ground state molecules. So the excimer formation depends directly upon the pyrene concentration. The decrease of excimer contribution indicates the concentration gradient of pyrene molecules from interface layer to the surface layer.

Depth profile of pyrene concentration is experimentally determined from the angular spectrum of TIRF spectroscopy as mentioned above. Figure 5a shows the angular spectrum and a gray curve is the fitting results of the uniform distribution of pyrene in a PHST film using the equation 2. There are three fitting parameters; refractive index of sample, background of noise intenisty, and a proportional constant in the least-squares method of simplex minimizaiton. The fitting is not good assuming a homogeneous depth distribution. Thus, the experimental data is fitted using a polynonimal function as an inhomogeneous distribution. The depth profile determined by equation 2 is shown in Figure 5b. Zero depth means the interface between a PHST film and the TIR substrate. It shows the concentration gradient from the interface, bulk to surface layers. It agrees with the results by analysis of the excimer formation. Note the large inhomogeneity of doped pyrene in a PHST film.

There are some possibilities to elucidate the inhomogeneous distribution. One is the effects of the casting solvent. The prebaking process is baking for ten min. at 80 °C. This is not sufficient to evaporate all solvents and some remain in the film, especially near the TIR substrate. The inhomogeneous distribution of casting solvents might cause the preferential incorporation of pyrene. Another possibility is the inhomogeneous structure of PHST. Chemical properties such as micropolarity and oxidation are different within each sublayer. These chemical properties affect the incorporation of pyrene into a PHST film. Further experiments are necessary to clarify

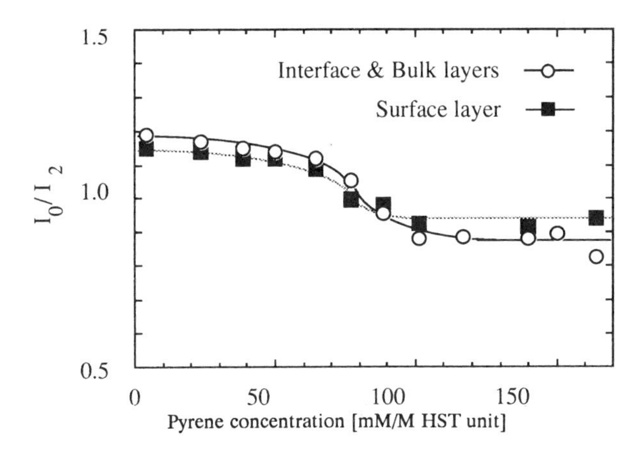

Figure 3. Vibrational intensity ratio dependence on pyrene concentration in the surface (closed square) and interface and bulk (open circle) layers.

Table I. Vibronic band intensity ratio

Solvent	I_0/I_2
1. Benzyl alcohol	1.22
2. Benzene	1.14
3. Toluene	1.11
4. Cumen	1.02
5. p-Xylene	1.00

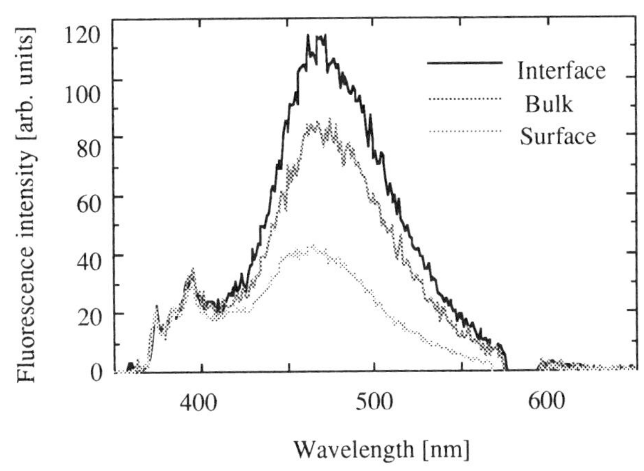

Figure 4. Fluorescence spectra of pyrene-doped PHST film of 140 mM / HST unit M. The spectra are normalized to the third vibronic band, I_2 in the monomer fluorescence.

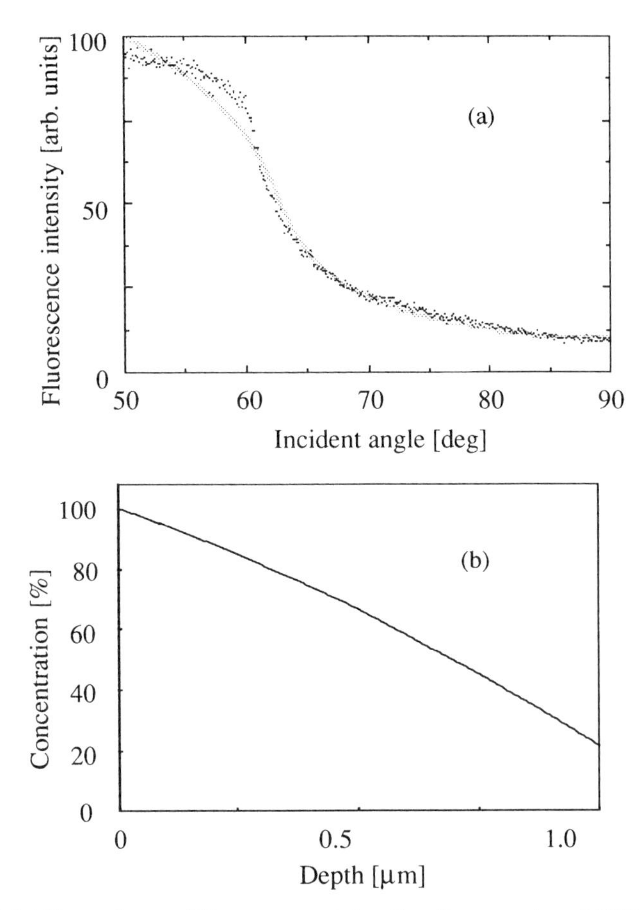

Figure 5. (a) Dependence of fluorescence intensity of pyrene-doped PHST film on incident angle and a fitting curve of the uniform depth profile (gray curve), (b) depth profile of pyrene concentration determined from angular spectrum (a).

the relationship between the chemical characteristics of a PHST film and the inhomogeneity of pyrene depth profile.

Rise and decay analysis of excimer fluorescence. The rise and decay curves of pyrene monomer and excimer fluorescence in the PHST films are measured. The emission maximum intensity of excimer is observed immediately after the laser excitation pulse. This is not the case of dynamic excimer formation or the diffusional migration of two pyrene molecules. For dynamic formation the excimer fluorescence intensity should increase with time with a delay after the laser excitation pulse. So the experimental results indicate static excimer formation from ground-state dimers of pyrene and no efficient movement in a PHST system. The fast rise time of the excimer fluorescence may be explained by the free volumes which provide sufficiently large voids for pyrene molecules to form excimer configurations. This static excimer formation is observed in all layers.

Rise and decay analysis of monomer fluorescence. Figure 6 shows the typical decay curves of monomer fluorescence of pyrene in the interface and bulk layers. The fluorescence decays faster in the interface layer than in the bulk layer. This is reasonable, because of the higher concentration of pyrene in the interface layer. A slow component is observed in the bulk layers as indicated by the dashed line. The contribution of the slowest component, defined as the intensity ratio of the component at time zero to that at peak intensity, is about 5 % and does not show obvious dependence on the concentration. The lifetime of this component does not depend upon the pyrene concentration and is equal to 250 ns, which is close to the monomer fluorescence lifetime in solvents as shown in Table II. The slowest component should be attributed to an isolated pyrene molecule in the bulk layer. And the lifetime in PHST matrix is between non-polar and polar solvents. The difference in lifetime of monomer fluorescence can be ascribed to environmental polarity. It is suggested that the polarity enhancement by solvents may be accompanied by a decrease in the radiative lifetime (*21*). So the observed value of the lifetime is reasonable from this point of view.

Baking effect in nitrogen environment. Baking is the one of the indispensable processes for preparing thin films but many problems remain to be solved. The decay curves of monomer fluorescence of films baked in the air and nitrogen gas are measured. The decay curves for nitrogen baking show a longer lifetime than those for air baking. Baking in nitrogen prevents PHST oxidation, producing a quinone-like structure. This is evidence for the existence of the oxide in the PHST film which is easily produced by air baking and quenches excited pyrenes effectively. This is the experimental data of the interface layer. Note that nitrogen baking has an effect even in the bulk and interface layers which are not directly exposed to the environmental gas.

Figure 7 shows fluorescence spectra in the bulk layers for nitrogen and air baking. The peak of excimer fluorescence is blue-shifted by 530 cm^{-1} or 66 meV for nitrogen baking but the spectral shape does not depend on the gas. This may be due to a different structure of PHST matrix. Excimers are less stable in a PHST film baked in nitrogen. The blue-shift of excimer fluorescence might be attributed to the formation of microcrystals.

Summary

Total-internal-reflection fluorescence spectroscopy has been shown to be a powerful technique for showing the structural properties of thin films, such as the inhomogeneous distribution of small molecules and the hydrophobicity of the interface layer. This inhomogeneity reflects the variation in the static microstructure and the

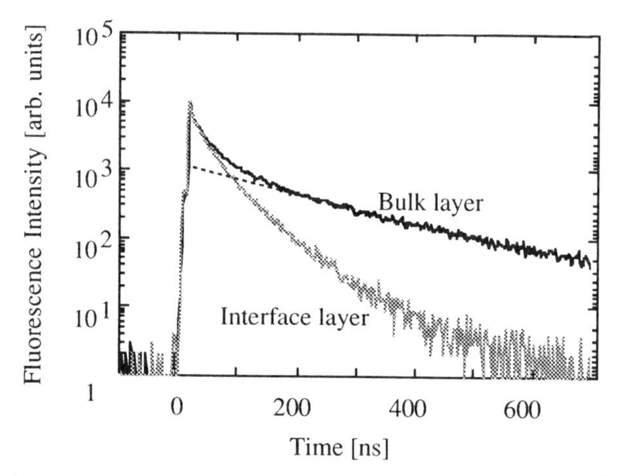

Figure 6. Decay curves of monomer fluorescence in the interface (gray curve) and bulk (solid curve) layers. The dashed line is a single-exponential function applied to the bulk fluorescence long after the excitation pulse.

Table II. Pyrene monomer lifetime

Solvent	lifetime
1. PHST	250 ns
2. cyclohexane	338 ns
3. 1,2-Dichloroethane	170 ns

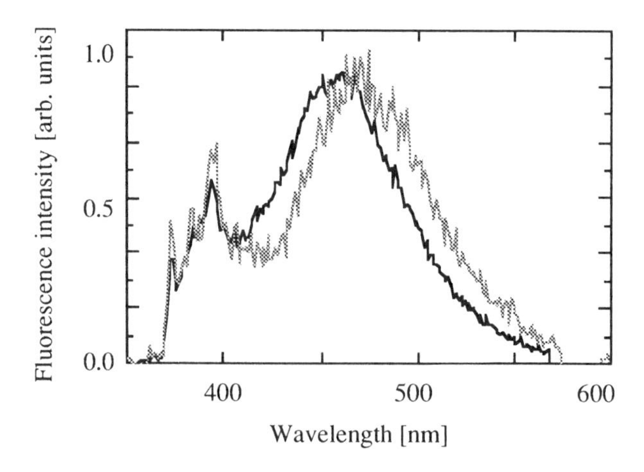

Figure 7. Fluorescence spectra of pyrene in bulk layers after nitrogen baking (solid curve) and air baking (gray curve).

dynamic local motions of molecules, such as the chemical and physical structure, aggregation, interaction between polymer and substrate, and conditions of preparation. Inhomogeneity is the intrinsic structure along the depth direction in the order of microns. TIRF spectroscopy with variable-angle equipment can give important information on the depth structure. Such information is important not only in fundamental research but also in applications to controlling and creating the desired surfaces and thin film properties in fields such as microelectronics, optical computers, molecular assembly, imaging and display media, in-vivo biosensors, and drug delivery systems.

Acknowledgements

The authors wish to thank Dr. Naoto Tamai for a helpful discussion on pyrene photochemistry.

References

1. Willson, C. G. *Introduction to Microlithography; ACS Sym. Ser.* **219**; Thompson, L. F.; Willson, C. G.; Bowden, M.J., Eds.; American Chemical Society, Washington, D. C., 1983; pp.87-159.
2. Masuhara, H; Mataga, N.; Tazuke, S.;Murao, T.; Yamazaki, I. *Chem. Phys. Lett.* **100,** 415 (1983).
3. Masuhara, H.; Tazuke, S.; Tamai, N.; Yamazaki, I. *J. Phys. Chem.* **90,** 5830 (1986).
4. Itaya, A.; Yamada, T.; Tokuda, K.; Masuhara, H. *Polymer J.* **22,** 697 (1990).
5. Toriumi, M.; Yanagimachi, M.; Masuhara, H. *MRS Extended abstracts;* Drake, C.; Klafter, J.; Kopelman, R., Eds.; 1990; pp.133-136.
6. Yanagimachi, M.; Toriumi, M.; Masuhara, H. *Chem. Mater.* 1991, **3,** pp. 413-418.
7. Toriumi, M.; Yanagimachi, M.; Masuhara, H. *Proc. SPIE Advances in Resist Technology and Processing VIII,* 1991, **1466,** pp. 458-468.
8. Toriumi, M.; Hayashi, N.;Hashimoto, M.; Nonogaki, S.; Ueno, T.; Iwayanagi, T. *Polym. Eng. Sci.* 1990, **29,** pp. 868-873.
9. Nonogaki, S.; Toriumi, M. *Makromol. Chem., Macromol. Symp.* 1990, **33,** pp. 233-241.
10. Daly, R. C.; DoMinh, T.; Arcus, R. A.; Hanrahan, M. J., ACS Sym. Ser. **346**; Bowden, M. J.; Turner, S. R., Eds.;1987; pp.237-249.
11. Miura, K.; Ochiai, K. M. T.; Kameyama, Y.; Kashi, C.; Uoya, S.; Nakajima, M; Kawai, A.; Kishimura, S. *Proc. SPIE Advances in Resist Technology and Processing V,* 1988, **920,** pp. 134-141.
12. Trefonas III, P.; Daniels, B.K.; Fischer, Jr., R. L. K. *Solid State Technol.,* August 1987, **30,** pp. 131-137.
13. Reiser, A. *Photoreactive polymers,* John Wiley and Sons, New York, N. Y.,1989.
14. Huangm, J. H.; Wei, T.; Reiser, A. *Macromolecules* 1989, **22,** pp. 4106-4112.
15. Okuda, Y.; Takashima, Y.; Miyai, Y.; Inoue, M. *Proc. SPIE Advances in Resist Technology and Processing IV,* 1987, **771,** pp. 61-68.
16. Ogawa, S.; Uoya, S.; Kimura, H.; Nagata, H.; *J. Photopolym. Sci. Technol.* 1989, **2,** pp. 375-382.
17. Coopmans, F.; Roland, B. *Proc. SPIE Advances in Resist Technology and Processing III,* 1986, **631,** pp. 34-39.
18. Kosbar, L. L; Kuan, S. W.; Frank, C. W.; Pease, R. F. W. *ACS Sym. Ser.*

381; Reichmanis, E.; O'Donnell, J. H. Eds.; American Chemical Society, Washington, D. C., 1989; pp.95-111.
19. Kosbar, L. L.; C. W. Frank, *Proc. SPIE Advances in Resist Technology and Processing V,* 1988, **920,** pp. 142-144.
20. Kuan, S. W. J.; Fu, C. C.; Pease, R. F. W.; Frank, C. W.M. *Proc. SPIE Advances in Resist Technology and Processing V,* 1988, **920,** pp. 414-418.
21. Nakajima, A. *Bull. Chem. Soc. Jpn.* 1971, **44,** pp. 3272-3277.
22. Dong, D. C.; Winnik, D. C. *Photochem. Photobiol.* 1982, **35,** pp. 17-21.
23. Dong, D. C.; Winnik, M. A. *Can. J. Chem.* 1984, **62,** pp. 2560-2565.
24. Toriumi, M.,; Masuhara, H. *Spectrochimica Acta Rev.* 1991, **14,** pp.353-377.
25. Meyerhofer,D.; *IEEE Trans. Elect. Devices.* 1980, **ED27,** pp.921-

RECEIVED September 1, 1992

Chapter 13

X-ray and Deep UV Radiation Response of *t*-Butoxycarbonyl-Protected 4-Hydroxystyrene–Sulfone Copolymers

A. E. Novembre, J. E. Hanson, J. M. Kometani, W. W. Tai, Elsa Reichmanis, and L. F. Thompson

AT&T Bell Laboratories, Room 1A–261, 600 Mountain Avenue, Murray Hill, NJ 07974

The x-ray and deep UV response of copolymers comprised of 4-tert-butoxycarbonyloxystyrene (TBS) and sulfur dioxide (SO_2) have been studied, and the results were utilized in the design of a lithographically useful chemically amplified positive acting resist material. The radiation sensitivity of the copolymer was a function of composition and increased with increasing SO_2 content in the material. These resists have been shown to respond most favorably to x-rays centered at $\lambda = 1.4$ nm and less efficiently to harder x-ray ($\lambda = 0.8$ nm) and 248 nm radiation. Arylmethyl sulfone model compounds have been investigated as a means of improving the harder x-ray and deep UV response of poly(4-tert-butoxycarbonyloxystyrene-co-sulfur dioxide) (PTBSS). The results obtained from studying the response of resists containing PTBSS and various sulfones have identified moieties which if incorporated into the structure of PTBSS should greatly improve the absorption properties at 0.8 and 248 nm, and consequently the lithographic sensitivity of the copolymer. A mechanism describing the radiation response of PTBSS and the model sulfone compounds is postulated to involve cleavage of the carbon-sulfur bond followed by hydrogen abstraction by the sulfonyl radical produced at the scission site. The acid species generated are then used to catalyze the removal of the t-butoxycarbonyl(t-BOC) protecting group of PTBSS during a subsequent heating step. Experimental data obtained from γ-radiolysis, IR and molecular weight studies provide evidence for the proposed mechanism. The deep UV quantum efficiency for acid formation for the investigated arylmethyl sulfones is discussed, and the results are used to identify those materials which can be used as an acid generating species in two component chemically amplified resists.

0097–6156/93/0527–0179$06.00/0

Resist systems based on chemical amplification have provided materials which exhibit both high sensitivity and resolution when used in conjunction with photon and electron-beam lithography(*1-3*). The overwhelming majority of these resists are multi-component, and the radiation induced chemistry is substantially different from that which occurs in standard novolac-diazonaphthoquinone photoresist formulations. Central to each chemically amplified resist formulation is an acid generating species which can be ionic or nonionic in nature, and is the component that initially responds to the radiation used to expose the resist(*1,4*). Additional components act as dissolution inhibitors(*5,6*), crosslinking agents(*7,8*) and binder resins(*4*). In non-crosslinking systems, presence of the acid labile group on a dissolution inhibitor generally dictates that three species must be present, whereas, attachment of this group to a binder or matrix resin reduces the number of components to two. Recently, it has been shown that copolymers of 4-tert-butoxycarbonyloxystyrene (TBS) and sulfur dioxide (SO_2), in the absence of an added acid generating species, act as sensitive x-ray (λ = 1.4 nm), and weakly sensitive harder x-ray (λ = 0.8 nm) and deep UV chemically amplified resists(*9*). Exposure of these copolymers results in main chain scission, whereby, acidic species such as sulfinic acids are produced at the scission sites. The radiation generated acid serves to catalyze the thermolysis of the tert-butoxycarbonyl (t-BOC) groups present on the copolymer and yields hydroxystyrene-sulfone copolymers (PHSS). The mechanism of this reaction was further elucidated using arylmethyl sulfone model compounds. Alcohol solutions of these compounds have been shown to form sulfinic acids when photolyzed at 254 nm(*10*), and it is postulated that the mechanism of radiolytic acid formation is the same for these sulfone derivatives and PTBSS. A series of structurally symmetric and nonsymmetric arylmethyl sulfone derivatives were additionally studied, and their application in chemically amplified resist formulations is reported. Experimental data obtained from γ-radiolysis, IR and molecular weight studies are presented to provide evidence for the proposed radiation induced reaction mechanism occurring in PTBSS and arylmethyl sulfones.

Experimental

Material Synthesis and Characterization. Ply(4-tert-butoxycarbonyloxystyrene-co-sulfur dioxide) PTBSS and poly(4-tert-butoxycarbonyloxystyrene) PTBS were prepared and characterized as described in the literature(*9,11*). Arylmethyl sulfones were purchased commercially or synthesized from arylmethyl chlorides and sodium hydrosulfite ($Na_2S_2O_4$) using literature procedures(*12*). All sulfones were purified by recrystallization from toluene.

Sulfur concentration in the copolymers was used to determine the composition of each material. The weight percent of sulfur was determined by x-ray fluorescence as described previously(*13*), and by elemental analysis performed by Galbraith Laboratories, Knoxville, TN. The glass transition temperatures (T_g) in N_2 of the polymers and the thermal decomposition temperature (T_d) and melting points of the polymers and arylmethyl sulfones when applicable were determined using a Perkin-Elmer DSC-7 differential scanning calorimeter and TGS-7 thermogravimetric analyzer, respectively.

X-ray and Deep UV Lithographic Analysis. PTBS and PTBSS resist solutions (5 to 20 w/v%) were prepared by dissolving the polymer in cyclohexanone, chlorobenzene or 2-methoxyethyl acetate. Resist solutions containing an arylmethyl sulfone were prepared by adding 3-20 mole% of the sulfone to the dissolved polymer. The solutions were filtered and then used to coat 0.4 to 1.5 μm thick resist films onto hexamethyldisilazane primed 4 and 5" silicon and quartz substrates. The substrates were subsequently baked at 105 ±2°C for 2 min. on a hot plate equipped with a vacuum hold-down chuck. Film thicknesses were measured using a Tencor Alpha Step 200 profilometer or a Nanometrics Nanospec/AFT film thickness gauge.

X-ray (λ = 0.8-2.0 nm centered @ 1.4 nm) exposures in helium were performed using a Hampshire Instruments Series 5000 point source proximity print stepper. A nylon-based radiochromic film (Far West Technologies, Inc.) was used for x-ray flux calibration of the pulsed x-ray source(*14*). Harder x-ray (λ = 0.8 nm) exposures were performed using the x-ray lithography station ES-0 of Alladin, the Synchrotron Radiation Center located at the University of Wisconsin (Madison)(*15*). Alladin was operated at 800 MeV with injected beam currents up to 230 mA and flux of 0.20 mW/mA·cm. Deep UV exposures (λ = 248 nm) were performed using a Süss Model MA56A contact aligner equipped with a Lambda Physik KrF excimer laser operating at 100 Hz. The laser putput was measured using a calibrated thin-film thermopile, and the incident dose per pulse ranged from 0.02 to 0.10 mJ/cm^2.

The resist films were postexposure baked (PEB) on a hot plate equipped with a vacuum hold-down chuck at 140°C for 2.5 minutes. Film thickness loss resulting from exposure and baking was measured using a profilometer. The exposed and baked films were immersion developed in a 0.17 N tetramethylammonium hydroxide (TMAH) solution for 30 seconds and rinsed for 20 seconds in de-ionized water. The minimum dose (D_s) necessary to completely develop a >100 μm^2 exposed area was used as the reported sensitivity for the PTBS and PTBSS polymers and for these polymers containing an arylmethyl sulfone.

Mechanistic Studies.

Conversion of PTBSS to Poly(4-hydroxystyrene sulfone). The extent of the loss of the tert-butoxycarbonyl group as a function of x-ray dose was calculated by recording FT-IR spectra of exposed and baked PTBSS and PTBSS plus sulfone films spun onto silicon wafers which had been polished on both sides. FT-IR spectra were obtained using a Mattson Instruments Inc., Galaxy Series model 8020 spectrometer.

Chain Scission Efficiency. Powdered samples of PTBSS were irradiated at room temperature in air with cobalt-60 γ-radiation. The dose rate was 0.45 Mrad/hr and the molecular weight of the exposed materials was measured by size exclusion chromatography (SEC). The change in the number average molecular weight ($\overline{M}_n$) as a function of γ-irradiation dose was then used to determine G(s) which represents the number of scission events for every 100 eV of energy absorbed by the irradiated material.

Identification of the Radiation Generated Acidic Species / Relative Deep UV Quantum Efficiency Determination. A 1.4 mg powdered sample of the 2.1:1 TBS:SO_2 copolymer was γ-irradiated (dose = 3 Mrad), and 1-2 mg of PTBSS plus an arylmethyl sulfone film which was deep UV exposed at a specified dose were dissolved in cyclohexanone containing the sodium salt of tetrabromophenol blue (TBPB) acid/base indicator. UV-visible absorption spectra of the TBPB solutions with and without the irradiated material were taken on a Hewlett-Packard 8452A diode array spectrophotometer. The measured change in absorbance of TBPB at 610 nm recorded from the two spectra was used in the calculation yielding the amount of acid present in the irradiated sample. The concentration of acid was determined by a comparison to a calibration plot which was established by measuring the TBPB absorbance @ 610 nm as a function of the amount of added hydrochloric acid titrant(*16*). The relative deep UV quantum efficiency was determined from knowledge of the concentration of acid produced in the exposed sample, the 248 nm absorption of a film of the material prior to exposure, and the incident exposure dose.

Results and Discussion

Materials Properties. The weight average molecular weight ($\overline{M}_w$), molecular weight distribution ($\overline{M}_w/\overline{M}_n$), and thermal properties of PTBS and PTBSS polymers used as single component resists, and as t-BOC protected matrix resins in resist systems containing arylmethyl sulfone derivatives are listed in Table I.

Table I. PTBS and PTBSS Material Properties

Composition (TBS:SO_2)	$\overline{M}_w \times 10^{-5}$ (g/mole)	$\overline{M}_w/\overline{M}_n$	T_g °C	T_d °C
1.75:1	3.53	2.70	168	172
2.0:1	5.75	1.54	164	165
#2.1:1	4.74	1.88	165	170
2.8:1	2.14	1.89	159	169
3.0:1	1.70	1.59	157	169
3.75:1	2.50	1.98	151	174
1.0:0	1.57	1.44	127	190

#4% of the t-BOC groups have been removed from the copolymer

These copolymers were prepared to evaluate the effect of material composition on the x-ray and deep UV radiation response. The thermal properties of the copolymers listed in Table I indicate that PTBSS process baking conditions using temperatures up to 140°C should not deteriorate resist performance, since this temperature is below the measured T_d.

Arylmethyl sulfones which were not available commercially were prepared by a one-step synthesis consisting of the appropriate arylmethyl chloride and sodium hydrosulfite. The structures and thermal properties of the sulfones studied are given in Table II. These sulfones were chosen to provide materials which would improve the 248 nm and x-ray absorption properties of resist materials formulated with PTBSS. Additionally, the series should provide information on how the structure and absorption properties of the sulfone affect the radiation induced acid formation efficiency.

Table II. Material Properties of Selected Arylmethyl Sulfones

Arylmethyl sulfone	Structure	Mp °C	T_d °C
Dibenzyl sulfone (DBS)	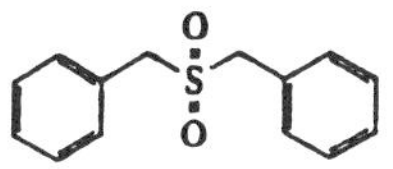	155	223
(2-Methylbenzyl)-phenyl sulfone (MBPS)	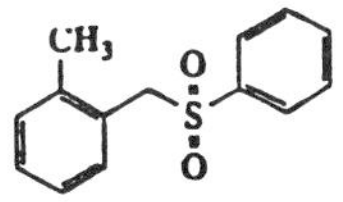	85	300
Bis(3-methoxybenzyl) sulfone (MEOBS)	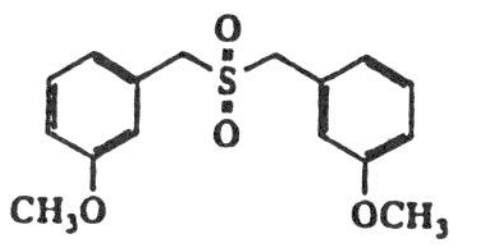	125	225
Bis(naphthomethyl) sulfone (NMS)	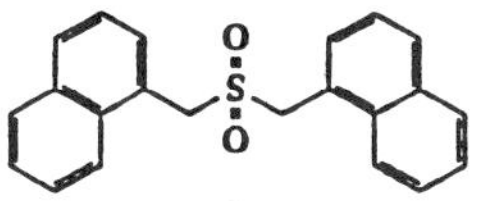	220	310
Bis(3,4-dichlorobenzyl) sulfone (DCBS)	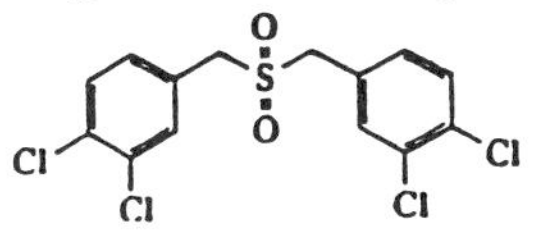	177	300
Bis(4-fluorobenzyl) sulfone (FBS)	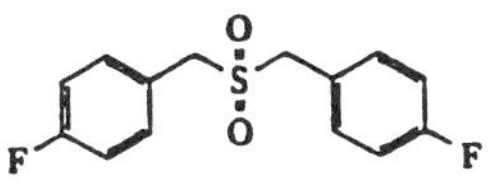	193	240

Lithographic Performance.

Single Component Resist Formulations. The x-ray (λ = 1.4 nm) exposure response curves for the TBS homopolymer and for PTBSS having a composition ranging from 1.75:1 to 3.75:1 TBS:SO_2 are shown in Figure 1. The initial film thickness was 0.5 μm and postexposure vacuum hot plate bake conditions were 140°C for 2.5 minutes. A four fold increase in the minimum dose necessary to dissolve a >100 μm^2 exposed area in the 0.17 N TMAH developer was observed as the ratio of TBS:SO_2 in the copolymers increased from 1.75:1 to 3.75:1. The former exhibited a D_s value of 10 mJ/cm^2, whereas, the latter had a value of 40 mJ/cm^2. These results differ from what was observed for the TBS homopolymer where no change in film thickness after PEB and development was observed using x-ray exposure doses up to 120 mJ/cm^2. The slight thickness change that was observed for PTBS resulted solely from the exposure. At a dose of 120 mJ/cm^2 approximately 3 percent of the initial film thickness was lost. This loss is a result of the direct removal of the tert-butoxycarbonyl group during the exposure. The resulting small percentage of hydroxyl groups formed in the homopolymer were ineffective in catalyzing the deprotection reaction during the 2.5 min. at 140°C postexposure bake step. Increasing both the bake time and temperature may, however, provide for conditions in which further deprotection is observed. Thus, it is clear that the presence of sulfur dioxide in the copolymer, and the ability to postexposure bake at elevated temperatures affords the observed high x-ray (λ = 1.4 nm) resist sensitivity.

Figure 2 represents a plot of D_s as a function of mole fraction of TBS (X_{TBS}) in the copolymer. This figure can be used to project what the x-ray sensitivity for copolymers having a X_{TBS} < 0.64 will be. For the case when X_{TBS} = 0.60 (1.5:1 TBS:SO_2), the extrapolated value for D_s is 3 mJ/cm^2. Preparation of a 1:1 TBS:SO_2 copolymer would provide for the highest possible sensitivity, but the decrease in solubility observed for copolymers having $X_{TBS} \leq 0.66$ may result in materials having limited process latitude.

Two Component Resist Formulations. In order to ascertain whether the same radiation induced chemistry that occurs in PTBSS also takes place with the arylmethyl sulfone derivatives, a film containing the 3.75:1 TBS:SO_2 copolymer and dibenzyl sulfone was exposed on the Hampshire Inst. Series 5000 x-ray proximity print stepper. DBS was chosen for this initial study due to the structural similarity to the backbone of PTBSS. DBS (19.3 mole%) was added to PTBSS, which yielded a formulation with an overall composition equivalent to a 2.2:1 TBS:SO_2 copolymer. The value obtained for this two-component resist after postexposure baking at 140°C for 2.5 min. was 23 mJ/cm^2. This point is represented in Figure 2, which shows that addition of DBS to the 3.75:1 TBS:SO_2 resist provides a sensitivity comparable to that which should be observed for a single component resist having an equivalent SO_2 content. These results strongly suggest that sulfones such as DBS exhibit an x-ray (λ = 1.4 nm) response analogous to that of PTBSS, and should function as sensitive photoacid generators for use in chemically amplified resists.

The synchrotron x-ray (λ = 0.8 nm) exposure of a 2:1 TBS:SO_2 copolymer, and formulations consisting of this material plus 12 mole% of DBS, MBPS, FBS and

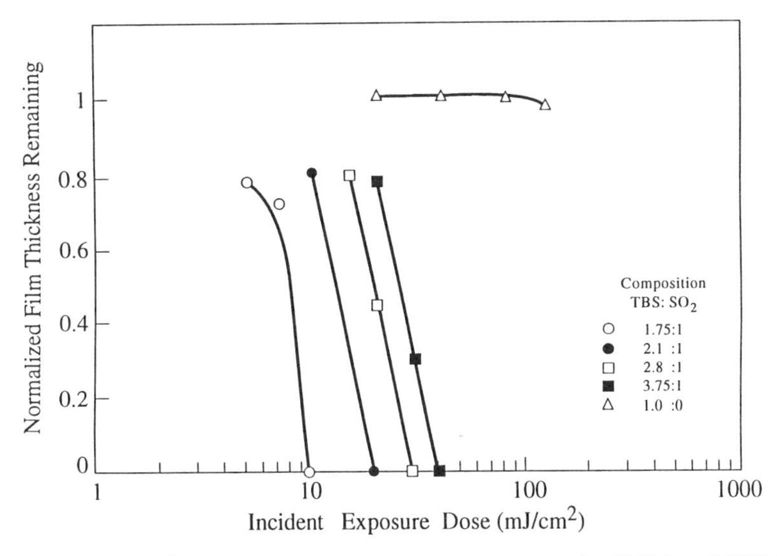

Figure 1. X-ray (λ = 1.4 nm) exposure response curves for PTBS and PTBSS resists of different composition.

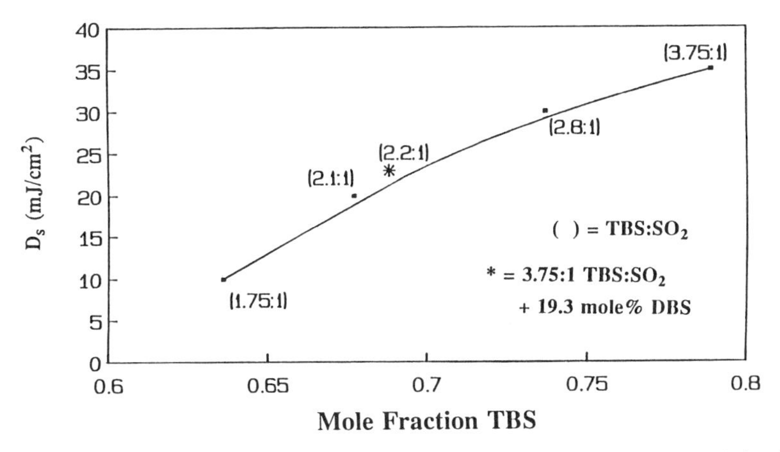

Figure 2. Plot of D_s taken from Figure 2 versus mole fraction of TBS in the copolymer and D_s resulting from addition of DBS to PTBSS.

DCBS photoacid generators are shown in Figure 3. In contrast to the sensitivity obtained for the 2:1 TBS:SO_2 resist when exposed at $\lambda = 1.4$ nm ($D_s = 20$ mJ/cm^2), this resist exhibits poor sensitivity using the higher energy x-rays. The D_s value was > 375 mJ/cm^2 using a PEB condition of 140°C for 2.5 min. This dose represented the highest value in the exposure dose array, and increasing the sensitivity by using more aggressive PEB conditions was not viewed as a viable approach. Increasing the copolymer SO_2 content was also not expected to provide a material having an x-ray sensitivity comparable to chemically-amplified resist systems specifically developed for use with 0.8 nm x-ray synchrotron radiation(*2*). The decreased x-ray sensitivity of the 2:1 TBS:SO_2 resist is attributed to the sharp decrease in the absorption of the resist at the shorter wavelength. Improvement in the sensitivity at this wavelength is observed when any of the sulfones are added to the 2:1 TBS:SO_2 resist. The improvement observed with the addition of DBS results from the increased SO_2 content of the total formulation. At a DBS loading of 12 mole% the TBS:SO_2 ratio in the film is 1.6:1, and as a result of the higher SO_2 content, D_s was reduced slightly to ~350 mJ/cm^2. In contrast, incorporation of MBPS, FBS, or DCBS into the PTBSS resist formulation was found to have a dramatic effect on the harder x-ray exposure response characteristics. For a resist formulation containing MBPS, the D_s value was reduced by 50% when compared to the formulation containing DBS. Although the x-ray absorption properties of DBS and MBPS are similar, the observed greater x-ray sensitivity using MPBS is attributed to the methyl group located at the ortho ring position. This methyl group provides a source of hydrogens which can be abstracted by the radiation generated sulfonyl radical.

The D_s values observed for resists containing FBS and DCBS were 150 and 125 mJ/cm^2 respectively, and represented a further improvement in sensitivity. The enhanced sensitivity observed with DCBS and FBS is a consequence of the increased x-ray absorption of these sulfones at 0.8 nm and, presumably, the generation of stronger acidic species which more effectively catalyze the thermal deprotection of PTBSS. In the case of DCBS, a 65% improvement in sensitivity is observed as compared to the resist containing DBS.

Deep UV Exposure Response Characteristics. The deep UV exposure response of a 3:1 TBS:SO_2 resist, and this resist containing 7.5 mole% of MBPS, NMS, MEOBS, FBS and DCBS is shown in Figure 4. The corresponding D_s values are also listed in Figure 4. The starting film thickness in all cases was 1.0 µm, and PEB conditions of 140°C for 2 min. were used. Table III gives the O.D./µm @248 nm for films containing the sulfones in the 3:1 TBS:SO_2 copolymer and are listed in the relative order of increasing quantum efficiency (Φ_{248nm}) for acid formation. For this relative basis FBS and MBPS consistently exhibited the lowest and highest values for Φ_{248nm}, respectively. Absolute values for Φ_{248nm} are not given because of the observed inconsistencies in the calculated amount of acid formed from the exposures. This result is believed to be a consequence of performing the experiments outside a purified air environment and not rigorously holding the time constant between exposure and when the samples were titrated. With the exception of FBS, addition of the sulfones improved the deep UV sensitivity of the 3:1 TBS:SO_2 copolymer by a factor of ≥ 6. The sensitivity improvement observed for NMS and

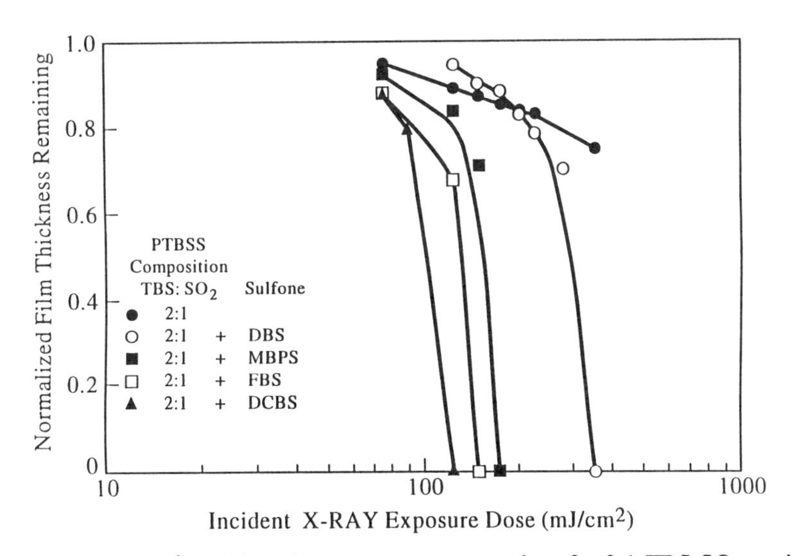

Figure 3. X-ray (λ = 0.8 nm) exposure response plot of a 2:1 TBS:SO_2 resist with and without a sulfone photosensitizer.

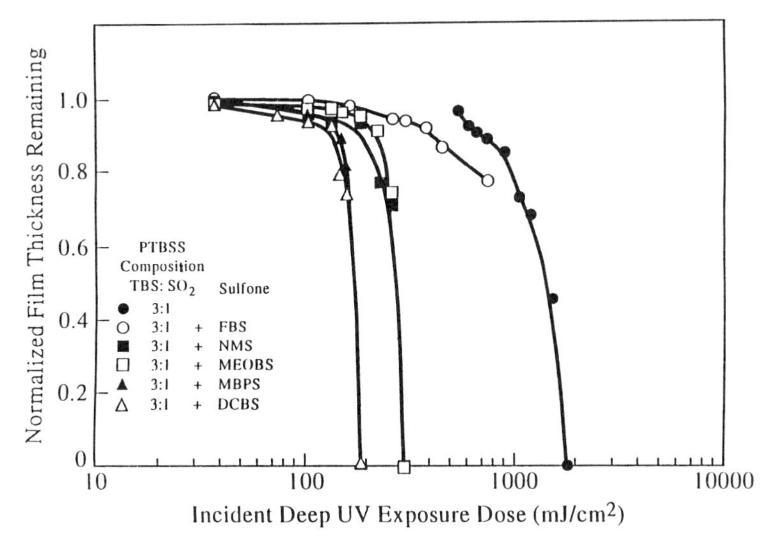

Figure 4. Deep UV exposure response curves of a 3:1 TBS:SO_2 resist with and without a sulfone photosensitizer.

MEOBS is most likely a consequence of the higher O.D./μm @248nm of films containing these sulfones vs the copolymer by itself. As shown in Table III, the absorption characteristics of these two component resist films showed a > 200% increase in O.D. over the copolymer. Addition of 7.5 mole% DCBS and MBPS to the 3:1 $TBS:SO_2$ resulted in the highest measured sensitivities. For DCBS, this results from a combination of improved absorption and quantum yield, and presumably, the greater strength of the radiation generated acid. In the case of MBPS, high sensitivity is derived mostly from the relatively improved quantum yield which results from the structural features discussed in the section on synchrotron x-ray response. FBS is shown to provide only a negligible increase in absorption which, in conjunction with a relatively low quantum yield for acid formation, causes FBS containing resists to require a dose in excess of 750 mJ/cm^2 for imaging purposes.

Table III: Deep UV Absorption Properties in Order of Increasing Φ_{248nm} for Acid Formation for the Arylmethyl Sulfones, listed in Figure 4

Sulfone	$^+$O.D./μm) @248 nm
FBS	0.120
MEOBS	0.330
NMS	0.300
DCBS	0.220
MBPS	0.157

$^+$Sulfone content = 7.5 mole% in a 3:1 $TBS:SO_2$ copolymer, the O.D./μm @248 nm for the 3:1 $TBS:SO_2$ resist = 0.10

Radiation Induced Reaction Mechanism. Figure 5 depicts the proposed radiation induced reaction mechanism for PTBSS and arylmethyl sulfones (i.e. DCBS). Exposure results in homolytic cleavage of the carbon-sulfur bond present in both materials which in turn produces benzylic and sulfonyl radical species. The latter species provides the source for the formation of acidic moieties since a certain percentage of the radicals abstract a hydrogen from the solid matrix to form a sulfinic acid. In the case of PTBSS, the acidic moiety represents the end group of the fragmented polymer chain. The remaining percentage of the sulfur containing radicals can undergo desulfonylation generating sulfur dioxide which may then react with a proton source in the film to produce sulfurous or sulfuric acid. In a subsequent heating step the acidic moieties catalyze the thermolysis of the tert-butoxycarbonyl group to form the aqueous base soluble polymer, poly(4-hydroxystyrene-sulfone). The occurrence of main chain scission and the evolution of gaseous products resulting from γ-radiolysis of olefin-sulfones have been previously reported by Brown and O'Donnell(*13,14*) and

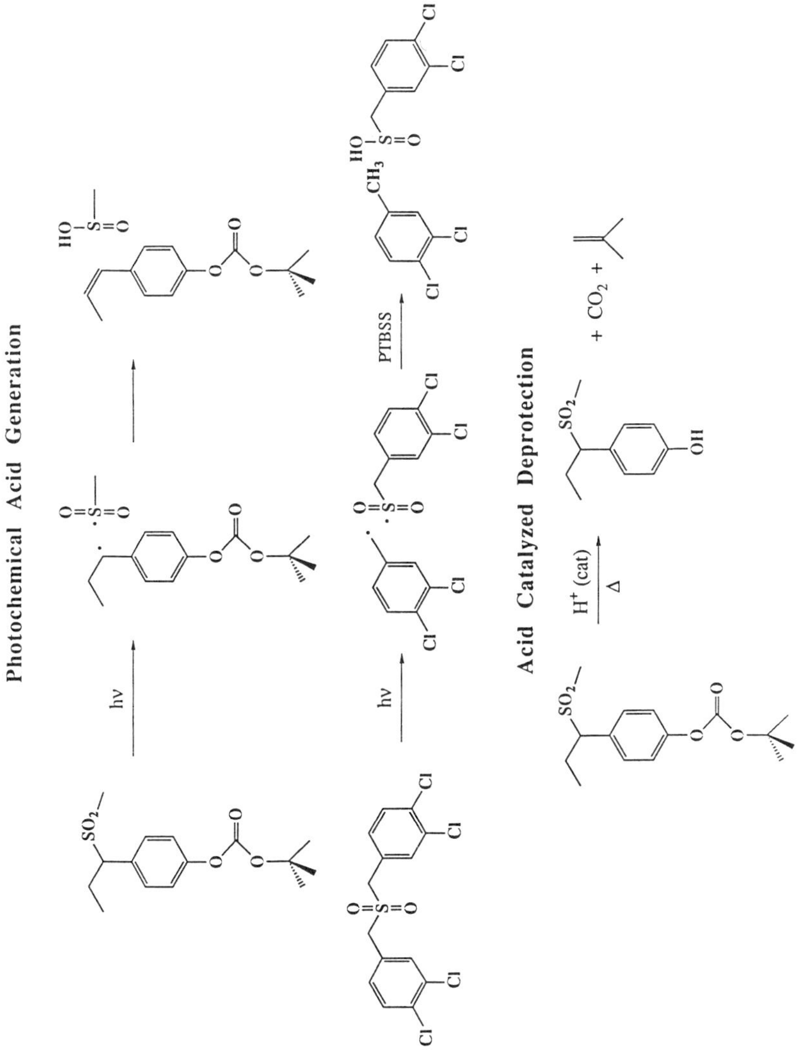
Photochemical Acid Generation
hν
hν
PTBSS
Acid Catalyzed Deprotection
H⁺ (cat)
Δ
+ CO₂ +

Figure 5. Schematic depicting the chemistry resulting from exposing and baking PTBSS and PTBSS plus arylmethyl sulfone containing films.

these results provide justification for the proposed PTBSS and arylmethyl sulfone radiation induced degradation mechanism.

Evidence for PTBSS Main Chain Scission. Direct evidence for main chain scission was obtained from γ-radiolysis of solid samples of a 2.1:1 TBS:SO_2 copolymer. The molecular weight data obtained for the nonirradiated and irradiated (3-21 Mrads) samples are summarized in Table IV. An approximate seven fold decrease in $\overline{M}_n$ can be seen for the sample irradiated at 3 Mrads. Increasing the dose to 21 Mrads further decreased the value of $\overline{M}_n$ by 68%. In order to ascertain that the change in $\overline{M}_n$ was not due to the partial conversion of PTBSS to PHSS upon exposure, IR spectra in the region of 2600-3800 cm^{-1} for the sample irradiated at 3 Mrads and for a nonirradiated sample are compared in Figure 6. No difference in absorption intensity at 3500 cm^{-1} between the two samples was observed, confirming that the decrease in molecular weight is soley due to the polymer backbone fragmentation.

The value of G(s) was obtained by determining the linear relationship of the reciprocal of $\overline{M}_n$ as a function of absorbed dose. The G(s) was taken to be proportional to the calculated slope, and for the 2.1:1 TBS:SO_2 copolymer was equal to 3.6.

Table IV: Change in Molecular Weight of a 2.1:1 TBS:SO_2 Copolymer as a Function of γ-Radiation Dose

DOSE (Mrad)	$\overline{M}_n \times 10^{-5}$ (g/mole)
0	2.52
3	.38
11	.23
21	.12

Presence of Acidic Moieties. The amount of acid produced in a γ-irradiated 2.1:1 TBS:SO_2 copolymer sample, and a deep UV exposed sample containing a 3:1 TBS:SO_2 copolymer + 7.5 mole% MBPS was determined by spectrophotometrically following the protonation of the sodium salt of tetrabromophenol blue present in the solutions containing the irradiated polymers. A calibration curve was established by plotting the absorbance of TBPB @610 nm as a function of varying amounts of titrant (10^{-4} M HCl) added. Figures 7 and 8 portray the absorption spectra in the region of 300 to 800 nm for the TBPB solution in cyclohexanone before and after the addition of the γ-irradiated 2.1:1 TBS:SO_2 copolymer and the deep UV exposed MBPS + 3:1 TBS:SO_2 copolymer sample, respectively. From Figure 7, the absorbance value after addition of the γ-irradiated material was 0.26 and this value

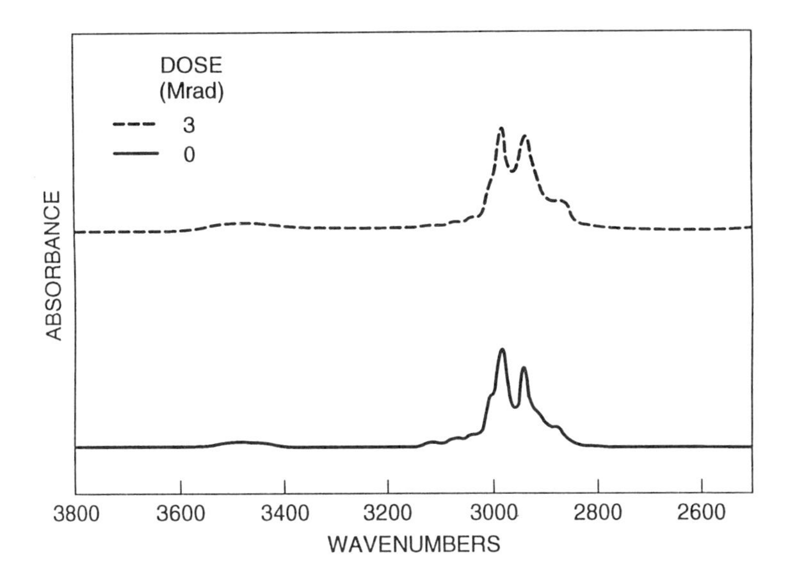

Figure 6. IR spectra of a γ-irradiated (3 Mrad) and nonirradiated powdered samples of the 2.1:1 TBS:SO_2 copolymer.

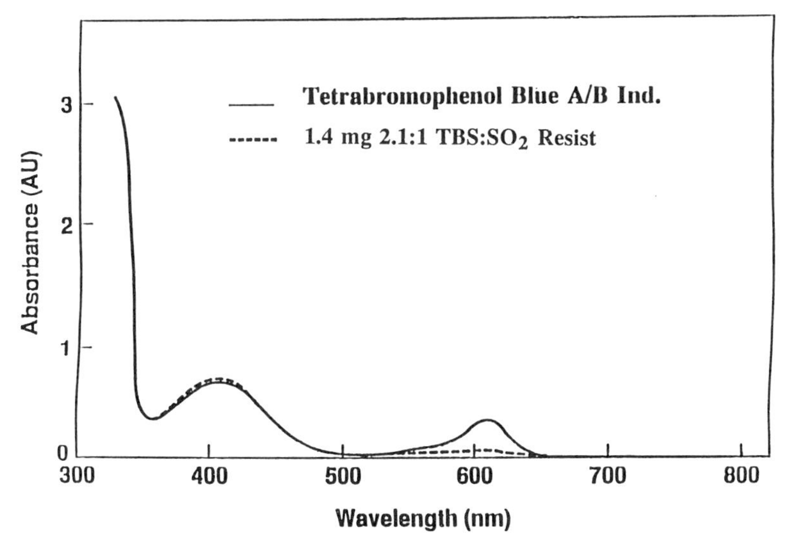

Figure 7. UV-visible absorption spectra of a TBPB cyclohexanone solution before (—) and after (----) the addition of γ-irradiated 2.1:1 TBS:SO_2 resist.

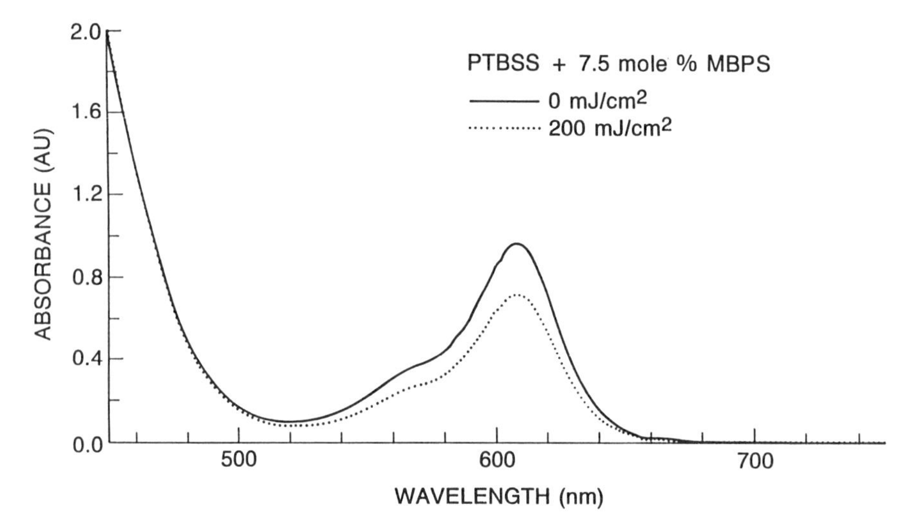

Figure 8. UV-visible absorption spectra of a TBPB cyclohexanone solution containing nonirradiated (—) and irradiated (----) sample of a PTBSS + MBPS resist.

corresponded to the production of 7.0 x 10^{-6} mmol of acidic moieties. If it is assumed that the amount of acid produced per unit absorbed dose is constant, the G-value for acid formation G(acid), which is the total number of acidic moieties produced for every 100 eV of energy absorbed by the irradiated copolymer, was calculated to be 1.6. The ratio of G(acid) to G(scission) can, therefore, be used as a measure of the efficiency of acid generation at each scission site. For the 2.1:1 TBS:SO_2 copolymer, the ratio was calculated to be 0.45, indicating nearly 50% efficiency. The amount of acid generated in the two-component resist was used to determine the relative quantum yields for acid formation of the sulfones as discussed in a previous section.

Evidence for the Conversion of PTBSS to PHSS. The formation of PHSS was confirmed by analyzing IR spectra taken of x-ray exposed regions of a thin film of a 2.8:1 TBS:SO_2 copolymer and a deep UV exposed 3:1 TBS:SO_2 copolymer containing 10 mole% MBPS. Both samples were postexposure baked at 140°C for 2.5 min. Peak areas corresponding to the carbonyl absorption at 1676-1876 cm^{-1} were integrated, and normalized to the areas of the carbonyl absorption peak obtained from a nonexposed region of the same film. The ratio was used to determine the percentage of t-BOC groups removed from each of the exposed and baked areas. Table V lists the percentage of t-BOC groups removed from each sample as a function of exposure dose.

Table V: X-ray and Deep UV Exposure Results for a 2.8:1 and 3:1 TBS:SO_2 Copolymer Containing 10 Mole% MBPS

Exposure Wavelength (nm)	Exposure Dose (mJ/cm^2)	% t-BOC Groups Removed
1.4	10	26
1.4	20	81
1.4	40	97
248	50	25
248	80	80
248	190	97

Figure 9 represents the IR spectra of the x-ray exposed and nonexposed regions of the 2:8:1 TBS:SO_2 resist film. The appearance and increase in absorption at 3500 cm^{-1} with increasing exposure dose is indicative of the presence of PHSS. Similar spectra were obtained from the deep UV exposed two-component PTBSS plus MBPS sample.

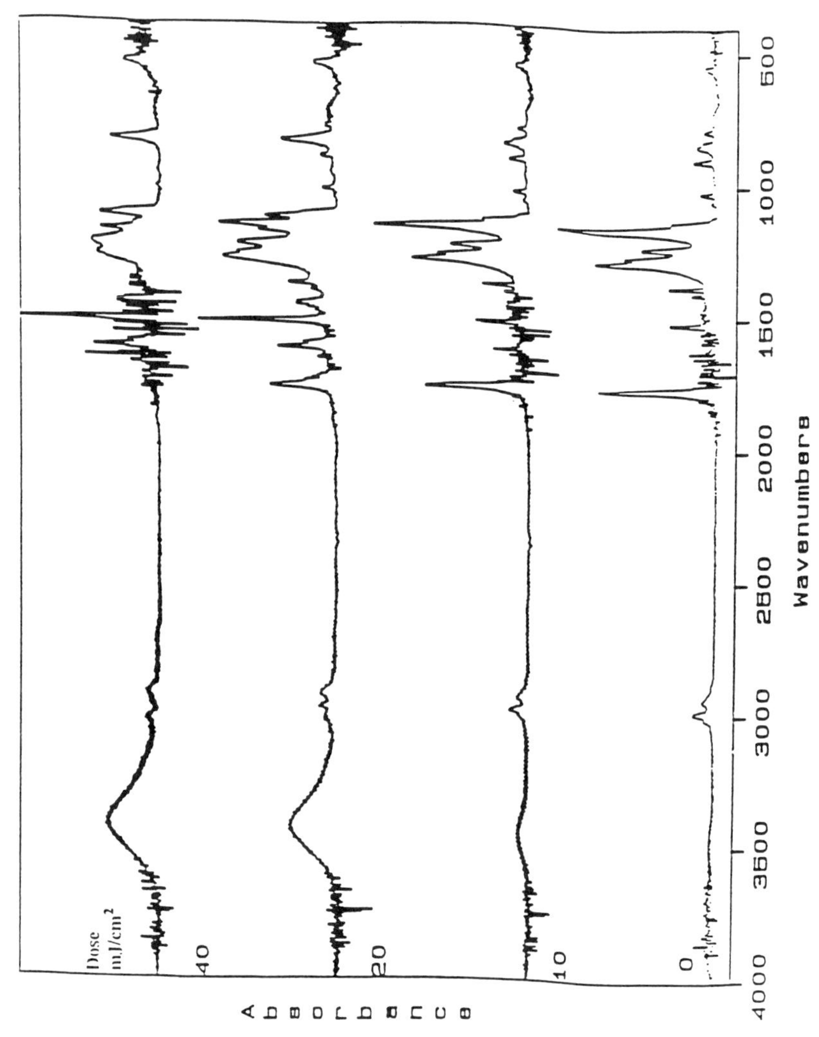

Figure 9. IR spectra of x-ray exposed and nonexposed regions of a 2.8:1 TBS:SO_2 resist film.

Conclusions

Resists containing PTBSS and an arylmethyl sulfone photoacid generator have been shown to act as sensitive deep UV (λ = 248 nm) and x-ray (λ = 1.4, 0.8 nm) resists. PTBSS alone has been shown to exhibit high sensitivity when exposed to the softer x-rays (λ = 1.4 nm). Addition of a sulfone to the copolymer is necessary to achieve high deep UV and harder x-ray resist sensitivity. For example, incorporation of a chlorinated sulfone to a 2:1 and 3:1 TBS:SO_2 copolymer greatly enhanced the resist sensitivity as does placement of a methyl group at the ortho ring position.

The radiation induced reaction mechanism in PTBSS and arylmethyl sulfones is proposed to involve cleavage of the carbon-sulfur bond producing radical species. These species in turn are responsible for the formation of acidic moieties. Such moieties in the case of PTBSS may exist as sulfinic acid end groups of the fragmented polymer chain. The radiation generated acid formed from PTBSS or the sulfone is then used to catalyze the deprotection of PTBSS during the process postexposure baking step. The efficiency of the acid formation process in PTBSS was measured by taking the ratio of G(acid) to that for G(scission) and has shown that for the 2.1:1 TBS:SO_2 copolymer approximately 50% of all scission sites produce an acid moiety.

Acknowledgements

The authors would like to thank Paula Trevor for x-ray fluorescence measurements; Dave Mixon, Mike Bohrer for preparation of some of the materials; Om Nalamasu, Gary Taylor and Ed Chandross for many helpful discussions; Hampshire Instruments technical and engineering support staff for x-ray exposures and resist process development; Professors Franco Cerrina and James Taylor and all the staff members who were instrumental in obtaining results using synchrotron radiation.

References

[1] a) Ito, H., Willson, C. G., In "Polymers in Electronics", *ACS Symposium Series* **242**, Davidson, T., Eds., ACS, Washington, D.C., 1984, pp. 11-23.
b)Frechet, J. M. J., Ito, H., Willson, C. G., "Proc. Microcircuit Engineering 82", Grenoble, France, 1982, pp. 260.
c) Crivello, J. V., In "Polymers in Electronics", *ACS Symposium Series* **242**, Davidson, T., Ed., ACS, Washington, D.C. 1984, pp. 3-10.

[2] Dössel, K.-F., Huber, H. L. and Oertel, H. *Microelectr. Eng.*, 1986, **5**, 97.

[3] Ito, H., Pederson, L. A., Chiong, K. N., Sonchik, S. and Tsai, C. *Proc. SPIE*, 1989, **1086**, 11.

[4] Reichmanis, E., Houlihan, F. M., Nalamasu, O. and Neenan, T. X., *Chemistry of Materials,* 1991, **3**, 394.

[5] O'Brien, M. J., *Polym. Eng. Sci.*, 1989, **29**, 846.

[6] Willson, C. G., Ito, H., Frechet, J. M. J., Tessier, T. G. and Houlihan, F., *J. Electrochem Soc.,* 1986, **133**, 181.

[7] Feely, W. E., Imhof, J. C. and Stein, C. M., *Polym. Eng. Sci.,* 1986, **26**, 1101.

[8] Dammel, R., Dössel, K.-F., Lingnau, J. and Theis, J., *Microelectr. Eng.,* 1989, **9**, 575.

[9] Novembre, A. E., Tai, W. W., Kometani, J. M., Hanson, J. E., Nalamasu, O., Taylor, G. N., Reichmanis, E. and Thompson, L. F., *Proc. SPIE,* 1991, **1466**, 89.

[10] Langler, R. F., Marini, Z. A. and Pincock, J. A., *Can. J. Chem.,* 1978, **56**, 903.

[11] Tarascon, R. G., Reichmanis, E., Houlihan, F. M., Shugard, A. and Thompson. L. F., *Polym. Eng. Sci.,* 1989, **29**, 850.

[12] Sandler, S. R. and Karo, W., Organic Functional Group Preparations, 2nd Edition, New York, Academic Press, 1983.

[13] Kanga, R. S., Kometani, J. M., Reichmanis, E., Hanson, J. E., Nalamasu, O., Thompson, L. F., Heffner, S. A., Tai, W. W. and Trevor, P., *Chemistry of Materials,* 1990, **3**, 660.

[14] Peters, D. W., Dardinski, B. J. and Kelly, D. R., *J. Vac. Sci. Techn.,* 1990, **B8**, 1624.

[15] Lai, B., Mitchell, G., Well, G. M. and Cerrina, F., *Nucl. Instr. Meth,* 1986, **A246**, 681.

[16] Pawlowski, G., Dammel, R., Lindley, C. R., Merrem, H., Röshert, H. and Lingnau, J., *Proc. SPIE,* 1990, **1262**, 16.

[17] Brown, J. R. and O'Donnell, J. A., *Macromolecules,* 1970, **3**, 265.

[18] Brown, J. R. and O'Donnell, J. H., *ibid.,* 1972, **5**, 109.

RECEIVED October 7, 1992

Chapter 14

Polarity Change for the Design of Chemical Amplification Resists

Hiroshi Ito

Research Division, Almaden Research Center, IBM Corporation, 650 Harry Road, San Jose, California 95120–6099

High resolution lithographic technologies such as deep UV (<300 nm), electron beam, and x-ray exposure techniques demand extremely high resist sensitivities for their economical operation in the production environments. We have proposed "chemical amplification" as a method to dramatically increase the resist sensitivity. In chemically amplified resist systems, an initial radiation event induces an avalanche of subsequent chemical transformations. We have focused our attention on the use of acid as a catalytic species. In this paper are described chemical amplification resist systems based on a change of polarity (solubility). Three modes of the polarity alteration are discussed; i) a polarity change from a nonpolar to a polar state, ii) a polarity reversal, and iii) a reverse polarity change from a polar to a nonpolar state. The polarity change (i) utilizes acid-catalyzed deprotection or deesterification of polymer pendant groups. Examples of such polymers are poly(4-*t*-butoxycarbonyloxystyrene), poly(*t*-butyl 4-vinylbenzoate), and an alternating copolymer of α,α-dimethylbenzyl methacrylate with α-methylstyrene. The polarity reversal (ii) is based on thermal deesterification and acid-catalyzed rearrangement of a dimethyl cyclopropyl carbinol ester. The reverse polarity change (iii) can be built on the classical pinacol-pinacolone rearrangement, wherein *vic*-diols (pinacols) are converted to ketones or aldehydes. These acid-catalyzed polarity change mechanisms provide highly sensitive, swelling-free, high resolution resist systems.

The past decade witnessed a remarkably rapid advancement of the microelectronics technology. The improvements of integrated circuit devices have been accomplished by increasing the number of components per chip, which has been made possible by reduction of the minimum feature size on the chip. The trend continues; the feature size continues to shrink. Four-megabit

0097–6156/93/0527–0197$07.75/0

dynamic random access memory (DRAM) devices are currently in production with minimum features in the 0.8 μm range. High numerical aperture (NA) step-and-repeat reduction projection tools operating at the conventional g line (436 nm) of the Hg arc lamp are used to produce these devices. Furthermore, fabrication of 16 Mbit DRAM's has recently entered into production with features as small as 0.5 μm. The state-of-the-art devices are manufactured typically with conventional diazonaphthoquinone/novolac resists in conjunction with high NA i line (365 nm) steppers. However, a new lithographic technology will be required for further reduction in feature size.

The resist resolution is inversely proportional to NA of the lens and directly proportional to the exposure wavelength. A higher resolution is achieved at a shorter wavelength (deep UV<300 nm, electron beam, or x-ray). However, the conventional diazonaphthoquinone resists do not perform adequately in the high resolution, short wavelength lithographic technologies. The economy of operating these high resolution but low power technologies in the production environments dictates that resists must possess extremely high sensitivities. In order to meet the sensitivity requirement, we have sought a chemical method and proposed "chemical amplification" *(1-3)*. In chemically amplified resist systems, an initial radiation or photochemical event triggers a cascade of subsequent chemical reactions. We have chosen "acid catalysis" of acid-labile polymer matrices to induce an avalanche of desired reactions using acidic species generated by irradiation *(4-6)*.

Resist materials, chemistries responsible for imaging, and process conditions also affect the resist resolution. Crosslinking negative resist systems tend to suffer from a limited resolution due to swelling during development with organic solvents. However, negative resists based on crosslinking of phenolic resins are typically devoid of swelling *(7-9)*. Another approach to the design of swelling-free negative resist systems is radiation-induced alteration of polarity or solubility of the resist resin. In this paper are described three modes of the polarity change for the design of chemically amplified resist systems (Scheme I); i) a polarity change from a nonpolar to a polar state, which is based on acid-catalyzed deprotection (deesterification) of polymer pendant groups *(1,2,5,10-14)*, ii) a polarity reversal, which can be accomplished by thermal deprotection and acid-catalyzed rearrangement *(15)*, and iii) a reverse polarity change from a polar to a nonpolar state, which can be induced by acid-catalyzed pinacol-pinacolone rearrangement *(16,17)*.

Experimental

Materials. Poly(4-*t*-butoxycarbonyloxystyrene) (PBOCST) was prepared by radical polymerization of 4-*t*-butoxycarbonyloxystyrene (BOCST) with 2,2-azobis(isobutyronitrile) (AIBN) or benzoyl peroxide (BPO) in toluene *(18)* or by reacting commercial poly(4-hydroxystyrene) (Maruzen Oil, Lyncur PHM) with di-*t*-butyl dicarbonate *(12)*. *t*-Butyl 4-vinylbenzoate (TBVB) was obtained by treating 4-vinylbenzoyl chloride with lithium *t*-butoxide or by reacting 4-vinylphenylmagnesium chloride with di-*t*-butyl dicarbonate and then subjected to radical polymerization *(10,15)*. α,α-Dimethylbenzyl methacrylate (DMBZMA) was synthesized by reaction between methacryloyl chloride and lithium α,α-dimethylbenzyl alkoxide and copolymerized with α-methylstyrene (MST) using AIBN in toluene at 60 °C *(14)*.

2-Cyclopropyl-2-propyl 4-vinylbenzoate was synthesized by reacting 4-vinylbenzoyl chloride with Li alkoxide of dimethyl cyclopropyl carbinol which had been prepared from cyclopropyl methyl ketone and methylmagnesium bromide *(15)*. Radical polymerization afforded the desired benzoate polymer. The polymeric pinacol, poly[3-methyl-2-(4-vinylphenyl)-2,3-butanediol], was prepared as follows *(16,17)*. Commercial 4-chlorostyrene was converted to a Grignard reagent, which was treated with a half equivalent of 3-hydroxy-3-methyl-2-butanone in THF at 0 °C. The monomer was purified after workup by column chromatography and by recrystallization from hot hexanes and subjected to radical polymerization with AIBN in THF at 60 °C. The polymerization mixture was chromatographed to remove unreacted monomer and the polymer was isolated by precipitation in water.

Benzopinacole and *meso*-hydrobenzoin were purchased from Aldrich and 3-methyl-2-phenyl-2,3-butanediol was synthesized by reacting two equivalents of phenylmagnesium bromide with 3-hydroxy-3-methyl-2-butanone *(16,17)*. The phenolic resins used as the matrix polymers for the pinacol rearrangement were *p*- and *m*-isomers of poly(hydroxystyrene) (PHOST) and a cresol-formaldehyde novolac resin. The PHOST's were prepared by heating a powder of corresponding poly(*t*-butoxycarbonyloxystyrenes).

The acid generators employed in these studies were triphenylsulfonium hexafluoroantimonate (TPS) and 4-thiophenoxyphenyl(diphenyl)sulfonium hexafluoroantimonate (TPPDPS), which had been synthesized according to the literature *(19,20)*. The resists were formulated using cyclohexanone or propylene glycol monomethyl ether acetate as a casting solvent. Diglyme was also used as a casting solvent for IR studies.

Lithographic Imaging. Spin-cast resist films were baked on a hot plate at 90-130 °C for 1.5-10 min. The e-beam exposures were performed on an IBM VS-1 (20 keV) or VS-4 (25 keV) or on a Hontas-4 (25 keV) system. The x-ray exposures were carried out at Stanford Synchrotron Radiation Laboratory and also at Brookhaven National Laboratory. Preliminary contact lithography was carried out using an Optical Associate Inc. exposure station, a 254-nm bandpass filter, and a quartz mask. A Perkin Elmer Micralign 500 was also used in the UV2 mode to generate a sensitivity curve. Projection printing was performed on a Canon KrF excimer laser step-and-repeat tool or on a Ultratech X-248E 1x stepper. The exposed wafers were postbaked on a hot plate at 90-160 °C for 1-2 min. The development of images was achieved with use of an appropriate solvent.

Measurements. Molecular weight determination was made by gel permeation chromatography (GPC) using a Waters Model 150 chromatograph equipped with 6 μStyragel columns at 30 °C in THF or by membrane osmometry using a Wescan 230 recording membrane osmometer in toluene at 30 °C. Thermal analyses were performed on a Du Pont 1090 thermal analyzer or a Perkin Elmer TGS-2 at a heating rate of 5 °C/min for thermogravimetric analysis (TGA) and 10 °C/min for differential scanning calorimetry under nitrogen atmosphere. IR spectra of the resist films were obtained with an IBM IR/32 FT spectrometer using undoped Si wafers (for e-beam exposure) or 1-mm thick NaCl discs as substrates. UV spectra were recorded on a Hewlett-Packard Model 8450A UV/VIS spectrometer using thin films cast on quartz plates. NMR spectra were obtained on an IBM Instrument NR-250/AF spectrometer. Film thickness was measured on a Tencor alpha-step 200.

Polarity Change from a Nonpolar to a Polar State by Acid-Catalyzed Deprotection

Poly(4-*t*-butoxycarbonyloxystyrene) (PBOCST). PBOCST can be prepared by radical polymerization (or cationic polymerization in liquid sulfur dioxide) of monomer *(18)* or from PHOST *(12)* (Scheme II). This polymer is converted upon postbake at ca. 100 °C to PHOST by reaction with strong Bronsted acid produced by irradiation of acid generators such as onium salts, quantitatively releasing CO_2 and isobutene to regenerate a proton (Scheme III), which is rather a classical example of chemical amplification resists *(1,2,5,11,12)*. Figure 1 shows IR spectra of a film of PBOCST (made from PHOST) containing 7.7 wt% (3.0 mole% to polymer repeat unit) of TPPDPS. Deprotection is insignificant at 0.5 $\mu C/cm^2$ (20 keV) (Figure 1b) with conversion and film thinning amounting to only ca. 6 and 3.3 %, respectively. Exposure to 2.5 $\mu C/cm^2$ followed by postbake at 130 °C (Figure 1c) results in 95 % conversion and 38 % thinning. The deprotection reaction results in conversion of a lipophilic polymer to a hydrophilic and acidic polymer, providing either positive or negative imaging simply by changing the polarity of the developer solvent (Scheme III). The resist is capable of resolving sub-half-micrometer features by e-beam or x-ray irradiation *(11,12)*. The incident dose required for 100 % final thickness in the case of the negative development with anisole of the tBOC resist (1.1 μm thick film containing 15 wt% of TPS) was 13 mJ/cm^2 at 1.0-3.5 keV of x-ray radiation with a contrast (γ) of 15 *(11)* Sub-half-micrometer negative images (0.3/0.3 and 0.25/0.5 μm line/space) delineated in a tBOC resist containing 15 wt% of TPPDPS by e-beam exposure of 3 $\mu C/cm^2$ at 25 keV are presented in Figure 2. Figure 3 exhibits the e-beam sensitivity curve for the tBOC resist (ca. 0.4 μm thick) consisting of PBOCST made from PHOST and 15 wt% of TPPDPS. The remaining thickness was measured after postbake at 100 °C for 2 min as well as after development in anisole. The shrinkage that occurs during postbake is a good measure of conversion of PBOCST to PHOST. The resist exhibits a maximum thickness loss of 37 % at 2 $\mu C/cm^2$ upon postbake due to loss of isobutene and carbon dioxide. The sensitivity defined as a dose at which 50 % of the film thickness remains after development is 1.4-2 $\mu C/cm^2$ with γ of 2.3.

As the scanning electron micrographs in Figure 2 demonstrate, the resist does not suffer from swelling in the negative development with anisole because the developer solvent has little affinity toward PHOST. In contrast, conventional negative resist systems based on crosslinking tend to exhibit a limited resolution due to swelling during development with organic solvents, which arises from the fact that the organic solvent that dissolves the uncrosslinked (unexposed) regions has an affinity toward the crosslinked (exposed) areas. E-beam nanolithography with the tBOC resist has provided 19 nm resolution, indicating that chemical amplification and high resolution are not mutually exclusive *(21)*. One proton has been shown to cleave 1000 tBOC groups under standard processing conditions, which corresponds to the acid diffusion distance of ca. 50 Å *(22)*.

1. polarity change : nonpolar → polar

2. polarity reversal : nonpolar → less polar
   ↓
   more polar

3. reverse polarity change polar → nonpolar

Scheme I. Three modes of polarity alteration.

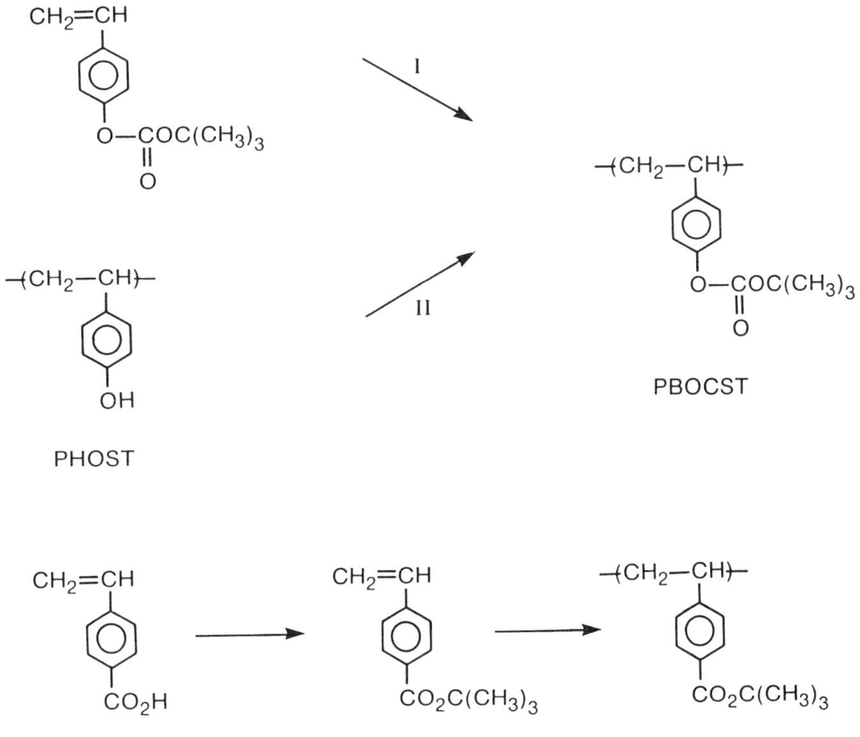

$CH_2{=}CH$–C_6H_4–$O{-}C(=O)OC(CH_3)_3$ —I→

$-(CH_2{-}CH)-$ with C_6H_4–OH (PHOST) —II→

$-(CH_2{-}CH)-$ with C_6H_4–$O{-}C(=O)OC(CH_3)_3$ (PBOCST)

$CH_2{=}CH$–C_6H_4–CO_2H → $CH_2{=}CH$–C_6H_4–$CO_2C(CH_3)_3$ → $-(CH_2{-}CH)-$ with C_6H_4–$CO_2C(CH_3)_3$ (PTBVB)

Scheme II. Preparation of PBOCST and PTBVB.

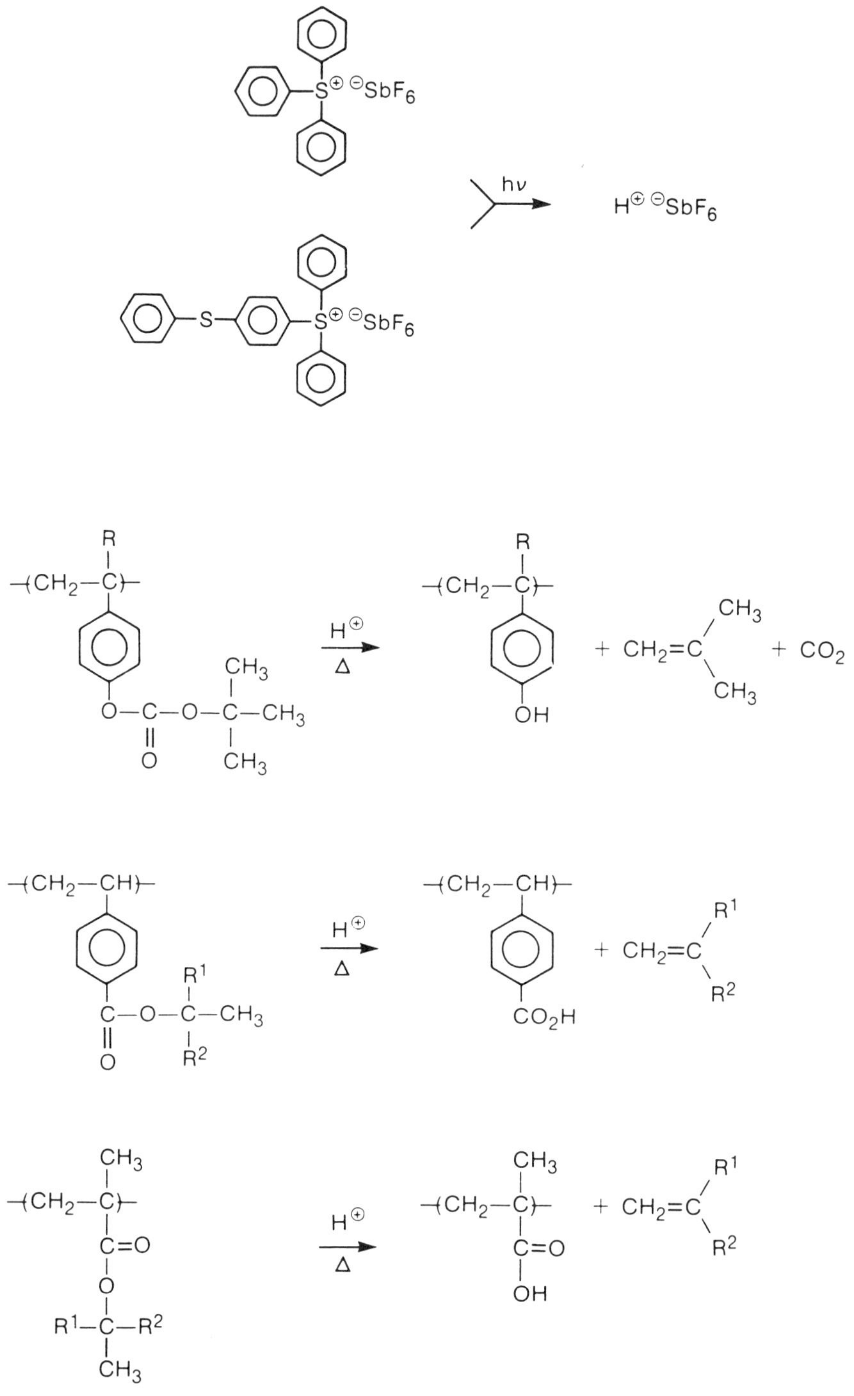

$S^{\oplus}$ $^{\ominus}SbF_6$

S $S^{\oplus}$ $^{\ominus}SbF_6$

$h\nu$

$H^{\oplus}$ $^{\ominus}SbF_6$

R

$-(CH_2-C)-$

CH_3

$O-C-O-C-CH_3$

O CH_3

$H^{\oplus}$

Δ

R

$-(CH_2-C)-$

OH

$+ CH_2=C(CH_3)_2 + CO_2$

$-(CH_2-CH)-$

R^1

$C-O-C-CH_3$

O R^2

$H^{\oplus}$

Δ

$-(CH_2-CH)-$

CO_2H

$+ CH_2=C(R^1)(R^2)$

CH_3

$-(CH_2-C)-$

$C=O$

O

R^1-C-R^2

CH_3

$H^{\oplus}$

Δ

CH_3

$-(CH_2-C)-$

$C=O$

OH

$+ CH_2=C(R^1)(R^2)$

Scheme III. Radiation-induced acid-catalyzed deprotection.

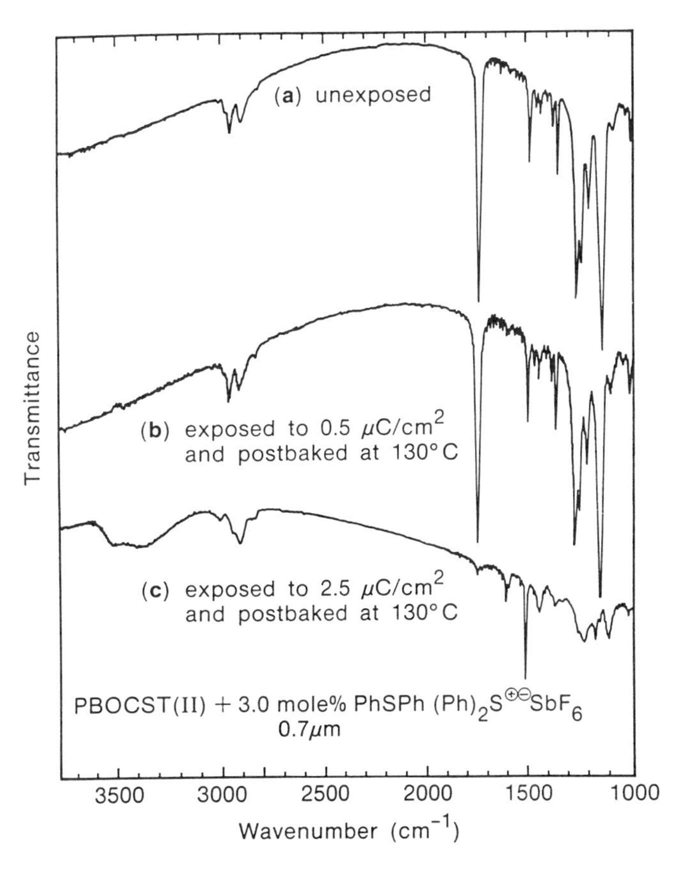

Figure 1. IR spectra of tBOC resist (0.7 μm) containing 3.0 mol% TPPDPS. (Reproduced with permission from ref. 12. Copyright 1989 Society of Photo-Optical Instrumentation Engineers.)

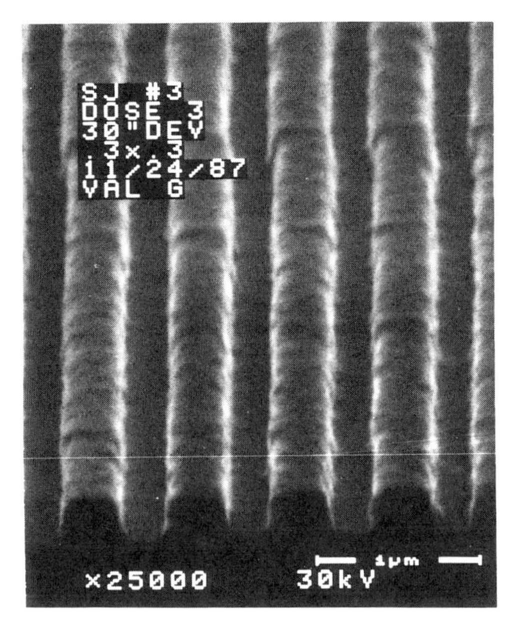

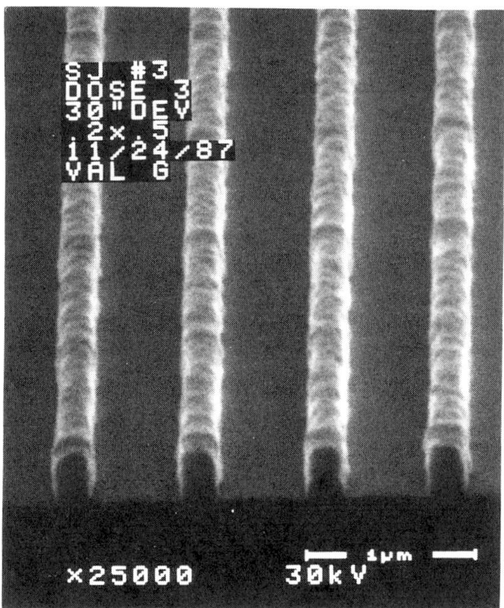

Figure 2. Negative images of tBOC resist printed by e-beam exposure (3 $\mu C/cm^2$ at 25 keV). (Reproduced with permission from ref. 12. Copyright 1989 Society of Photo-Optical Instrumentation Engineers.)

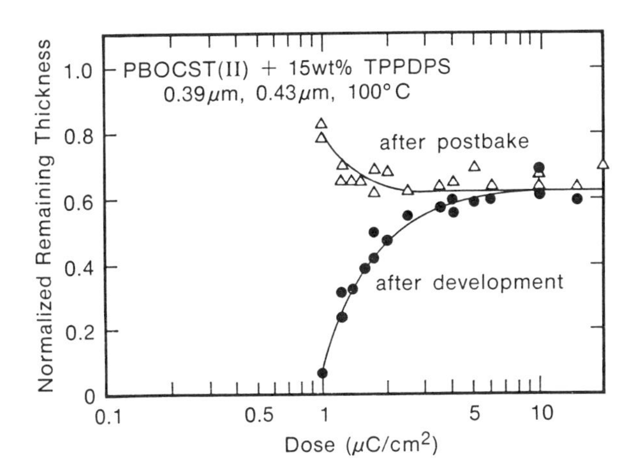

Figure 3. E-beam sensitivity curve of tBOC resist at 20 keV. (Reproduced with permission from ref. 12. Copyright 1989 Society of Photo-Optical Instrumentation Engineers.)

Poly(*t*-butyl 4-Vinylbenzoate) (PTBVB). PTBVB can be readily afforded by radical polymerization of TBVB (Scheme II) *(10,12,15)*. This lipophilic polymer is converted upon postbake to poly(4-vinylbenzoic acid) (PVBA) by reaction with an acid generated by radiation, quantitatively liberating isobutene and a proton (Scheme III). IR spectra of a PTBVB resist (0.68 μm thick) containing 8.2 wt% (3.0 mol% to polymer repeat unit) of TPPDPS are presented in Figure 4. The resist shows ca. 45 % conversion and 27 % thinning at 0.5 $\mu C/cm^2$ upon postbake at 130 °C (Figure 4b). Exposure to 2.5 $\mu C/cm^2$ induces ca. 65 % deprotection and 27 % thinning (Figure 4c). Development with anisole results in negative imaging and use of aqueous base as a developer provides positive imaging *(8)*. In contrast to the tBOC resist, the PTBVB resist is not useful as a single layer deep UV resist due to a very high absorption of PTBVB and PVBA below 300 nm. As the e-beam sensitivity curve of the PTBVB resist in Figure 5 indicates, however, the system containing 8.2 wt% of TPPDPS is a highly sensitive e-beam resist, providing 0.5-0.65 $\mu C/cm^2$ (20 keV) sensitivity for 50 % film retention and γ of 3.8 in the case of the film thickness of ca. 0.7 μm.[10] The typical imaging dose is ca. 0.7 $\mu C/cm^2$ *(10)*. Liberation of isobutene upon postbake (130 °C in this case) results in film shrinkage. No further thickness loss is observed during development with anisole once the maximum shrinkage is attained during postbake.

In order to directly compare the intrinsic sensitivity of the *t*-butyl carbonate and the *t*-butyl ester, a copolymer of BOCST (42 %) with TBVB (58 %) was prepared by radical polymerization, mixed with 3.0 mol% of TPPDPS, exposed to 0.5 and 2.5 $\mu C/cm^2$, postbaked at 130 °C, and subjected to IR analysis. The unexposed film exhibits two carbonyl absorptions at 1755 (carbonate) and 1715 cm^{-1} (benzoate). The 0.67-μm thick film showed ca. 90 % tBOC loss and ca. 50 % t-butyl ester deprotection when exposed to 0.5 $\mu C/cm^2$ followed by postbake at 130 °C, indicating that the *t*-butyl carbonate is more susceptible to acidolysis than the *t*-butyl benzoate. The tBOC deprotection is completed and the *t*-butyl ester conversion is very high at 2.5 $\mu C/cm^2$. The lithographic sensitivity of the PTBVB resist is higher than that of the tBOC resist whereas the radiation-induced acidolysis is more facile on the tBOC group than on the *t*-butyl ester group, suggesting that conversion of PTBVB to PVBA provides a more pronounced solubility differentiation than the conversion of PBOCST to PHOST.

The PTBVB resist is characterized by its high thermal stability *(10,12)*. Its negative images are devoid of thermal flow at 230°C owing to the high glass transition temperature (T_g=250°C) of PVBA.

Alternating Copolymers of α,α-Dimethylbenzyl Methacrylate with α-Methylstyrene. Poly(methyl methacrylate) (PMMA) has been known for its very high resolution ever since it was employed as the very first polymeric e-beam and deep UV resist material. Its sensitivity is, however, very low. The sensitivity of the methacrylate-based resists can be readily increased by incorporating chemical amplification, especially by selecting such an ester group as the α,α-dimethylbenzyl ester which is highly susceptible to A_{AL}-1 acidolysis *(13)*. In order to render the methacrylate resists more attractive in device manufacturing, however, two more problems have to be solved; low thermal stability and low dry etch resistance. A sensitive deep UV resist was designed by copolymerizing DMBZMA with MST by radical initiation such

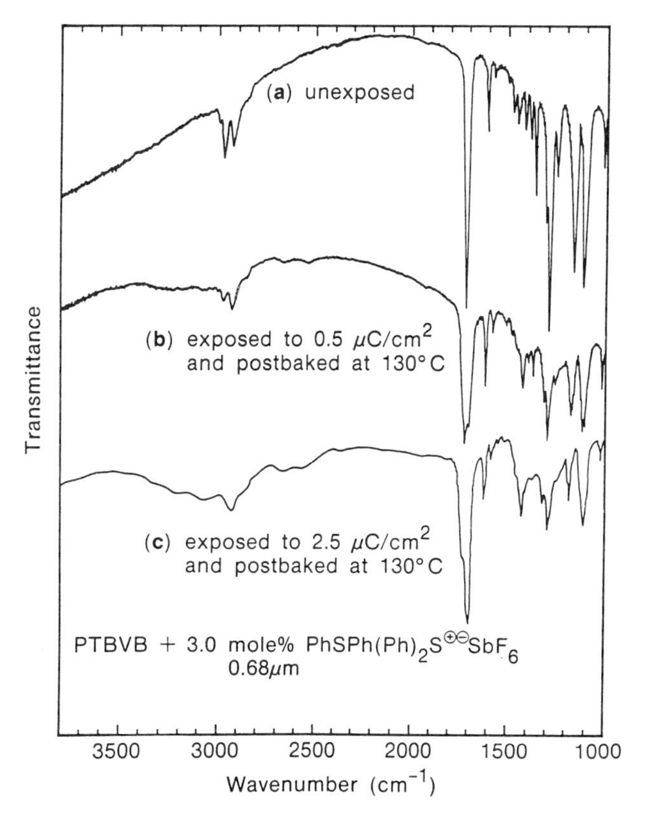

Figure 4. IR spectra of PTBVB resist (0.68 μm) containing 3.0 mol% TPPDPS. (Reproduced with permission from ref. 12. Copyright 1989 Society of Photo-Optical Instrumentation Engineers.)

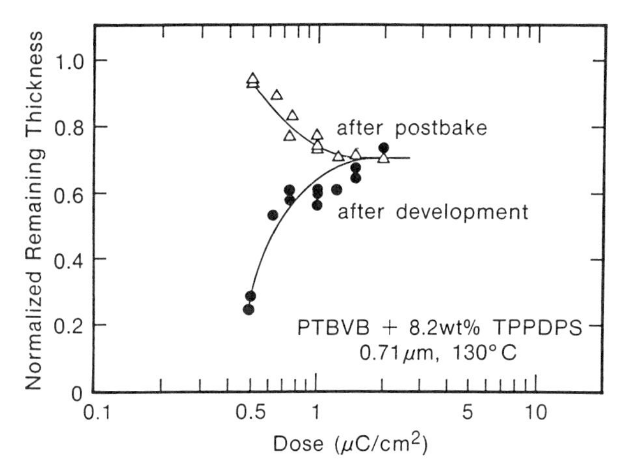

Figure 5. E-beam sensitivity curve of PTBVB resist at 20 keV. (Reproduced with permission from ref. 12. Copyright 1989 Society of Photo-Optical Instrumentation Engineers.)

that each component provides specific functions (Scheme IV) *(14)*. The methacrylate unit in the polymer chain provides good UV transmission (an optical density, OD, of the copolymer is 0.10/μm at 248 nm). The α,α-dimethylbenzyl ester moiety offers facile acidolysis and therefore a high sensitivity as well as a polarity change for dual tone imaging. The MST unit in the polymer chain offers dry etch durability and high thermal stability. The negative images of the copolymer resist are devoid of thermal flow up to 220 °C. The high thermal stability is the result of minimizing the intramolecular dehydration through isolation of the methacrylate unit in the copolymer by alternating copolymerization, which allows to take advantage of the high T_g of poly(methacrylic acid) (T_g=228 °C) and of poly(α-methylstyrene) (T_g=168 °C) *(14)*.

The copolymer resist (0.4 μm thick) containing 9.3 wt% of TPPDPS retains 50 and 100 % of its thickness at 0.3 and 0.7 $\mu C/cm^2$, respectively, of 20 keV e-beam radiation when developed with anisole after postbake at 100°C (Figure 6). In this system, the dual tone imaging is still possible, although the concentration of the polar group, poly(methacrylic acid), generated in the film is always <50 %. The copolymer resist containing 4.7 wt% of TPS (OD=0.24/μm at 248 nm) is completely soluble in a KOH solution when exposed to 2 mJ/cm^2 of 254 nm radiation and postbaked at 130 °C. IR spectra of the copolymer resist before and after exposure to 4.8 mJ/cm^2 and postbake at 120 °C for 2 min are presented in Figure 7, which indicates that the conversion of the ester to the acid is 70 %. Thus, generation of <35 % of the polar unit solubilizes the copolymer in aqueous base, which is in sharp contrast with the tBOC resist that requires >90 % conversion for base solubility. The resist film begins to become insoluble in anisole at an extremely low dose of 0.4 mJ/cm^2 and exhibits ca. 38 % conversion at 0.48 mJ/cm^2 when postbaked at 130 °C for 2 min. Thus, the polar methacrylic acid unit at a concentration as low as 19 mol% is sufficient to insolubilize the copolymer in anisole. As discussed earlier, in addition to ease of acidolysis, the structure of the protecting groups affects the solubility differentiation, which in turn governs the resist sensitivity as well. Conversion of esters to carboxylic acids offers greater solubility differentiation than deprotection of carbonate to phenol.

Hatada and co-workers have shown that poly(DMBZMA) is deesterified to a certain extent upon e-beam exposure and functions as a sensitive, single-component, positive resist when developed with an alcoholic solution of alkoxide *(23)*.

The acid-catalyzed deprotection reaction was utilized in the form of the negative tBOC resist in manufacturing of 1 Mbit DRAM's on Perkin Elmer mirror projection scanners operating in the deep UV mode *(24)*, has attracted a great deal of attention in the resist community, and is currently being pursued in many laboratories for the design of aqueous base developable, positive deep UV resists for KrF excimer laser lithography *(25-29)*.

Polarity Reversal by Thermal Deesterification and Acid-Catalyzed Rearrangement

We have studied the effects of the structure of the ester groups on the thermal deesterification temperature and the resist sensitivity for the

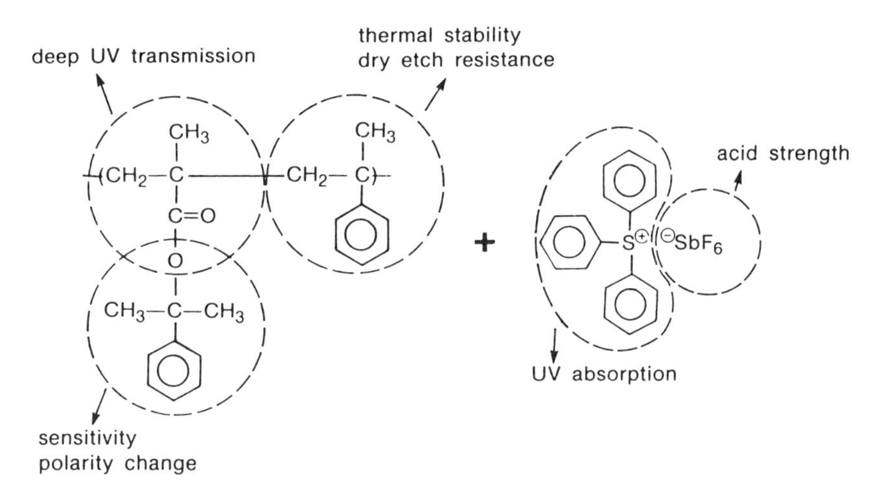

Scheme IV. Two-component copolymer resist.

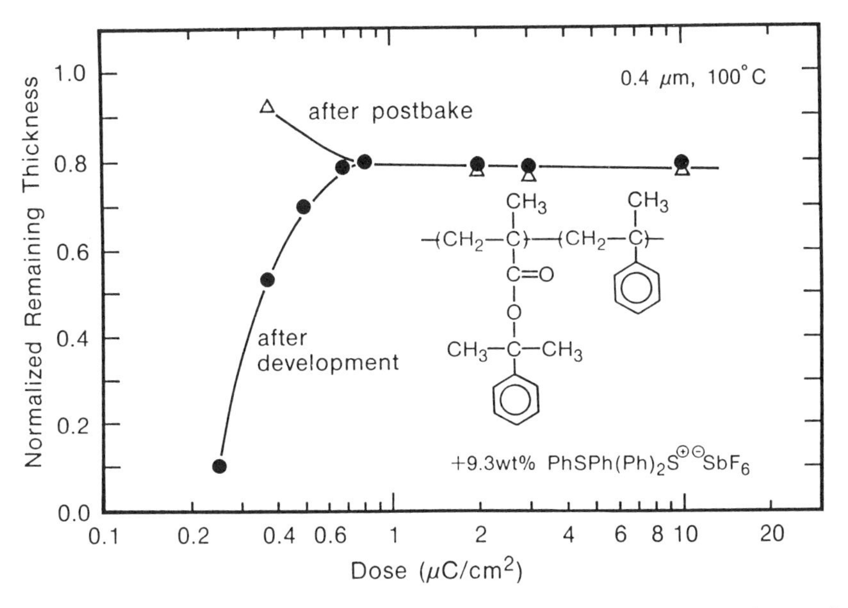

Figure 6. E-beam sensitivity curve of poly(DMBZMA-co-MST) (0.4 μm) resist at 20 keV.

polymethacrylates *(14)* and poly(4-vinylbenzoates) *(10)*. In Figure 8 is presented a TGA curve of poly(2-cyclopropyl-2-propyl 4-vinylbenzoate) (PCPPVB) *(15)* together with those of PTBVB and PBOCST. Replacement of one of the methyl groups of the *t*-butyl ester with a cyclopropyl group results in dramatic reduction in the deesterification temperature by ca. 80 °C. Thus, the dimethyl cyclopropyl carbinol ester is even more thermally labile than the tBOC group. However, whereas the thermal deesterification of PBOCST and PTBVB is quantitative, the weight loss (ca. 32 %) that occurs at 160 °C in the case of PCPPVB is smaller than the quantitative loss of 2-cyclopropylpropene (35.7 wt%) accounts for, which is due to concomitant rearrangement to a thermally stable 4-methyl-3-pentenyl ester (ca. 10 %), according to our authentic syntheses and spectroscopic studies (Scheme V) *(13)*. As mentioned earlier, since polyvinylbenzoates have strong UV absorptions below 300 nm, the TPPDPS salt which extends its absorption to 350 nm was used as the photochemical acid generator and the resist film was exposed to 313 nm radiation. The rearrangement of the cyclopropyl carbinol ester is much more pronounced in the presence of acid (Scheme V). The effects of exposure dose and postbake temperature on conversion in terms of remaining film thickness are presented in Figure 9. The PCPPVB film containing 9.4 wt% of TPPDPS is stable when heated without UV exposure at 100 or 130 °C for 5 min and shrinks more upon postbake at higher doses due to loss of more olefin. The 130 °C bake provides a higher conversion than the 100 °C bake at the same dose but the degree of shrinkage saturates at ca. 12 % at about 5 and 50 mJ/cm^2 when postbaked at 130 and 100 °C, respectively. Thus, the shrinkage never reaches the maximum value of 32 % expected from TGA or 35.7 % expected from the quantitative loss of 2-cyclopropylpropene. On the contrary, the 160 °C bake results in almost maximum thinning in the unexposed regions and the exposed film retains more thickness at higher doses, which is clearly due to more pronounced rearrangement in the presence of acid. The shrinkage saturates again at ca 12 % at about 20 mJ/cm^2 when postbake is carried out at 160 °C. The maximum degree of acid-catalyzed rearrangement is estimated to be about 66 % in the solid state.

The PCPPVB resist system allows negative imaging either with a nonpolar organic developer or with aqueous base, depending on the postbake temperature as illustrated in Scheme VI and Figure 10. Unexposed films are cleanly soluble in anisole but insoluble in aqueous base when heated below 130 °C because the film consists of the lipophilic PCPPVB. As the film is exposed, more benzoic acid units are generated upon postbake and the film becomes insoluble in anisole at about 3 mJ/cm^2 when postbaked at 130 °C (polarity change). The exposed films never become soluble in aqueous base presumably because the concentration of the vinylbenzoic acid units formed is not high enough (ca. 34 %), failing to provide positive imaging. When postbaked at 160 °C, the PCPPVB resist system behaves completely differently. The 160 °C postbake renders the unexposed film insoluble in anisole but soluble in aqueous base due to the predominant thermal deprotection to convert PCPPVB to PVBA containing only about 10 % of the rearrangement product. The exposed films are insoluble in anisole because of generation of the polar benzoic acid units and become insoluble in aqueous base at about 5 mJ/cm^2 when postbaked at 160 °C because the exposed area mainly consists of the nonpolar rearrangement product with only 34 % of the polar benzoic acid unit. Thus, the polarity is reversed by the high temperature postbake.

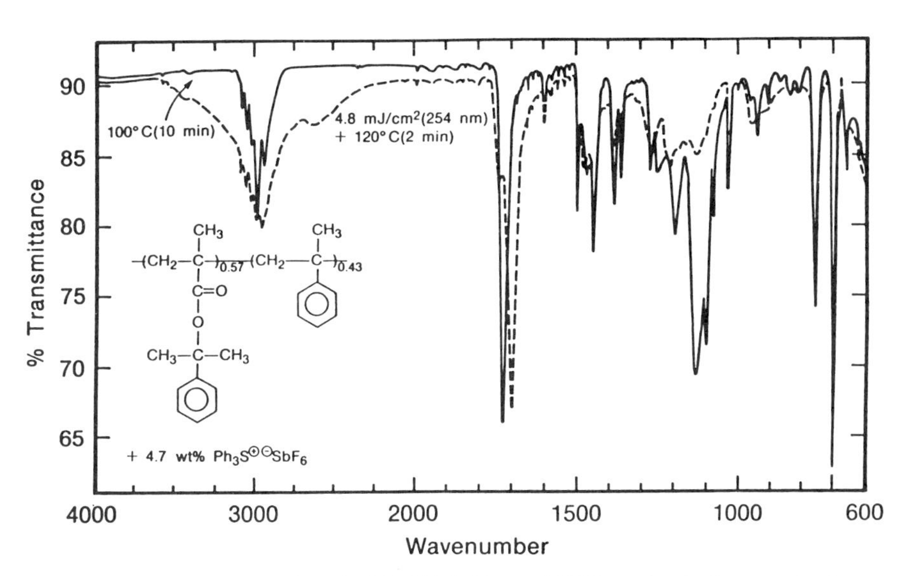

Figure 7. IR spectra of poly(DMBZMA-co-MST) resist before and after deep UV exposure. (Reproduced with permission from ref. 13. Copyright 1989 American Chemical Society.)

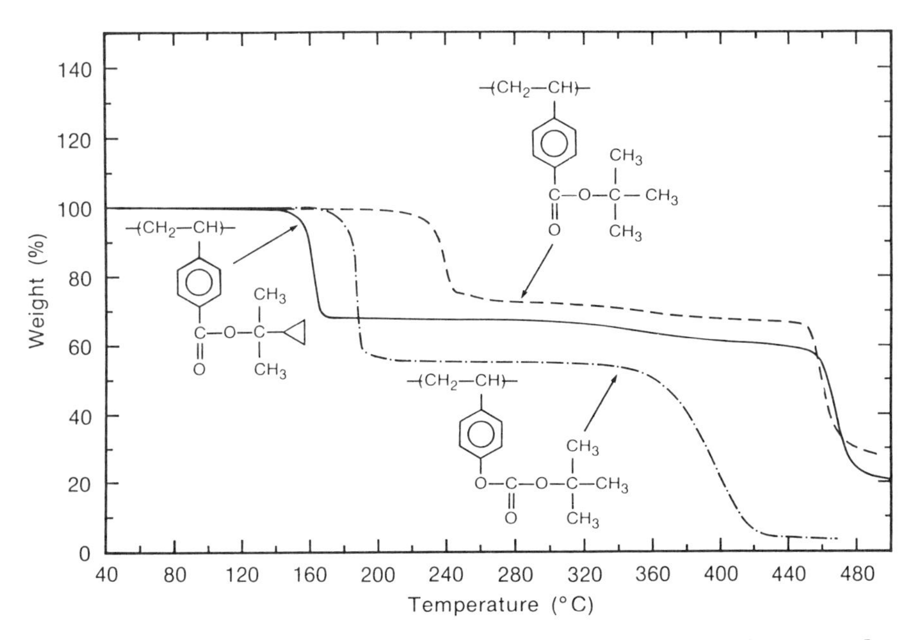

Figure 8. TGA of PCPPVB, PTBVB, and PBOCST (heating rate: 5 °C/min). (Reproduced with permission from ref. 15. Copyright 1990 American Chemical Society.)

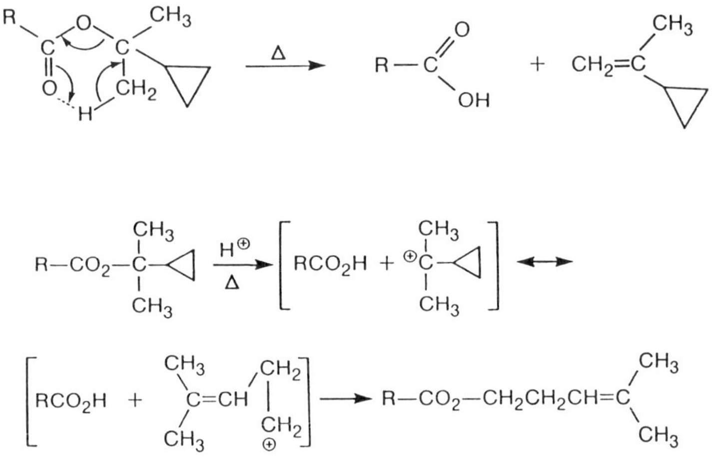

Scheme V. Thermal deesterification and acid-catalyzed rearrangement of cyclopropyl carbinol ester.

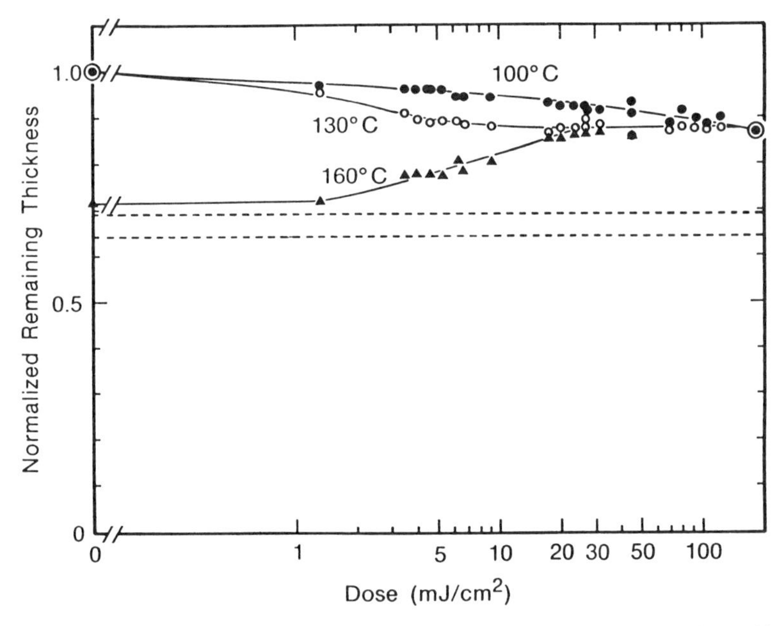

Figure 9. Shrinkage of PCPPVB resist film containing 9.4 wt% of TPPDPS as a function of 313 nm UV dose and postbake temperature. (Reproduced with permission from ref. 15. Copyright 1990 American Chemical Society.)

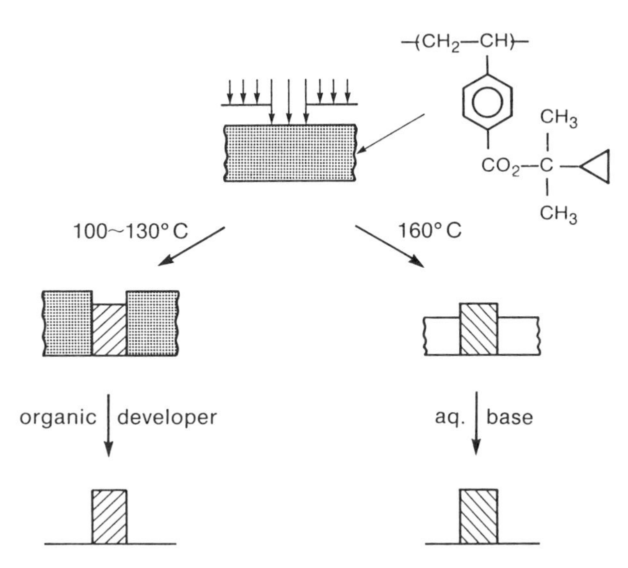

Scheme VI. Negative imaging of PCPPVB resist by polarity change and polarity reversal.

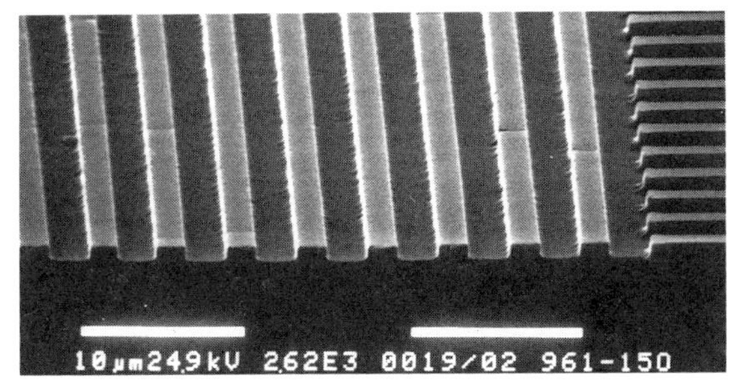

(**a**) negative imaging with organic developer

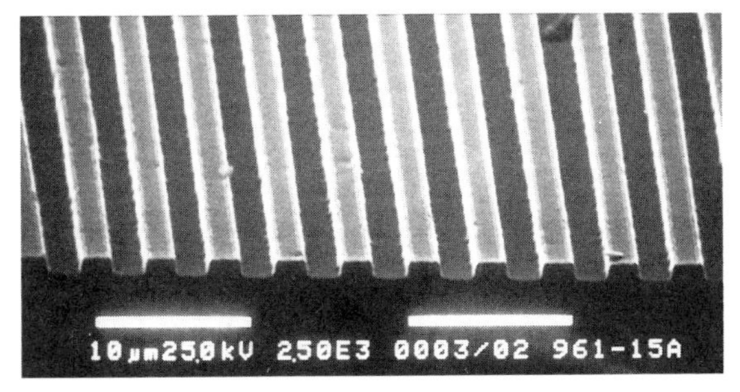

(**b**) negative imaging with aqueous base

Figure 10. Scanning electron micrographs of negative images delineated in PCPPVB resist by development with anisole and with aqueous base after postbake at 130 and 160 °C, respectively. (Reproduced with permission from ref. 15. Copyright 1990 American Chemical Society.)

Reverse Polarity Change from a Polar to a Nonpolar State by Pinacol Rearrangement

We have next turned our attention to a change from a polar to a nonpolar state as a third mode of polarity alteration. Two examples of the reverse polarity change that have been reported are photochemical transformation of pyridinium ylides to 1,2-diazepines *(30)* and acid-catalyzed silanol condensation *(31)*. To achieve the reverse polarity change, we selected the classical pinacol-pinacolone rearrangement *(16,17)*, wherein *vic*-diols (pinacols) are converted to ketones or aldehydes with acid as a catalyst (Scheme VII).

The pinacol polymer was prepared by radical polymerization and treated with trifluoromethanesulfonic (triflic) acid in methanol at 0 °C. The product was isolated by precipitation in water and characterized. NMR and IR spectra of the rearranged product confirmed quantitative and clean rearrangement to a single structure. A model reaction has also demonstrated that the unconjugated ketone is the sole rearrangement product. The molecular weights and distribution remained unchanged ($M_n = 27{,}000$, $M_w = 139{,}000 \rightarrow M_n = 31{,}400$, $M_w = 139{,}500$).

IR spectra of the pinacol polymer film containing 4.7 wt% of TPS are presented in Figure 11. The spectrum of the pinacol polymer exhibits a large OH absorption at ca. 3400 cm^{-1} (Figure 11a), which completely disappears upon postbake at 120 °C for 2 min after exposure to 5.5 mJ/cm^2 of 254 nm radiation (Figure 11b). The disappearance of the OH absorption is accompanied by appearance of a very intense C=O absorption at 1710 cm^{-1}, which is characteristic to nonconjugated ketones. This behavior is in sharp contrast with the tBOC resist chemistry; disappearance of the C=O absorption accompanied by appearance of the OH absorption on IR spectra (see Figure 1). The spectrum of the exposed and postbaked film (Figure 11b) is identical to that of the well-characterized solution acidolysis product mentioned earlier. Thus, the pinacol rearrangement facilely occurs also in a solid polymer with a photochemically-generated acid as a catalyst. In spite of its high sensitivity toward acidolysis, the pinacol polymer is extremely stable thermally as the IR spectrum of the film heated at 220 °C for 15 min (Figure 11c) indicates. The polymeric pinacol resist film containing 4.6 wt% TPS is very much transparent at 248 nm (OD=0.28/μm). The absorption of the film increases upon postbake due to conversion of diol to ketone.

Deep UV sensitivity curves for the pinacol resist (1.2 μm thick) containing 4.6 wt% TPS are presented in Figure 12. The film thickness was measured after postbake at 120 °C for 2 min (△) and after development with methanol (•). Postbake results in film shrinkage due to loss of water. The theoretical maximum shrinkage of 8.7 % is attained at ca. 2.5 mJ/cm^2. The exposed film begins to become insoluble in methanol at ca. 0.8 mJ/cm^2 and the full thickness retention is achieved at ca. 2.5 mJ/cm^2 with $\gamma = 14$.

Use of aqueous base as a developer is very attractive in the production environments. Certain alcohols could be dissolution promoters of phenolic resins while ketones and aldehydes could inhibit the phenolic resin dissolution in aqueous base (Scheme VIII). Then, the reverse polarity change in a phenolic matrix through the pinacol rearrangement could provide chemical amplification negative resist systems that can be developed with aqueous base. In fact, acid-catalyzed silanol condensation in a phenolic resin has been

Scheme VII. Pinacol rearrangement of polymeric *vic*-diol.

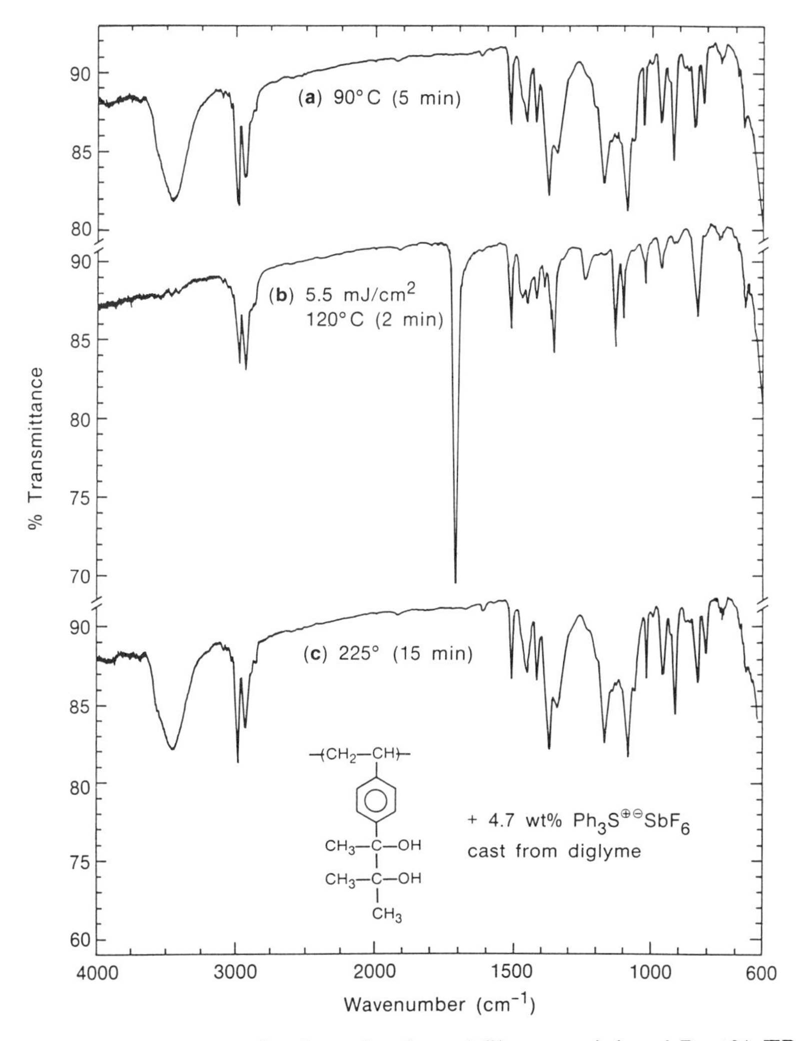

Figure 11. IR spectra of polymeric pinacol film containing 4.7 wt% TPS. (Reproduced with permission from ref. 16. Copyright 1991 Society of Photo-Optical Instrumentation Engineers.)

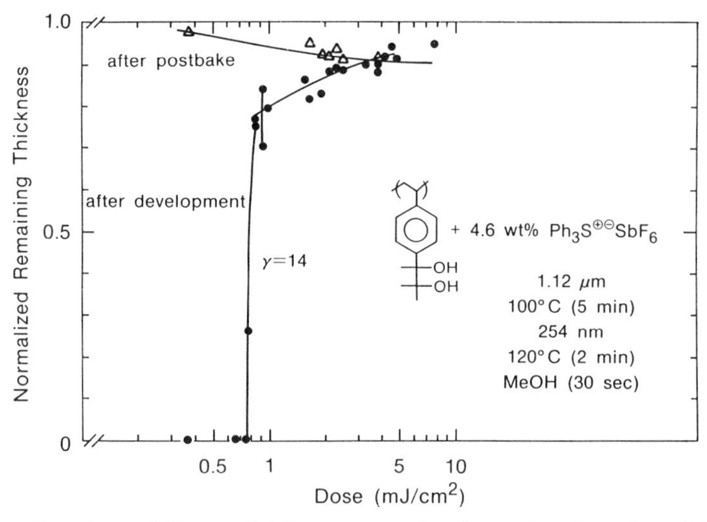

Figure 12. Deep UV sensitivity curve of polymeric pinacol resist containing 4.6 wt% TPS. (Reproduced with permission from ref. 16. Copyright 1991 Society of Photo-Optical Instrumentation Engineers.)

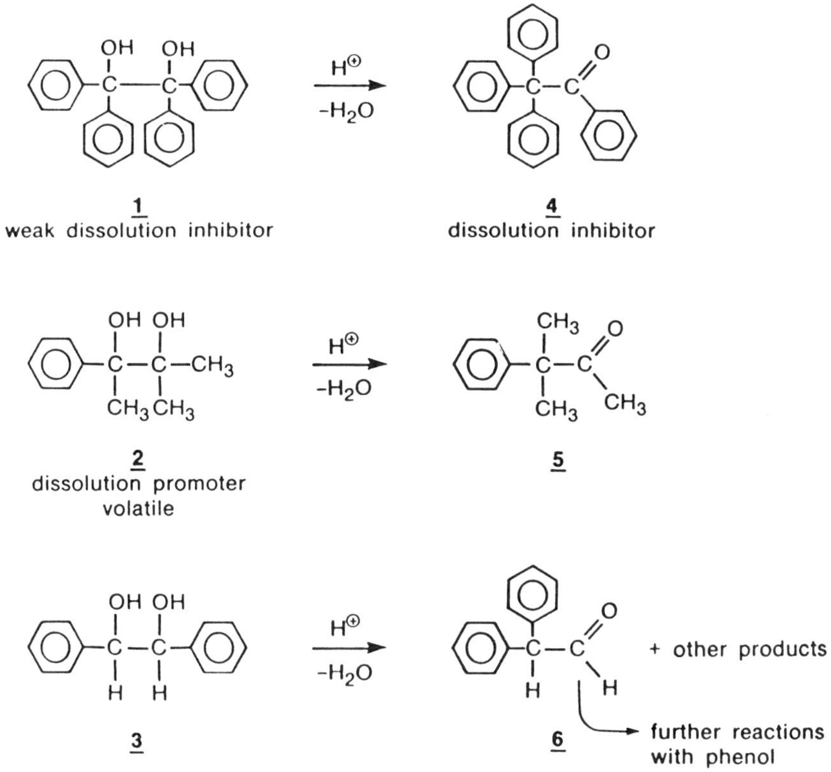

$H^{\oplus}$, $-H_2O$

1 weak dissolution inhibitor → **4** dissolution inhibitor

2 dissolution promoter volatile → **5**

3 → **6** + other products; further reactions with phenol

Scheme VIII . *vic*-Diols and their pinacol rearrangement products.

successfully applied to the design of such resist systems *(31)*. In this scheme, a dissolution promoting silanol undergoes acid-catalyzed condensation to form a dissolution inhibiting siloxane in a phenolic resin.

In Figure 13 are presented IR spectra of a PMMA film containing 32.3 wt% benzopinacole and 3.2 wt% TPS. The film prebaked at 100 °C for 1.5 min shows an absorption due to OH at ca. 3600 cm^{-1} (Figure 13a), which completely disappears at 2.0 mJ/cm^2 upon postbake at 100 °C for 2 min (Figure 13b). The exposed and postbaked film exhibits a C=O absorption of a conjugated ketone at 1680 cm^{-1}, which is well separated from the large ester carbonyl absorption of the matrix PMMA. The IR study demonstrates facile and complete rearrangement of a small diol in a polymer film. The rearrangement of benzopinacole was also investigated in phenolic matrices such as a novolac resin and *m*-PHOST. IR spectra of a 0.98-μm thick novolac resin film containing 16.1 wt% of benzopinacole and 4.1 wt% TPS are presented in Figure 14. The diol OH absorption is buried in the huge absorption of the phenolic OH. A new C=O absorption appears at 1666 cm^{-1} upon UV exposure (4.2 mJ/cm^2) and postbake (100 °C, 2 min). Similarly, the C=O absorption was also detected at 1667 cm^{-1} in an exposed and postbaked film of *m*-PHOST containing benzopinacole and TPS. Thus, benzopinacole is transformed to an aromatic ketone in the phenolic matrices. The carbonyl group of the rearranged aromatic ketone absorbs at a shorter wavenumber (1666-7 cm^{-1}) in phenolic resins than in polystyrene or PMMA films (1675-80 cm^{-1}), which is due to hydrogen-bonding between C=O of the rearrangement product and OH of the phenolic resin. It is well known that the hydrogen-bonding plays a vital role in the dissolution inhibition of phenolic resins in aqueous base.

Triphenylsulfonium salts are efficient dissolution inhibitors *(32,33)*. Incorporation of 4.8 wt% of TPS in a PHOST film results in several-fold decrease in its dissolution rate in aqueous base. The dissolution inhibition effect of the small pinacol compounds depends on their structure. Whereas benzopinacole retards the dissolution of PHOST, 3-methyl-2-phenyl-2,3-butanediol is a dissolution promoter but tends to evaporate out of the film during prebake. *meso*-Hydrobenzoin does not affect the dissolution of PHOST much but has been reported to increase the dissolution rate of a novolac resin film *(34)*. Although benzopinacole is a dissolution inhibitor of PHOST, its rearranged product, phenyl trityl ketone, is much more efficient in inhibiting the PHOST dissolution, allowing its use in a three-component resist formulation.

meso-Hydrobenzoin performs quite well when formulated with *p*-PHOST while a novolac resin is a better partner of benzopinacole. A three-component resist film consisting of *p*-PHOST, 16.1 wt% of hydrobenzoin, and 4.2 wt% of TPS with an OD of 0.33/μm at 248 nm becomes completely insoluble at 6 mJ/cm^2 with γ of 3.2 in MF319 (aqueous tetramethylammonium hydroxide) for at least 45 sec when postbaked at 90 °C for 2 min. A deep UV sensitivity curve of a negative resist composed of a novolac resin, 15 wt% of benzopinacole, and 2 wt% of TPS is presented in Figure 15. The 0.75-μm thick resist film had an OD of 0.53 at 248 nm. The resist exhibits a high γ of 4 and its maximum thickness retention of ca. 93 % at ca. 7 mJ/cm^2 on a Perkin Elmer 500 operating in the deep UV mode when postbaked at 90 °C for 90 sec and developed with Shipley 2401 CD208 (aqueous KOH) for 110 sec. Figure 16 presents scanning electron micro-

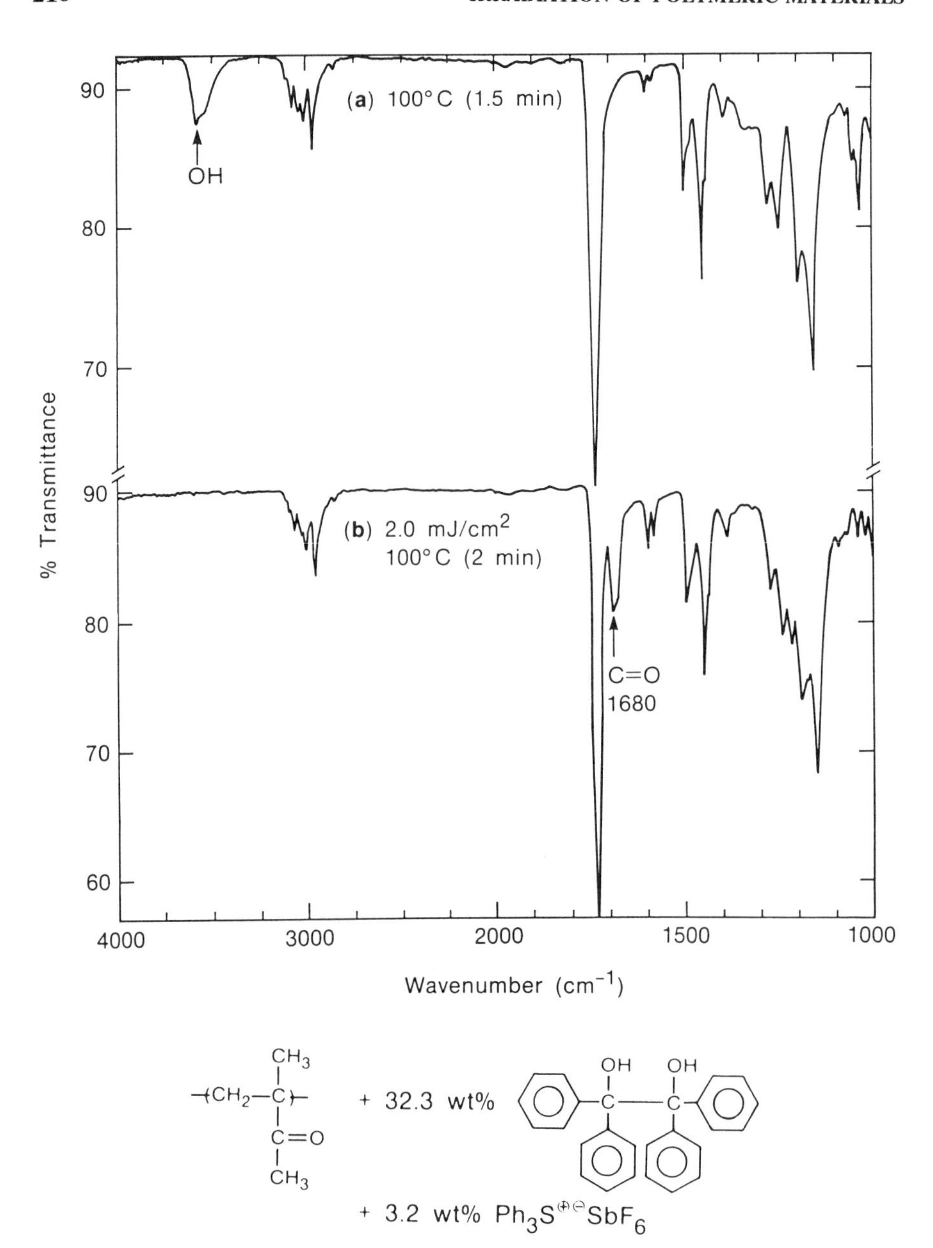

Figure 13. IR spectra of PMMA film containing 32.3 wt% benzopinacole and 3.2 wt% TPS.

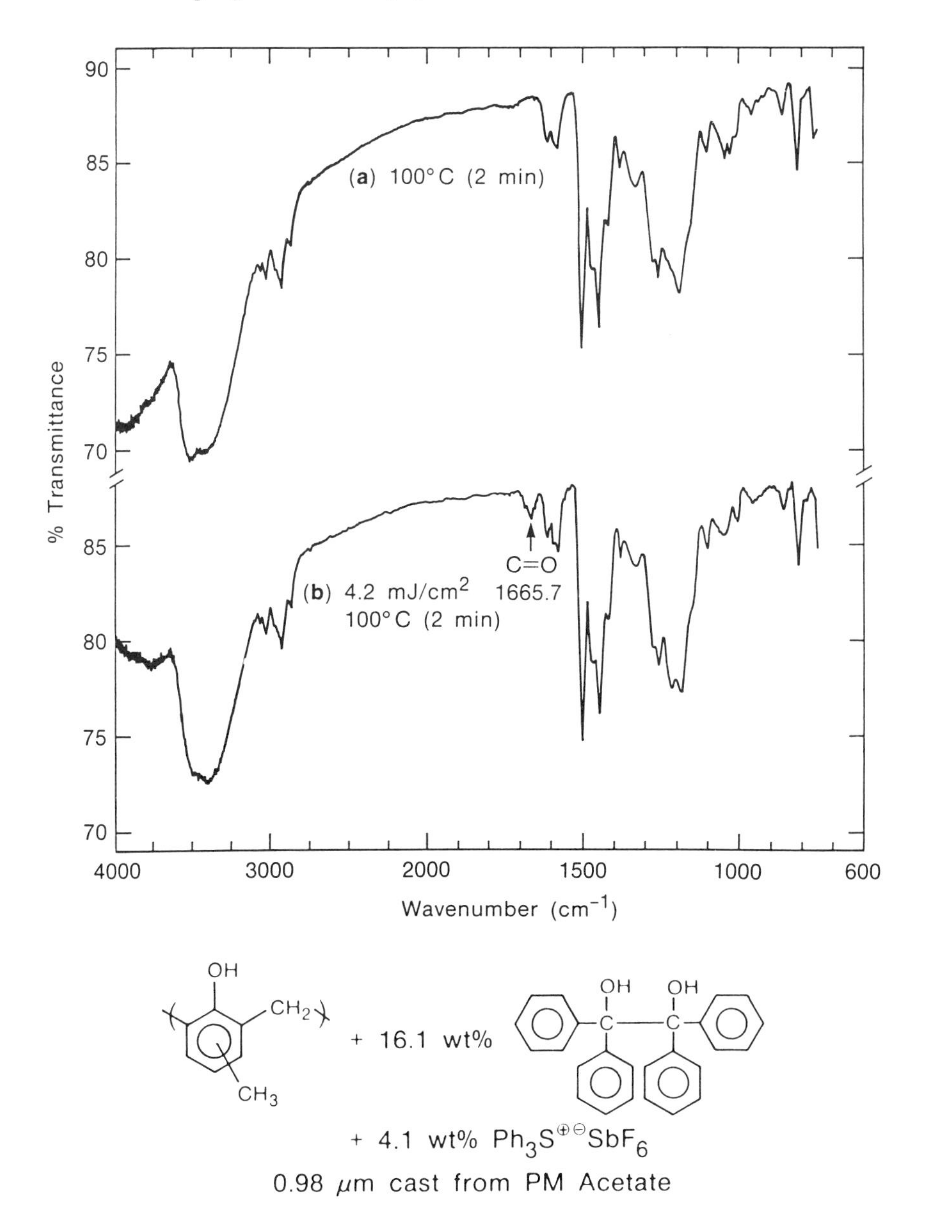

Figure 14. IR spectra of novolac film containing 16.1 wt% benzopinacole and 4.1 wt% TPS. (Reproduced with permission from ref. 16. Copyright 1991 Society of Photo-Optical Instrumentation Engineers.)

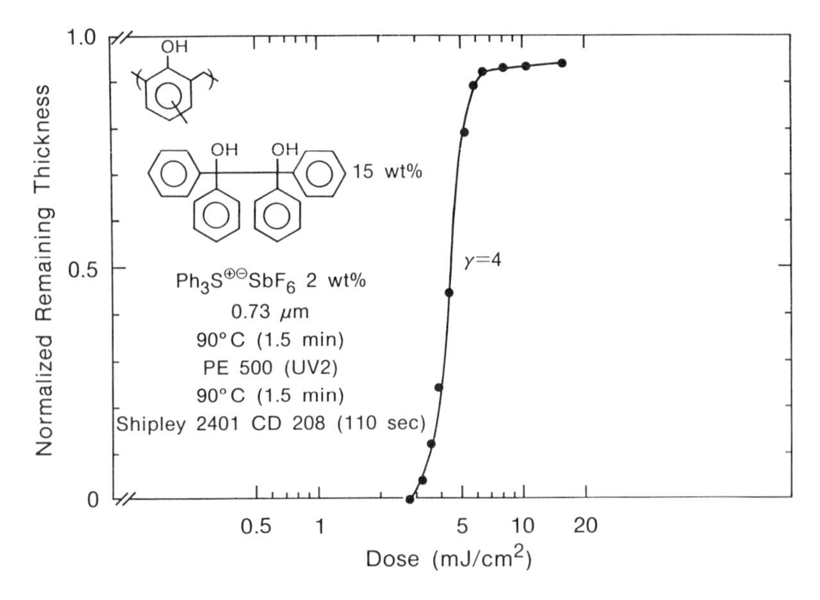

Figure 15. Deep UV sensitivity curve for aqueous base development of three-component negative resist consisting of novolac resin, 15 wt% benzopinacole, and 2 wt% TPS. (Reproduced with permission from ref. 16. Copyright 1991 Society of Photo-Optical Instrumentation Engineers.)

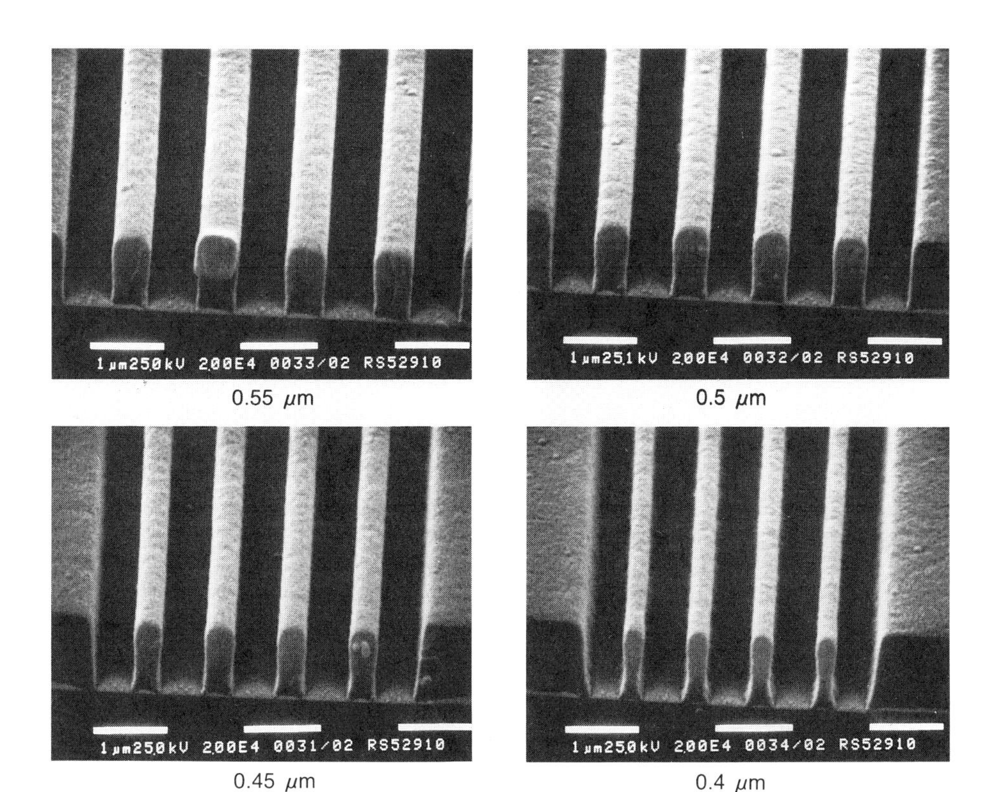

Figure 16. Scanning electron micrographs of sub-half-micrometer negative images projection-printed in three-component benzopinacole resist at 10 mJ/cm^2 on Canon KrF excimer laser stepper. (Reproduced with permission from ref. 16. Copyright 1991 Society of Photo-Optical Instrumentation Engineers.)

graphs of sub-half-micrometer negative images projection-printed in the benzopinacole/novolac resist at 10 mJ/cm^2 on a Canon KrF excimer laser stepper with a NA of 0.37.

In contrast to the *tertiary* diols which undergo clean pinacol rearrangement to ketones (Scheme VIII), the *secondary* diol provides several products including diphenylacetaldehyde, in the presence or absence of isopropylphenol *(16,17)*. High boiling aldehydes such as diphenylacetaldehyde (bp 315 °C) produced in the exposed regions can undergo further acid-catalyzed addition-condensation reactions with the surrounding phenol *(35)*, which contributes to the negative imaging mechanism.

Summary

The chemical amplification mechanisms based on the acid-catalyzed polarity change were described, which are useful for the design of sensitive resist systems that develop without suffering from swelling or allow aqueous base development. Three modes of the polarity alteration were discussed; i) a polarity change from a nonpolar to a polar state based on acid-catalyzed deprotection, ii) a polarity reversal induced by thermal deesterification and acid-catalyzed rearrangement, and iii) a reverse polarity change from a polar to a nonpolar state achieved by pinacol rearrangement.

Acknowledgment

The author thanks his co-workers whose names appear in the cited references for their contributions.

Literature Cited

1. Ito, H.; Willson, C. G.; Fréchet, J. M. J. *Digest of Technical Papers of 1982 Symposium on VLSI Technology* **1982**, 86.
2. Fréchet, J. M. J.; Ito. H.; Willson, C. G. *Proc. Microcircuit Eng.* **1982**, 260.
3. Ito, H.; Willson, C. G. *Technical Papers of SPE Regional Technical Conference on Photopolymers* **1982**, 331.
4. Ito, H.; Willson, C. G. *Polym. Eng. Sci.* **1983**, *23*, 1012.
5. Ito, H.; Willson, C. G. In *Polymers in Electronics* Symposium Series 242, Davidson, T. Ed.; American Chemical Society: Washington, D. C., 1984, p. 11.
6. Ito, H. *Proc. KTI Microelectronics Seminar* **1988**, 81.
7. Iwayanagi, T.; Kohashi, T.; Nonogaki, S.; Matsuzawa, T.; Douta, K.; Yanazawa, H. *IEEE Trans. Electron Devices* **1981**, *ED-28*, 1306.
8. Feely, W. E.; Imhof, J. C.; Stein, C. M. *Polym. Eng. Sci.* **1986**, *26*, 1101.
9. Reck, B.; Allen, R. D.; Twieg, R. J.; Willson, C. G.; Matuszczak, S.; Stover, H. D.; Li, N. H.; Fréchet, J. M. J. *Polym. Eng. Sci.* **1989**, *29*, 960.
10. Ito, H.; Willson, C. G.; Fréchet, J. M. J. *Proc. SPIE* **1987**, *771*, 24.
11. Seligson, D.; Ito, H.; Willson, C. G. *J. Vac. Sci. Technol.* **1988**, *B6(6)*, 2268.

12. Ito, H.; Pederson, L. A.; Chiong, K. N.; Sonchik, S.; Tsai, C. *Proc. SPIE* **1989**, *1086*, 11.
13. Ito, H.; Ueda, M.; Ebina, M. In *Polymers in Microlithography*; Reichmanis, E.; MacDonal, S. A.; Iwayanagi, T. Eds.; Symposium Series 412; American Chemical Society, Washington, D. C., 1989, p. 57.
14. Ito, H.; Ueda, M. *Macromolecules* **1988**, *21*, 1475.
15. Ito, H.; Ueda, M.; England, W. P. *Macromolecules* **1990**, *23*, 2589.
16. Sooriyakumaran, R.; Ito, H.; Mash, E. A. *Proc. SPIE* **1991**, *1466*, 419.
17. Ito, H.; Sooriyakumaran, R.; Mash, E. A. *J. Photopolym. Sci. Technol.* **1991**, *4*, 319.
18. Fréchet, J. M. J.; Eichler, E.; Ito, H.; Willson, C. G. *Polymer* **1983**, *24*, 995.
19. Crivello, J. V.; Lam, J. H. W. *J. Org. Chem.* **1978**, *43*, 3055.
20. Miller, R. D.; Renaldo, A. F.; Ito, H. *J. Org. Chem.* **1988**, *53*, 5571.
21. Umbach, C. P.; Broers, A. N.; Willson, C. G.; Koch, R.; Laibowitz, R. B. *J. Vac. Sci. Technol.* **1988**, *B6*, 319.
22. McKean, D. R.; Schaedeli, U.; MacDonald, S. A. *J. Polym. Sci., Part A, Polym. Chem.* **1989**, *27*, 3927.
23. Hatada, K.; Kitayama, T.; Danjo, S.; Tsubokura, Y.; Yuki, H.; Morikawa, K.; Aritome, H.; Namba, S. *Polym. Bull.* **1983**, *10*, 45.
24. Maltabes, J. G.; Holmes, S. J.; Morrow, J.; Barr, R. L.; Hakey, M.; Reynolds, G.; Brunsvold, W. R.; Willson, C. G.; Clecak, N. J.; MacDonald, S. A.; Ito, H. *Proc. SPIE* **1990**, *1262*, 2.
25. Tarascon, R. G.; Reichmanis, E.; Houlihan, F. M.; Shugard, A.; Thompson, L. F. *Polym. Eng. Sci.* **1989**, *29*, 850.
26. Yamaoka, T.; Nishiki, M.; Koseki, K. *Polym. Eng. Sci.* **1989**, *29*, 856.
27. Hayashi, N.; Hesp, S. M. A.; Ueno, T.; Toriumi, M.; Iwayanagi, T.; Nonogaki, S. *Proc. ACS Div. PMSE* **1989**, *61*, 417.
28. Kikuchi, H.; Kurata, N.; Hayashi, K. *J. Photopolym. Sci. Technol.* **1991**, *4*, 357.
29. Endo, M.; Tani, Y.; Yabu, T.; Okada, S.; Sasago, M.; Nomura, N. *J. Photopolym. Sci. Technol.* **1991**, *4*, 361.
30. Schwalm, R.; Böttcher, A.; Koch, H. *Proc. SPIE* **1988**, *920*, 21.
31. Ueno, T.; Shiraishi, H.; Hayashi, N.; Tadano, K.; Fukuma, E.; Iwayanagi, T. *Proc. SPIE* **1990**, *1262*, 26.
32. Ito, H. *Proc. SPIE* **1988**, *920*, 33.
33. Ito, H.; Flores, E. *J. Electrochem. Soc.* **1988**, *135*, 2322.
34. Uchino, S.; Iwayanagi, T.; Ueno, T.; Hayashi, N. *Proc. SPIE* **1991**, *1466*, 429.
35. Ito, H.; Schildknegt, K.; Mash, E. A. *Proc. SPIE* **1991**, *1466*, 408.

RECEIVED August 14, 1992

Chapter 15

Chemically Amplified X-ray Resists

J. W. Taylor[1-4], C. Babcock[1-3], and M. Sullivan[1,2]

[1]Center for X-ray Lithography, [2]Materials Science Program, [3]Engineering Research Center in Plasma-Aided Manufacturing, and [4]Department of Chemistry, University of Wisconsin at Madison, 3731 Schneider Drive, Stoughton, WI 53589–5397

Conventional photoresists employed to create microcircuits generally lack sufficient sensitivity to meet the desired manufacturing throughput when X-rays in the region of 0.8 to 1.0 nm are used for exposure. Chemical amplification of the exposure event does provide the sensitivity. Exposures of submicron features with these chemically-amplified resists and subsequent development in aqueous base solution demonstrate that several of the commercial negative-tone and positive-tone formulations are capable of high resolution provided the variables which produce the chemistry are understood and controlled. Examples of these variables and the limits of necessary control are demonstrated for two basic commercial formulations, Shipley (negative) XP-90104C and Hoechst (positive) AZ-PF. The variables examined include: preprocessing for adhesion, spin conditions for thickness, pre-baking conditions, exposure, post-bake, and development time, temperature, and normality. Each step has a chemical connection, and these connections are discussed. Finally, the etch selectivity of two chemically-amplified resists is examined, and the resists show similar selectivity to the conventional novolac-based photoresists but are capable of producing features at least to 0.30 microns.

The results of a three-wavelength ellipsometer approach to the measurement of surface roughness and resist swelling is also described. The negative tone chemically-amplified resist XP-90104C is shown to swell to less than 1% in aqueous base developer at the temperatures normally used for processing.

The initial requirement for the polymer formulation used as a resist by the semiconductor industry is the resolution of the image which is produced by the exposure agent. When processing throughput becomes a dominant concern, however, the sensitivity of the resist is a factor which must be optimized. Several additional factors, including the flow characteristics (which define the uniformity of the thin film produced by spinning), the thermal stability, the sensitivity to environmental conditions, the resistance to plasma etching, and the stability of the image are also critical to the performance of the resist.

0097–6156/93/0527–0224$06.25/0

Traditionally, the polymer chemist will rely on increasing the quantum efficiency of the exposure agent to achieve a sensitivity increase or will utilize functional groups in the base polymer system to produce a lower activation energy for reaction to yield products which react differently with the developing agent. This provides contrast between the exposed and unexposed resist. There is some variety in the differentiation approach depending on the method of development. With solution development, it is the differential solubility which provides the discrimination between the exposed and unexposed polymer. In the dry or plasma development, it is the differential reactions between the exposed and unexposed resist-coated surface to the radicals and ions striking the resist surface. Once the unwanted resist areas are removed, it is the differential reactions to the etching gas and plasma with the resist and the exposed wafer surface which provide the discrimination. In this latter situation, the resist may also be utilized as a mask for protecting areas where plasma etching is not desired. Polymer matrix selection becomes important in this situation to provide the necessary etch selectivity and protect the covered areas.

What then can be done when the traditional lines of chemical synthesis fail to produce the sensitivity that a photoresist system requires when X-ray photons are the exposure agent? One answer has been the development of a new strategy for increasing the effect of the photon absorption, or electron absorption in the case of e-beam exposure. In this new strategy, primarily pioneered for photoresists by research at IBM, AT&T, and in Germany (*1-8*),the photon event generates an acid, and the acid subsequently acts as a catalyst to promote the reaction producing the required differentiation. Because the catalyst is produced by one photo event but can participate in several cycles of chemical reactions, the new resists are called "chemically amplified". In the X-ray region these chemically-amplified resists are of critical importance because the absorption cross section is a function of wavelength and decreases substantially as the wavelength decreases. Short wavelength(0.8-1.4 nm) X-rays are the choice for synchrotron-based X-ray lithography to provide the lowest contribution from diffraction and thereby produce the highest image resolution. In like manner, the point source X-ray effort is primarily centered at the Fe emission line at 1.4 nm(*9,10*). At these short wavelengths, however, the sensitivity of most traditional resists is too poor to be used given the throughput requirement for the production of microcircuit devices by X-ray lithography. For example, as can be seen in Table I. where the results with synchrotron exposures are listed, poly(methyl-methacrylate - PMMA) - one of the highest resolution resists for X-ray lithography - has a sensitivity of 1500-2000 mJ/cm^2; the desired sensitivity for manufacturing is 100 mJ/cm^2 or better(*2*).

The modification of PMMA and related acrylic polymers(*11*)by halogenation, substitution, and copolymerization with other monomers has resulted in improvements in sensitivity to below 100 mJ/cm^2. However, the resolution of these modified resists is worse than that of PMMA, and they can not be developed in an aqueous developer. Aqueous base development is desired for photoresists because it tends to give better resolution and presents much fewer solvent disposal problems. The novolac/diazonaphthoquinone dissolution-rate inhibition resists, such as those used for UV lithography in most wafer fabrication lines, can be developed with aqueous solution. These resists also have greater process stability against plasma etching and heat, but they do not meet the sensitivity target value of 100 mJ/cm^2. For example as shown in Table I, an experimental high-sensitivity resist of this type (Olin-Hunt HEBR 242) tested at the University of Wisconsin Center for X-ray Lithography (CXrL) had a sensitivity of 460 mJ/cm^2. Other novolac/ diazonaphthoquinone resists are even less sensitive than this, but the matrix does provide etching selectivity.

TABLE I. SUMMARY OF MEASURED X-RAY RESIST PROPERTIES

RESIST (tone)	Developer	Sensitivity (mJ/cm^2)	Contrast	Resolution	Etch rate (nm/min)[1]
PMMA (pos)[2]	MIBK/IPA	1500	3.5	<0.05	34
HEBR-242 (pos)[3]	HPRD-434	460	2.7	0.25	
HiPR-6512 (pos)[3]					22
ECX-1029 (neg)[4]	MF-312	32	8.9	0.35	
XP-8933B (neg)[4]	MF-312	22	7.2	0.30	
XP-90104B (neg)[5]	MF-312	36	5.4	0.25	22
XP-90104C (neg)[5]	MF-312	21	5.1	0.25	
AZ-PF (pos)[6]	AZ	40-60	6.0	0.25	20

[1] Measured in CHF_3/O_2 RIE of SiO_2.

[2] Mol. wt. 950 K, 6% solution in $C_6H_4Cl_2$, supplied by KTI; other solvents used for mask replication.

[3] Supplied by OCG Microelectronics (formerly Olin Hunt).

[4] Supplied by Rohm and Haas/Shipley

[5] Supplied by Shipley Company

[6] Supplied by Hoechst Celanese

Chemically-Amplified X-ray Resists

The most sensitive aqueous developable X-ray resists today appear to be the acid-catalyzed chemically-amplified resists. Because the X-ray active component in this resist works as a catalyst after exposure to X-rays, it can cause many crosslinking reactions in a negative tone formulation in response to a single photon event. This leads to sensitivities as low as 10 mJ/cm^2 (Table I.). A chemically-amplified positive resist has been evaluated at the Berliner Speicherring für Synchrotronstrahlung (BESSY) with a sensitivity of 30 mJ/cm^2 at 754 MeV(*12*). The novolac resins used in these commercial formulations are composed of a cresol-formaldehyde condensation polymer. The cresol is actually a blend of ortho-, meta-, and para-cresols, as well as phenol(*13*). The relative amounts of each of these components and the positions of the methylene bond in the novolac chain have a large effect on the properties of the resist(*14*).

The advantages of novolac resins for use in resists include non-swelling aqueous development, proven processing technology, and generally good etch resistance. They are used as the matrix polymer in a number of the more important commercial chemically-amplified resists. The chemically-amplified resists contain radiation sensitive acid generating compounds to release strong acids when a photon is absorbed. The negative chemically-amplified resists contain a crosslinker which is catalyzed by the acid, crosslinking the polymer and lowering its dissolution rate. The positive resist contains a dissolution inhibitor which decomposes through an acid-catalyzed reaction, causing an increase in the dissolution rate of the exposed resist.

Components of Commercial Chemically-Amplified Negative X-ray Resists

There are several types of photoacid generators which have been utilized for chemically-amplified resists(3,*5*,*15*,*16*). A few of the more important ones are onium salts, trichloromethyl-substituted triazines, and halogenated (especially brominated) hydrocarbons. For example, Hoechst AG has patented compositions containing the substituted trichloromethyl triazine photoacid generator(*16*).

Onium salts were investigated by Crivello(*17*), and their application to chemically-amplified resists have been extensively investigated by Ito and Willson(*18*). An example used for an X-ray resist evaluation by Bruns et al.(*19*) is diphenyliodonium hexafluorophosphate ($(C_6H_5)_2I^+PF_6^-$). Other examples are triphenylsulfonium hexafluoroantimonate and triphenylsulfonium hexafluoroarsenate ($(C_6H_5)_3S^+SbF_6^-$ and $(C_6H_5)_3S^+ AsF_6^-$)(*5*).

Another type of photoacid generator is a halogenated hydrocarbon. One of the simplest examples is carbon tetrabromide, which has also been evaluated by Bruns et al., who found it gave lower sensitivity than diphenyliodonium hexafluorophosphate or trichloromethyl triazine(*19*). They have reported that photoacid generators based on dibrominated phenols yielded good sensitivity(*20,21*).

The crosslinking agent used in negative chemically-amplified resists is a polyfunctional compound with several groups capable of acid-catalyzed condensation reactions linking them to the novolac chain via the hydroxyl group of the cresol structure. For example, hexamethoxymethylmelamine has a functionality of six (the number of methoxymethyl groups) and can react with up to a maximum of six cresols or phenols.

Examples of commercial chemically-amplified negative X-ray resists which have been studied at CXrL are ECX-1029 from Rohm and Haas, and XP-8933B, XP-90104B, and XP-90104C from Shipley. All of these work on the basis of acid-catalyzed crosslinking. They consist of novolac resin, a crosslinker (such as hexamethoxymethyl-melamine) , and a photoacid generator (PAG). They are similar to SAL-601, an e-beam resist which has been described in detail in the literature(*22-25*). The unexposed photoresist is soluble in an aqueous base developer. Upon exposure to X-rays, the PAG is converted to an acid. In the presence of this acid, the melamine crosslinker reacts with the novolac resin to form a crosslinked network. The kinetics of the crosslinking reaction are very slow at room temperature and are accelerated by the use of a post-exposure bake (PEB) at 110-115°C for one minute. The resultant crosslinking increases the novolac molecular weight and serves to prevent dissolution of the photoresist in the exposed areas.

Components of a Commercial Chemically-Amplified Positive X-ray Resist

For positive chemically-amplified resists, a dissolution inhibitor can be used which undergoes an acid-catalyzed hydrolysis into more soluble compounds. The chemically-amplified positive X-ray resist produced by Hoechst as AZ-PF works by this mechanism. This resist is novolac-based and formulated with three components(*8*). The functions of dissolution inhibition and photosensitivity, which are combined in the diazonaphthoquinone(DNQ) of a conventional novolac/DNQ resist, are performed by two separate compounds in AZ-PF. One is an acetal dissolution inhibitor (RO-CHR'-OR"), which is sensitive to acid-catalyzed hydrolysis, and the other is a brominated phenolic compound which produces hydrobromic acid upon exposure to X-rays.

According to Eckes et al.(*8*)the hydrogen bonding causes the hydrophilic hydroxyl groups to be held together in tight clusters, while the hydrophobic methyl groups are exposed to the developer. This causes the hydroxide ions of the developer to be repelled from the novolac resin, thereby greatly reducing the dissolution rate of the resin. Exposure of the resist produces a strong acid in the exposed regions from reactions of the radiation-sensitive acid-generating compound. The acid catalyzes a hydrolysis of the acetal into two alcohols and an aldehyde, via an intermediate complex between the acid and the dissolution inhibitor. One would anticipate that this breakup removes the binding force from the novolac hydroxyl groups, which are then free to react with the hydroxide ions of the base developer solution. An increase in the dissolution rate of the novolac resin beyond its native value occurs, leading to removal of the exposed resist by the developer. The activation energy of the hydrolysis reaction is low enough that it takes place in about 20-30 minutes at room temperature, while a post-exposure bake at about 60°C will cause it to be completed in less than a minute(*8*).

Process Latitude Measurements on Chemically-Amplified X-ray Resists

One important definition of process latitude is the dose latitude for ±10% linewidth variation. The observed linewidth of resist features patterned lithographically is dependent on the exposure dose. In general, for negative resists, an increase in dose causes an increase in linewidth, while in positive resists, increasing the dose causes a decrease in linewidth. It is important to determine how accurately the dose delivered to a resist must be controlled to print the desired mask features within a given tolerance. The dose latitude is determined by varying the exposure dose and measuring linewidths of a given feature geometry as they change with dose. The linewidth is then plotted versus dose, and a determination can be made as to what range of doses resulted in linewidths within ±10% of the nominal size. Sometimes a straight line can be fitted to the linewidth vs. dose data; at other times, the relationship is not linear and a judgment must be made how to fit the data. The accuracy and precision of linewidth measurements is crucial to obtaining a good process latitude measurement. Careful control of the processing conditions is also essential, as any uncontrolled changes in the process can lead to linewidth changes independent of dose, affecting the results obtained for process latitude. Results reported herein are from SEM measurements, but instrumentation for electrical linewidth measurements, which are much more precise, has just been installed at CXrL.

The interaction of these key steps all work to provide the observed sensitivity, resolution, image shape, and etch resistance. The interrelation of various processing variables are now being explored, and the complexity of the interactions is providing a challenge in the utilization of this new class of resists. Some of the efforts at CXrL in understanding the relationships of the processing variables on performance for commercially available resists at CXrL are outlined in this chapter. An alternative scheme of deprotection as the result of chemical amplification is discussed in the subsequent chapter by authors from AT&T.

Experimental Procedure

Resist Samples. Resist samples were acquired from the following sources: XP-90104B and XP-90104C resists from Shipley Company, AZ-PF 114 from Hoechst Celanese, PMMA from KTI Electronic Chemicals, and experimental resists from Rohm and Haas.

Lithographic Process. All lithography was done at the University of Wisconsin Center for X-ray Lithography (CXrL). Exposures were made on the ES-0 and ES-1 exposure stations using a computer-controlled single-field exposure system. The exposures were made using the 800 MeV beam of the synchrotron, filtered with a 25 micron beryllium window. For each wafer exposure, the exposure chamber was first evacuated with a mechanical pump to a pressure less than 100 millitorr and was then filled with helium to a pressure of 20 torr to provide wafer and mask cooling during the synchrotron radiation exposure.

All resist processing was done in the process cleanroom facilities at CXrL. The lithographic processing was done according to the following procedures, which had been chosen for their ability to pattern 0.30 micron features. All wafers were new 4" test grade silicon wafers from Tygh Silicon (Pleasanton, CA 94566). They were used directly from the package and primed for improved adhesion with the following process:

Priming (adhesion promotion) process

1. Bake 5 minutes on vacuum hot plate at 160°C to dehydrate surface.
2. Place in bell jar with HMDS (hexamethyl disilazane) vapor for 5 minutes
3. Bake 5 minutes on a hot plate at 110°C to dry and remove excess HMDS.

After allowing the wafers to cool to room temperature, wafer priming was followed immediately by the spincoating application of the resists. The two resists were processed as follows:

AZ-PF process

1. Spincoat: 60 sec @ 5000 rpm, thickness ~1.0 μm
2. Softbake: 60 sec @ 110°C, vacuum hot plate
3. Mount w/mask:
   a. Mask=IBM R091889D (mask transmission ~50%)
   b. Mask clamped to wafer with ring mount
   c. Three spacing pads: 5 μm aluminum foils
4. Expose: Dose to mask: 125 mJ/cm^2
   Dose to resist: ~62 mJ/cm^2 ; Exposed in ES-0 using 800 MeV
5. Holding period of at least 15 minutes at room temperature
6. Develop: 4 min., AZ developer, 1:1; immersion with manual agitation

XP-90104B-23.5 process

1. Spincoat: 60 sec @ 4000 rpm, thickness ~1.0 μm
2. Softbake: 60 sec @ 110°C, vacuum hot plate
3. Mount w/mask:
   a. Mask=IBM R091889D (mask transmission ~50%)
   b. Mask clamped to wafer with ring mount
   c. Three spacing pads: 5 μm Al foils
4. Expose: Dose to mask: 150 mJ/cm^2
   Dose to resist: ~75 mJ/cm^2
   Exposed in ES-0 at CXrL using 800 MeV beam
5. PEB: 60 sec @ 110°C, vacuum hot plate
6. Develop: 3 min., MF-312 developer , 1:1 (0.27 N); immersion / manual agitation

Spincoating was done on a model 5110-C-T system from Solitec (Santa Clara, CA 95050). The vacuum hot plate was controlled to within ±0.3°C of the nominal temperature with a CEE Model 1000 (Cost-Effective Equipment, division of Brewer Science, Inc., Rolla, MO). After development, all wafer samples were rinsed with deionized water, dried under a stream of nitrogen, and inspected under an optical microscope to see if the patterns had been reproduced well.

Etching Process. For the measurement of etch resistance, the Plasmatherm model 2484 reactive ion etcher at the Engineering Research Center for Plasma-Aided Manufacturing was used. A CHF_3/O_2 (in the ratio of 10:1 as measured by mass flow) RF plasma operated at 13.5 Mhz was used for SiO_2 etching and was chosen for the basic etch process. The initial goal of the study was to find whether or not patterns of 0.5 micron size and smaller could be successfully transferred into the SiO_2 from the resist. This was later extended to dimensions of 0.35 and 0.30 microns.

Patterned wafers were cleaved into quarters for the etching step in order to conserve the supply. In order to reproduce as closely as possible the conditions of etching a whole wafer, an additional unpatterned three quarters of a wafer were fitted together with the patterned wafer on the etch electrode to make a "whole" wafer. The water-cooled etch electrode was also covered with a mylar film to reduce electrode sputtering. The initial etching conditions were: pressure, 40 millitorr; power, 500 W; CHF_3 flow, 90 sccm; O_2 flow, 9 sccm; the DC self bias was -600V.

The bulk etch rate of SiO_2 was measured by means of a He-Ne laser interferometer designed and interfaced to the Plasmatherm(*26*). The initial film thickness was measured by ellipsometry using an Auto-El ellipsometer from Rudolph Instruments. Etch rates could also be measured as a function of time during the etch, using the interferometer trace.

Measuring Etch Resistance. The rate of resist (*27*)erosion was determined by film thickness measurement with a profilometer or by an in-situ laser interferometer built onto the plasma etch chamber(*26*). It has been shown that the etch rate of one material in a plasma is significantly affected by adjacent materials. Therefore, a resist that is patterned into small features on a substrate will be etched differently than a homogeneous resist-coated water, due to the exposed substrate. Features were used in the range of size of actual technological interest - in this case, 0.5 μm and smaller.

The main goal of the etch resistance measurements was to define etching conditions where an anisotropic, selective etch is obtained without damage to the resist. This relates to the question of the resolution of the lithography/etch system as a whole, since the lithographic goal is to etch features with maximum resolution.

Process Latitude Measurements. All X-ray exposures were performed on exposure station ES-1 at CXrL in 20 torr helium, using the 800 MeV beam of the Aladdin synchrotron. The beam current ranged from approx. 50 to 200 mA, giving a flux from the beamline of approx. 33 to 130 mW/cm^2 (17 to 67 mW per horizontal cm). A chopper was used to decrease the flux to approximately 5% of this value, in order to allow finer dose control. X-ray masks were obtained from Perkin-Elmer and DuPont. These had been fabricated with a patterned gold absorber about 0.5 micron thick supported by a boron nitride membrane about 5 microns thick(*28*) The proximity gap was set by clamping mask and wafer together with a 25 micron thick Kapton film spacer around the circumference. The radiation spectrum from ES-1, as it was configured at the time, was a broad peak centered at about 1500 eV (8.25 Å).

In order to find good process conditions and to evaluate process latitudes, the following process parameters were varied using statistically-designed(*29,30*) experiments: exposure dose, developer concentration, development time, and post-exposure bake (PEB) temperature and time (for those resists requiring a post-exposure bake). For each resist, a central condition was defined for each of these process parameters. Each of the parameters was then changed to values above and below the central condition, while keeping the other parameters constant. The following processes describe the central conditions of the five parameters, as well as process parameters which were fixed.

For Shipley XP-90104C-20.5, the wafers were first subjected to a 5 minute, 160-200°C dehydration bake on a vacuum hot plate, then primed 5 minutes in HMDS (hexamethyl disilazane) vapor. The priming was followed by a 5 minute drying bake at 100-110°C. XP-90104C-20.5 was spun at 1750 rpm for 60 seconds to achieve a film thickness of 1 micron. The rest of the conditions were similar to those listed earlier for the B resist except for exposing at 45 mJ/cm^2. After 3 minutes immersion in developer, the wafer was removed and immediately rinsed in deionized water, then dried under a stream of filtered nitrogen. In addition to the standard processing conditions, X-ray dose was varied from 35 to 55 mJ/cm^2, PEB temperature from 104 to 119.5°C, PEB time from 40 to 120 seconds, development time from 2.5 to 4 minutes, and developer concentration from 0.243 to 0.297 N.

For Hoechst AZ-PF the conditions were similar to those previously described. Hoechst AZ developer contains a sodium salt of silicic acid ($NaOSi(OH)_3$) and trisodium phosphate; its normality is 0.30 N. In addition to the standard processing conditions, wafers were processed while changing X-ray dose and holding all other variables at a constant setting. Several doses were also evaluated using AZ developer diluted with a ratio of 3:2 AZ:H_2O.

All samples were coated with approximately 100 Å of gold before examination in the SEM. The gold coating was deposited with a "SPI-Sputter" DC sputtering system from SPI Supplies (West Chester, PA). An argon pressure of approximately 200 millitorr was used for sputtering, with applied DC current of 60 mA, and a total sputtering time of 60 seconds. In the SEM, samples were tilted at about 60° to show their edge profiles. Linewidths were measured on SEM photos with magnification of 20,000, giving an estimated measurement uncertainty of 0.026 micron (1 sigma). Linewidth was measured on the same line of the test pattern, as nearly as possible to the same position on the line, for all measurements in a given series, to try to minimize variations due to the mask pattern. Two SEMs were used for the measurements. One was a JEOL 848IC at CXrL, and the other was a Hitachi SEM at Shipley Company. The JEOL 35 SEM at U. W. Materials Science Center was used for some of the preliminary work leading to these experiments.

SEM examination of the mask and patterned wafers showed that patterns on the mask had a bias of about 0.1 to 0.2 micron. Because of this, patterns which were designed as 0.5 micron lines with 0.5 micron spaces between them were in fact approximately 0.6 to 0.7 microns wide, with spaces of 0.3 to 0.4 microns. Because of this mask bias, it was possible to make process latitude measurements at 0.3 micron dimensions for the negative resist, XP-90104C.

Results of Process Latitude Measurements on a Negative Tone Chemically-Amplified Resist: XP-90104C

Figure 1 shows the pattern which was printed for XP-90104C with 0.3 micron lines and 0.7 micron spaces using the following process conditions to give good profile and sensitivity using a 0.3 μm line/ 0.7 μm space pattern: prebake 60 seconds at 110°C, exposure dose 50 mJ/cm^2 onto the mask, post-exposure bake (PEB) 60 seconds at 110°C, develop 3 minutes at 0.27 N with Shipley MF-312 developer. Comparable results were obtained with 0.2 micron lines and 0.4 micron spaces. Initial process latitude studies showed that PEB temperature must be closely controlled, since variation of ±5°C leads to poor results.

Increasing the PEB time for the negative-tone resist to as much as 120 seconds still gives good sidewalls, but causes lines to become broader by about 0.1 micron. Small changes in the concentration of the developer also do not significantly affect the sidewalls, but do affect the linewidth. Doses between 40 and 55 mJ/cm^2 onto the mask give good sidewalls. The transmission of the X-ray mask was ~70%, so this corresponded to doses between ~28 and 38 mJ/cm^2 delivered to the resist.

From Figure 2 it is apparent that the linewidth of 0.3 micron nominal features in Shipley XP-90104C resist decreases as the developer concentration increases. The reason for the decrease of linewidth with increasing developer concentration is the increased dissolution rate of the resist. It has been demonstrated that the dissolution rate of many types of unexposed novolac resin obeys a power law dependence on the developer concentration(*30*). The power law has been shown to hold true above a critical concentration C_0, below which dissolution does not take place, i.e. the dissolution rate is described by $R=A'(C-C_0)^{n'}$ for developer concentrations greater than C_0. C_0, A', and n' all vary greatly depending on the type of resin used, in terms of its relative content of different types of novolac, dissolution inhibitors, and other additives(*15,32-33*).

Since the resist dissolution rate increases with increasing developer concentration, we expect the linewidth to decrease as the developer concentration increases(*34-35*). This can be viewed simply as the result of the evolving resist profile moving inward at a faster rate during development. If this is the case, then the results of increasing the developer concentration should be identical to increasing the development time. Note that the slope of linewidth vs. developer concentration in the region examined here is very high. It appears that this particular system of resist, exposure parameters, and developer concentration lies in a fairly steep region of the exponential curve of dissolution rate vs. developer concentration.

A measurement of linewidth variation for 0.3 micron lines while varying the X-ray exposure dose was accomplished, but the measurement precision was poor. The trend toward greater linewidth with increasing dose was expected and observed for the negative resist. The slope of a linear least-squares fit to the linewidth vs. dose plot was 0.0075 microns per mJ/cm^2. (The slope of this plot determines how large the dose latitude will be for given conditions, which can be expressed as the change in dose which yields a 10% change in linewidth.) In this case, using the linear fit gives an estimate of the latitude as 4 mJ/cm^2. As a percentage of the nominal dose of 45 mJ/cm^2, this gives a dose latitude of ~9%.

It is apparent from the graph that the measurement errors associated with the techniques used here make a quantitative assessment of dose-linewidth latitudes difficult. This situation will be improved with the electrical linewidth system currently being installed at CXrL. However, some information does come from this measurement. It is clear that there can be substantial variations in 0.3 micron nominal feature linewidths as a function of dose, and that careful study must be undertaken to

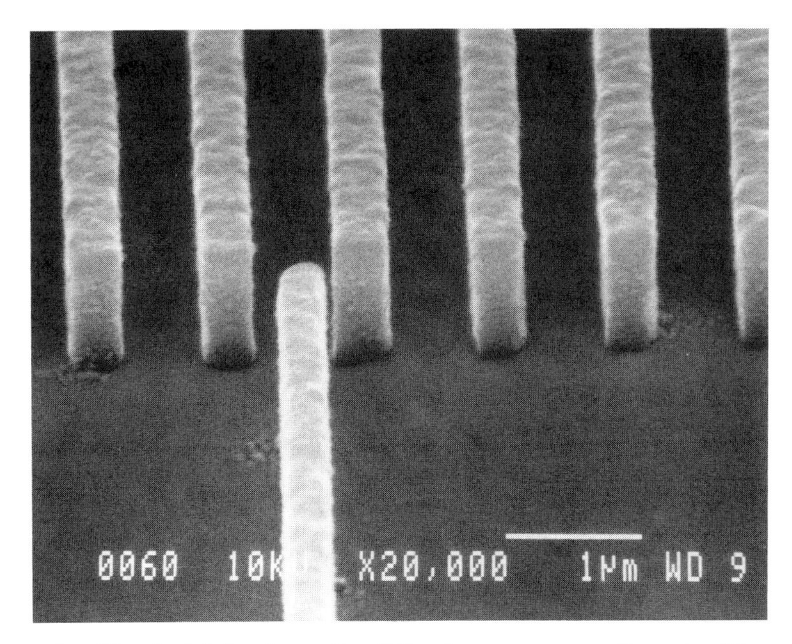

Figure 1. Nominal 0.3 micron line/0.7 micron spaces patterned with an exposure dose of 50 mJ/cm^2 in XP-90104C.

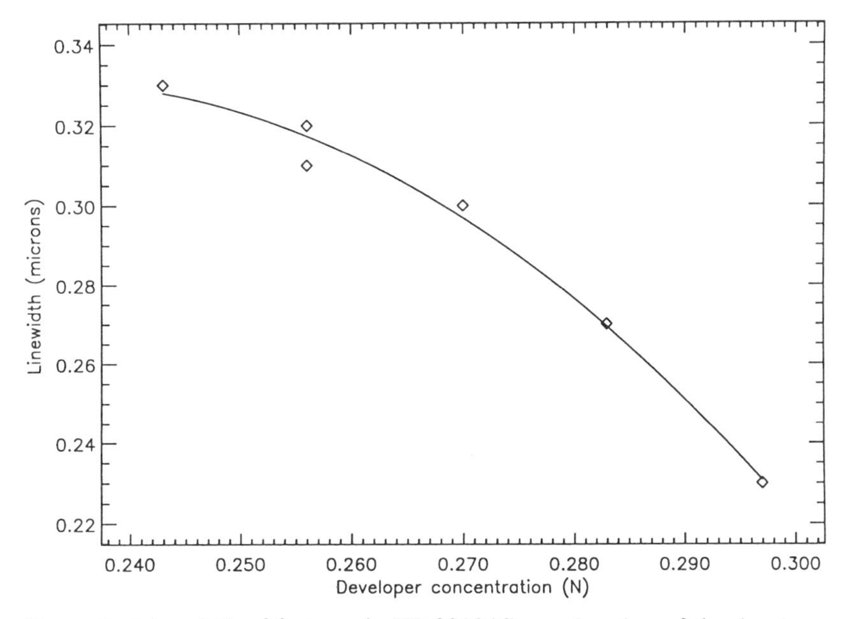

Figure 2. Linewidth of features in XP-90104C as a function of the developer concentration. The fit is to a quadratic and has a slope of -1.82 microns/(N).

minimize the effects of dose variation on linewidth, by means of process optimization, resist chemistry optimization, or other means. As has been shown recently by others(*36*)the superior dose latitude expected from X-ray lithography compared to optical lithography depends on processing and can not be taken for granted.

Figure 3 shows the linewidth of 0.3 micron nominal lines as a function of post-exposure bake time, which varied from 40 to 120 seconds. All other process variables were held at their central values. The relationship between the linewidth and the PEB time looks nearly linear between 60 and 120 seconds. A straight line fit to the data has a slope of 0.0025 microns/second. Expressed in terms of percentage changes from the standard values of 60 seconds and 0.30 microns, this gives a 20% latitude or 12 second change in PEB time. In terms of percentage changes in the parameters, these data show that PEB time does not change the linewidth as severely as several of the other process variables.

The linewidth variation measured while varying the post-exposure bake temperature showed the trend toward greater linewidth with increasing PEB temperature, as expected, since increasing the PEB temperature effectively increases the amount of chemical amplification. This should have the same effect as increasing the X-ray dose, i.e. it will increase the linewidth of the latent image. There were not sufficient points collected, however, to determine the functional dependence of linewidth on PEB temperature. A straight line fitted to the available data gives an estimate of the magnitude of the influence of PEB temperature on the linewidth and suggests that a 10% change corresponds to a temperature range of 1.4°C. Because the acid-catalyzed crosslinking reaction, which determines the relative dissolution rate of the resist, is thermally driven(*25,37*) and the gain of the chemical amplification has been reported to approximately double with each 10°C increase in the PEB temperature(*38*) the effect of PEB temperature on linewidth is expected to be large in comparison to the effects of most other process variables. From this it is clear that the vacuum hot plate PEB temperature is a critical variable for linewidth control, and must be controlled to within a fraction of a degree to keep from causing significant linewidth variations across the wafer as well as wafer to wafer.

Results on Process Latitude Measurements on a Positive Tone Chemically-Amplified Resist : AZ-PF

Figure 4 shows the pattern obtained for AZ-PF using nominal 0.6 micron lines and spaces. For this resist, the following process conditions were found to give vertical sidewalls, good resolution, and good sensitivity for the 0.6 micron features evaluated: prebake 60 seconds at 120°C, expose at 120 mJ/cm^2 dose to mask, wait at least 15 minutes with wafer at room temperature, develop 2 minutes with mild agitation in 1:1 AZ developer:H_2O. The dose latitude for good sidewalls as judged by the SEM images was from about 120 to 160 mJ/cm^2 onto the mask.

Figure 5 shows the results of linewidth variation as a function of X-ray dose. These data show that AZ-PF resist can reproduce nominal 0.6 μm line/ 0.4 μm space patterns over an approximate ±15% dose range (140±20 mJ/cm^2). The sidewalls produced by this resist look quite vertical for some process conditions; however, there was linewidth variation, and the profile of the lines became undercut and roughened as the dose increased above about 160 mJ/cm^2. While sidewall angle and clearing between features appeared to be acceptable under a wide range of process conditions, linewidth control appeared to be more difficult to quantify. Linewidth changes under ±10% were observed for only a few process conditions, and to separate the effects of run-to-run variation from the effects of dose when dealing with such high resolution will require a finer range of doses and more samples. The parameters investigated for

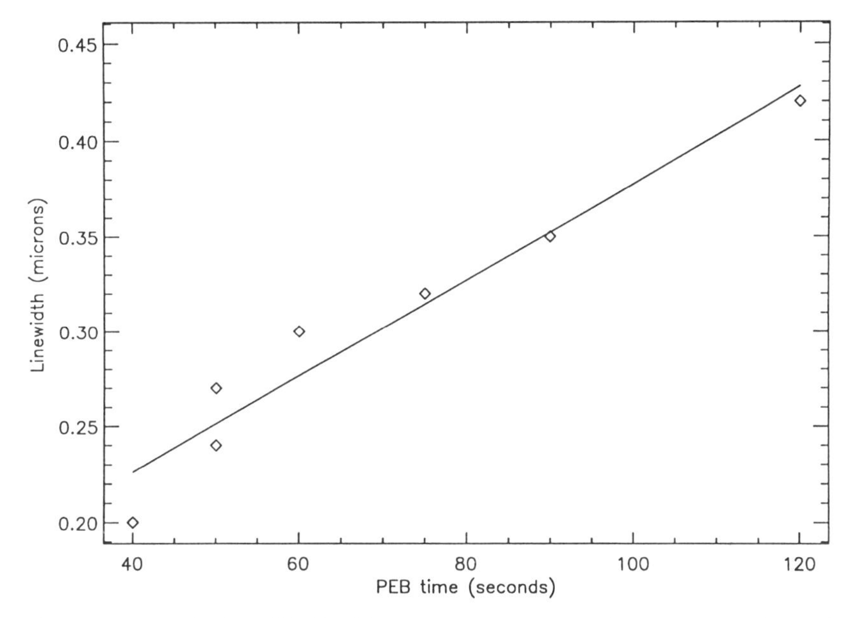

Figure 3. Linewidth of XP-90104C vs. post-exposure bake time. The slope of the line is 0.0025 microns/sec.

Figure 4. SEM of AZ-PF pattern of nominal 0.6 micron lines when exposed to X-ray dose of 120 mJ/cm^2 .

this resist were X-ray dose, developer concentration, and mask feature size. The AZ-PF process does not require a post-exposure bake; thus PEB time and temperature were not investigated in this study. The optional post-exposure bake has recently been examined in some detail by Hoechst(*8*); they have reported that a moderate PEB (about 60-65°C for 60-90 sec.) gave better results than no PEB. In particular, the PEB was reported to decrease the effects of the waiting time between exposure and development, which had led to problems with reproducibility.

Results on Plasma Etching Selectivity of the Chemically-Amplified Resists

Because both chemically-amplified resists studied utilize the novolac resist matrix, it was expected that they would show good etch resistance. For comparison, a non chemically-amplified conventional novolac resist, OCG Microelectonic Materials HiPR-6512 and poly(methylmethacrylate), PMMA (of molecular weight of 950 K from KTI Chemicals) were treated to the same etch conditions as the two chemically-amplified resists. For these experiments, however, XP-90104B was substituted for XP-90104C. The etch rates in nm/min. and selectivity compared to SiO_2 for the four resists, as measured in a CHF_3/O_2 RIE plasma at 500 watts were as follows: XP-90104B = 22 (1.5); AZ-PF = 20(1.7); HiPR-6512 = 22(1.5); and PMMA = 33(1.0). Under these conditions, the etch rate of SiO_2 = 33. The chemically-amplified resists, therefore, showed etch rates comparable to that for the conventional novolac resist, which was much better than that of PMMA. See Table I.

These data also suggest that the chemically-amplified resists can be used as etch masks over SiO_2 to protect it from etching. This concept was tested by preparing a 0.5 micron layer of SiO_2 on top of a silicon wafer, coating the wafer with 1.0 micron of resist, patterning the wafer with suitable mask features, and subjecting the wafer to RIE etching conditions. Figure 6 shows the results of patterning 0.30 micron features with XP-90104B. In this case 0.7 micron of resist remained on top of the masked layer of SiO_2 while all of the exposed SiO_2 was etched away in the exposed area. Because this was a negative resist, the resulting features from a hole pattern is a series of posts with feature sizes related to the hole pattern in the mask. A similar amount of resist was lost when AZ-PF was used as an etch mask for 0.30 micron features. This is shown in Figure 7.

Results on Surface and Edge Roughness Measurements on a Negative Tone Resist: XP-90104C

The printed resist image, in the ideal case, will have a sharp edge and show a 90 degree angle with respect to the surface of the wafer. In actual practice, the edge will reflect the mask features from the e-beam writing and can have an edge profile between 85-87 degrees and still be acceptable(*2*). Because of the solution development process, however, the surface and, therefore, the edge do show some roughness which is the consequence of the chosen processing parameters. It has been show experimentally and theoretically that ellipsometry can be very sensitive to surface roughness. Fenstermaker and McCracken(*39*) have shown that surface roughening can be accurately modelled using an effective medium approximation.

For these reasons, we have used(*40*) *in situ* ellipsometry to measure dissolution, swelling, and surface roughening of various chemically-amplified resists during development. In these experiments there are two parameters which are plotted and compared to theoretical plots. The first is Δ, which is the difference in phase between the reflected and the incident light. The second is Ø, which measures the relative change in magnitude of the s- and p- polarization upon reflection from the

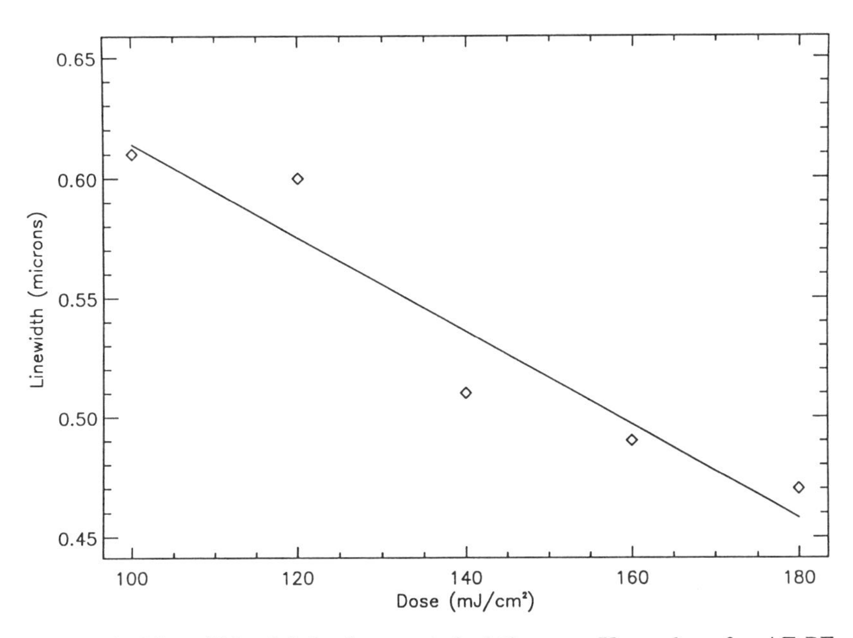

Figure 5. Linewidth of 0.6 micron nominal lines vs. X-ray dose for AZ-PF. The slope of the line is -0.002 microns/(mJ/cm^2).

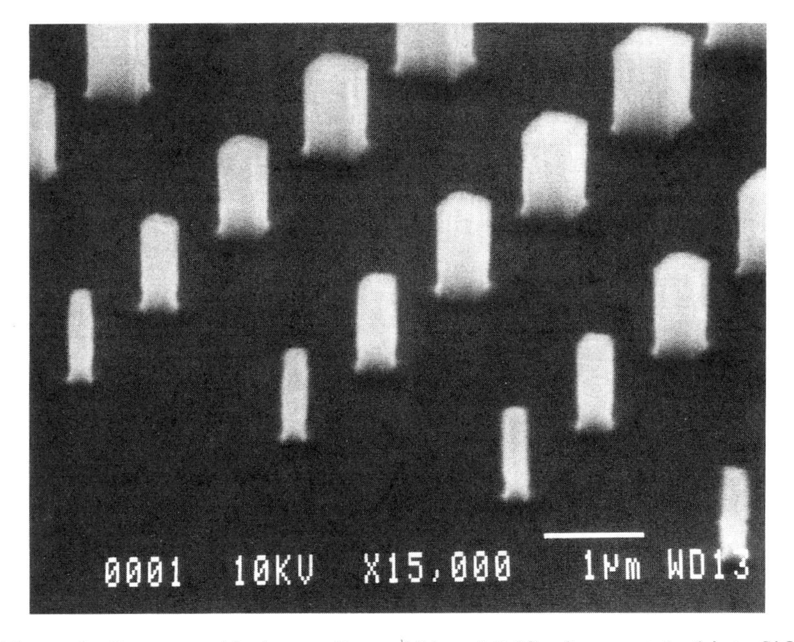

Figure 6. Patterns with the smallest widths of 0.30 microns etched into SiO_2 using Shipley XP-90104B as an etch mask.

thin film. (We have described the apparatus and the appropriate equations in a previous publication(*41*)).

The purpose of the theoretical model is to find a set of conditions which will fit the contours of the delta/psi plot which was obtained experimentally as the resist film is exposed to the development solvent. With the effective medium approximation (EMA) we have been able to model the effects of surface roughening by employing a multilayer film model. The assumption of the EMA is that the scale of the roughening is small enough so that most of the light is specularly reflected from the film-developer interface. The roughened surface layer of the resist film can be thought of as a mixture of the ambient (developer) and the bulk (resist) with a refractive index representative of such a mixture and determined by the Maxwell Garnett Theory of mixing(*42*.).

By using the EMA model and assuming that the surface layer increased linearly with time, it was possible to account for the departure of the experimental data from an ideal dissolution model (which assumes that the resist-developer interface is perfectly abrupt and smooth). Figure 8 shows the experimental data for the dissolution and subsequent roughening of a chemically-amplified Shipley XP-90104C resist in Shipley MF-312 developer. Included in the graph is the constant refractive index contour for the bulk film (n=1.620) and the EMA surface roughening model which best fits the data. The point marked "begin"was the film measured at the start of the development, whereas the point marked "end" was the measured value after a development time of 30 min. The resist had received 20 mJ/cm^2 exposure prior to development. The final surface roughness estimated from the fit was found to be 65.0 nm. The data were taken at a wavelength of 632.8 nm. Included in the figure are the results of a calculation using the ideal dissolution model and the EMA model. The experimental data for a similar run measured at a different wavelength (546.1 nm instead of 632.8 nm), show similar trends, and the roughness calculated from the best fit to the data in each case was almost identical.

In order to corroborate the results of the ellipsometry model, resist samples were exposed at low levels of exposure between 6,7 and 8 mJ/cm^2, developed, and then coated with a very thin gold film (on the order of 25.0 nm) for examination in the SEM. The decrease of roughening with increasing exposure can be seen in Figure 9. Surface roughening was also found to increase with development time for a given exposure using the same technique. Because a metal overcoating is required for SEM measurement and charging effects occur with very thin Au films, the true surface roughness of the sample could not be obtained for very smooth film surfaces by the SEM measurements. While ellipsometry may be capable of detecting roughening below 10.0 nm, it is difficult to confirm roughening on this scale. Work is now in progress to use a STM to measure the surface of these films after development.

Measurement of resist swelling was also performed using in situ ellipsometry. These measurements were undertaken to determine the regimes in which non-ideal dissolution behavior occurs for this chemically-amplified photoresist. For instance, at high exposures the resist becomes crosslinked, and penetration by solvent is difficult. If such a film is placed in the normal developer solution, nearly pure swelling could result only over a long time period or at elevated temperature. After several hours at 25°C, for example, the film swelling for XP-90104C appeared to be linear with time. Because of the linearity, the data shown as Figure 10 were fit to a Case II swelling model in which a swollen surface layer grows into a glassy underlayer with time. While the equilibrium solvent volume in the film is high (the final calculated film thickness is projected to be twice the initial thickness), the diffusion of solvent into the film rather slow. However, for a typical development time of 3 minutes used under normal processing, this film would increase in thickness by less than 1%. This is negligible under semiconductor manufacturing requirement standards.

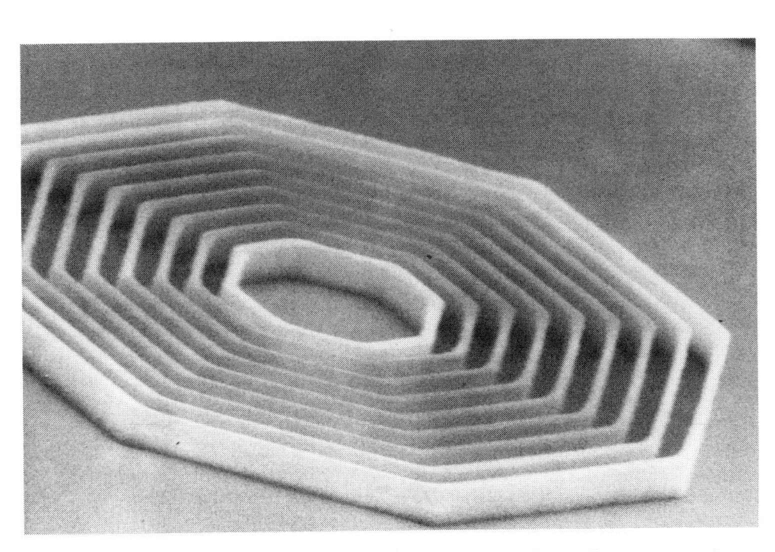

Figure 7. Test pattern etched into SiO_2 using AZ-PF as an etch mask. Linewidth of etched pattern is 0.30 microns.

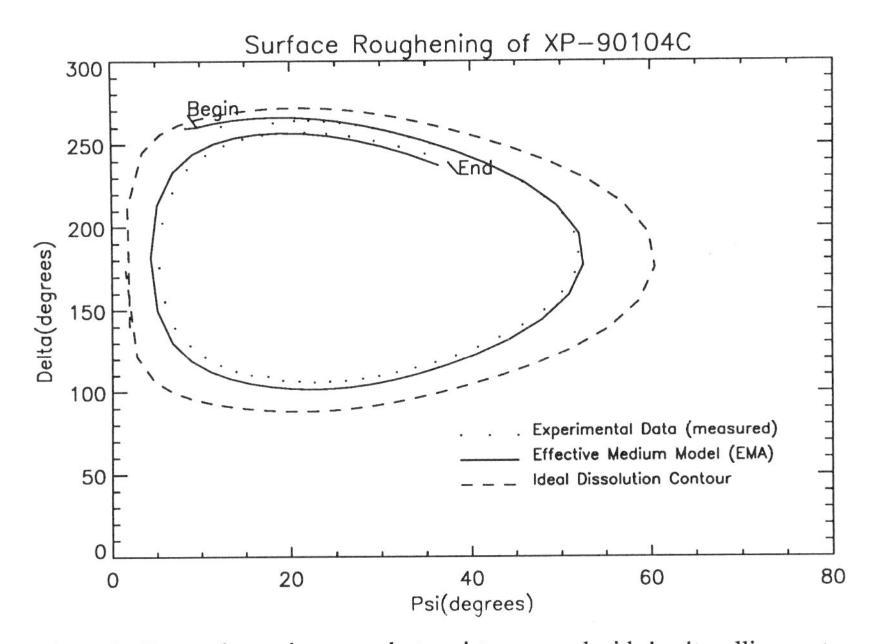

Figure 8. Exposed negative tone photoresist measured with *in situ* ellipsometry during development in MF-312 developer.

Figure 9. SEM measurements of surface roughening with exposure levels as shown.

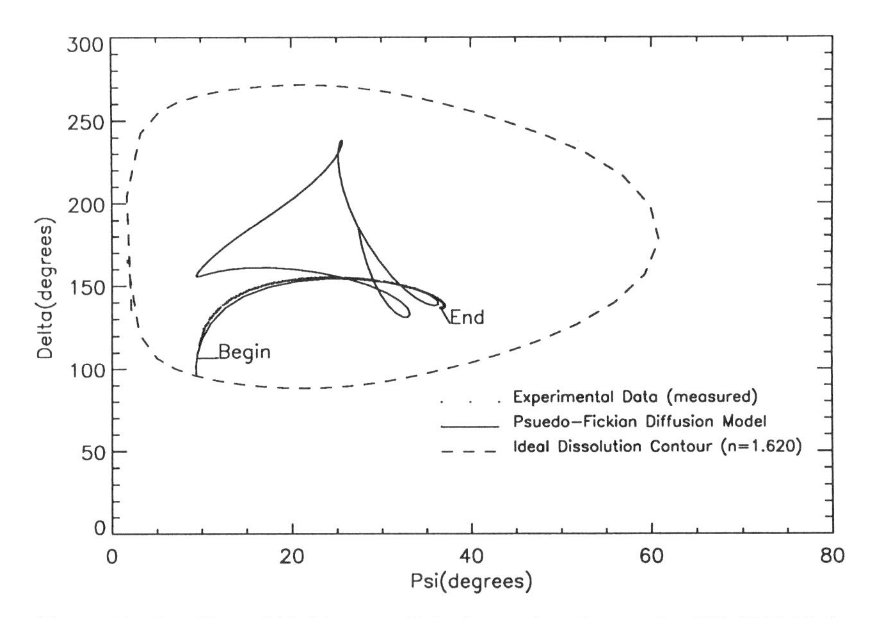

Figure 10. Swelling of highly cross-linked negative photoresist, XP-90104C, in MF-312 developer at 25°C. The model incorporates a mixture of Case II and Fickian diffusion. Resist was exposed to 70 mJ/cm^2 prior to development.

The swelling of the negative tone chemically-amplified Shipley XP-90104C resist can be accelerated at elevated temperature as shown in Figure 11 where the development is at 45°C for a highly exposed resist sample developed with MF-312 developer. The data points were recorded in 30 sec intervals. The rate of swelling is several times greater at 45°C than at 25°C. It is clear that the inflection points of the data from both experimental runs fall well within the bulk film contour. The swelling is not pure Case II, which is evidenced by the fact that the inflection points in the data do not lie on the ideal dissolution contour, and this means that a more sophisticated modelling will be required to reproduce the experimental data. The key point, however, is that neither the swelling nor the roughening of the chemically-amplified resist is large enough under these conditions to cause problems in the utilization of the resist for X-ray lithography.

Conclusions

Several resists have been evaluated for their sensitivity for synchrotron X-ray lithography. Several experimental resists show promising sensitivity. Shipley XP-90104C and Hoechst AZ-PF have been evaluated for their process latitude and it was found that they can reproduce features with good side walls over a range of process conditions. Linewidth control of both resists requires further study, but the variation of XP-90104C linewidth with process parameters investigated here behaved in a fairly predictable way and chances appear good that further studies could yield a process with good exposure latitude.

The results of these studies show that Shipley XP-90104C can reproduce 0.2 μm line/ 0.6 μm space patterns for some process conditions, and 0.3 μm line/ 0.7 μm space patterns for a substantial range of conditions. XP-90104C reproduced 0.3 μm line/ 0.7 μm space patterns with vertical side walls under a range of doses of ±15% (47±7 mJ/cm^2), with linewidth variation depending on the dose. The dose delivered to the X-ray mask can be controlled to ±3%, which is well within this ±15% range.

Results with AZ-PF so far have not been as good as with the negative chemically-amplified resists. From the present study, it was apparent that moderate overexposure or overdevelopment of AZ-PF led to drastic changes in the side wall profile, as well as causing problems with incomplete clearing of the resist. The use of a metal-ion-free developer, particularly TMAH (tetramethylammonium hydroxide), is reported to yield improved profile control(*8*). This may be because of the larger size of the tetramethylammonium cation compared to the sodium cation in the AZ developer. It is possible that the higher diffusion rate of the sodium cation causes the base developer to diffuse into the resin more rapidly than the resin surface can be dissolved away, causing stresses to build up and prompting cracking and uneven dissolution. Uneven dissolution was observed for the AZ-PF samples in the present study, in which the resist sometimes appeared to be removed in pieces. Development with TMAH is recommended to gain better control over the resist profile.

The electrical linewidth measurement system is the preferred choice for making measurements of linewidth and process latitude because of its reputed precision and speed. CXrL is now characterizing a Prometrix electrical linewidth measurement system for future linewidth measurements and process latitude studies. For this initial study to define the critical variables, however, the initial tools were satisfactory at the mask resolutions available at the time this study was done. It should be noted that our evaluation of the resolution ultimately achievable by these resists with synchrotron X-ray lithography is currently limited by the resolution of available X-ray masks. So far, it appears that these two resists, as well as other resists we have evaluated, can reproduce all features of our X-ray masks down to

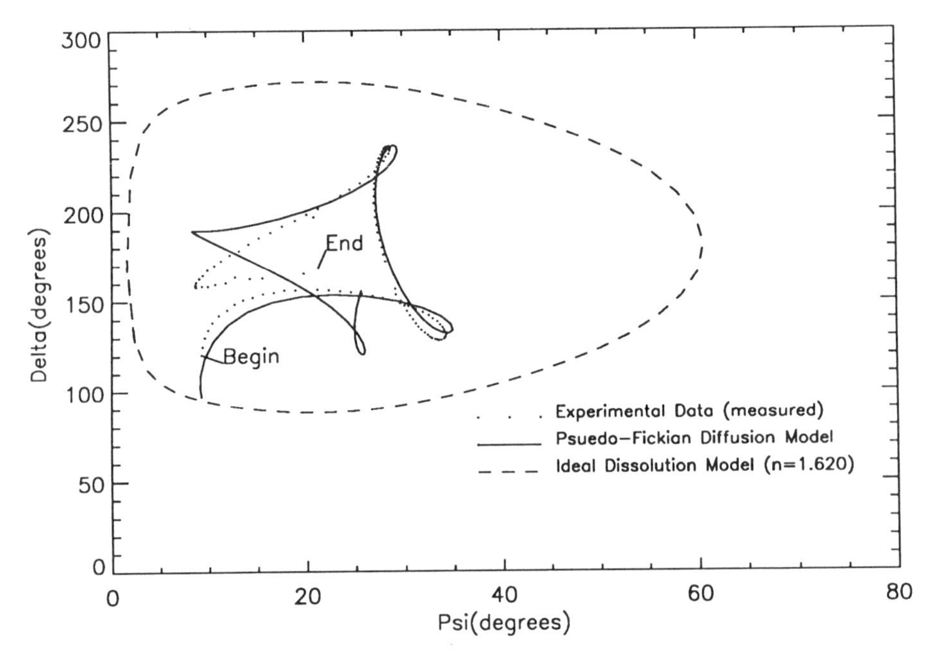

Figure 11. Swelling of XP-90104C in MF 312 developer at 45 °C after receiving 70 mJ/cm^2. A pseudo-Fickian model was used to analyze the data. Points were taken at 30 sec intervals.

dimensions as small as 0.2 μm line/ 0.4 μm space for negative-tone resist and 0.3 μm line/ 0.3 μm space for the positive-tone resist images.

Acknowledgments

The Center for X-ray Lithography(CXrL) is supported by the SEMATECH Centers of Excellence Program of the Semiconductor Research Corporation under Contract 88-MC-507, and the Department of Defense(DARPA) supplied funding for facilities support through Navy grant number N00014-91-J-1876. Support for C.B. and access to the plasma etching instrumentation was provided from the Engineering Research Center through funding from the National Science Foundation under grant number ECD-8721545. The Synchrotron Radiation Center is operated with support of the National Science Foundation under grant number DMR 881625. The authors acknowledge the substantial help from the Rohm and Haas Company, the Shipley Company, and the Hoechst Company in supplying resist samples.

Literature Cited

1. Ito, H. In *Radiation Effects on Polymers,*, Clough, R. L. and Shalaby, S., Eds. ACS Symposium Series 475, American Chemical Society, Washington, D. C.,1991; pp. 326-342.
2. Taylor, J. W.; Babcock, C. , Sullivan, M.; Suh, D.; Plumb, D.; ibid. ,pp. 310-325, and references therein.
3. *Polymers in Microlithography,* Reichmanis, E.; MacDonald, S. A.; Iwayanagi, T.; Eds. ACS Symposium Series 412, American Chemical Society, Washington, D.C., 1989.
4. Novembre, A.; Tai, W. W.; Kometani, J. M.; Hanson, J. E.; Nalamasu, O.; Taylor, G. N. Reichmanis, E.; Thompson, L. F.; *Proc. SPIE,* **1991**,1466, 89-97.
5. Reichmanis, E.; Houlihan, F. M. ; Nalamasu, O.; Neenan, T. X., *Chemistry of Materials*, **1991**, 3, 394-407.
6. Lignau, J.; Dammel, R.; Theis, J.; *Solid State Technol.,* **1989**, 32, 105-111.
7. Dammel, R.; Lindley, C. R.; Pawlowski, G.; Scheunemann, U.; Theis, J.; ibid. 378-391.
8. Eckes, C.;Pawlowski, G.; Przybilla.K.; Meier, W.; Madore, M.; Dammel, R.; *Proc. SPIE,* **1991**, 1466, 394-407.
9. Guo, J. Z. Y.; Cerrina, F.;*Proc. SPIE,* **1991**,1465, 330-337.
10. Peters, D. W.; Frankel, R. D.; *Solid State Technology,* March, 1989, 77-81.
11. Miller, L.; Brault, R.; Granger, D; Jensen, J.; van Ast, C.; Lewis, M.; *J. Vac.Technol.* , **1989**, B 7, 68-72.
12. Doessel, K.-F.; Huber, H. L.; Oertel, H.;*Microelectronic Engineering*, **1986**, 5, 97-104.
13. Toukhy, M.; Beauchemin, B.; *Proc. SPIE,* **1989**, 1086, 374-387.
14. Hanabata, M.; Uetani, Y.; Furuta, A.; *J. Vac. Soc.Technol.*, **1989**, 7(4), 640-650.
15. Neenan, T. X.; Houlihan, F. M.; Reichmanis, E.; Kometani, J. M., Backman, B. J. ; Thompson, L. F.; *Proc. SPIE,* **1989**, 1086, 2-9.
16. Buhr, G.; patent US 4,189,323.
17. Crivello, J. V., Lee, J. L. ; *Macromolecules*, **1981**, 14, 1141-1147, and references therein.
18. Ito, H. ; Willson, C. G. In *Polymers in Electronics*, Davidson, T. Ed. , American Chemical Society Symposium Series 242, American Chemical Society, Washington, D. C., 1984, pp. 11-23.
19. Bruns, A.; Luethje, H.; Vollenbroek, F.A.;Spiertz, E.; *Microelectronic Eng.*, **1987**, 6, 467-471.

20. Buhr, G. ; Dammel R.; Lindley, A.; *Polymeric Mater. ,Sci.. & Eng.*, **1989,** 61, 269-277.
21. Doessel, K.-F., EP-A0312751.
22. Liu, H.; deGrandpre, M.; Feely, W.; *J. Vac. Sci. Technol.*,**1988**, B6, 379-383.
23. Blum, L.; Perkins, M.; Liu, H.; *ibid.*, 2280-2285.
24. deGrandpre, M.; Graziano, K.; Thompson, S. D.; Liu, H.; Blum, L.; *Proc. SPIE*, **1988**, 923, 158-168.,
25. Seligson, D.; Das, S.; Gaw, H.; Pianetta, P.;*J. Vac. Sci. Technol.* , **1988**,B6, 2303-2307.
26. Mishurda, H.; Hershkowitz, N.; *Proc. SPIE*, **1991**, 1392, 563-569.
27. Fedynyshyn, T. H., Grynkewich, G. W., Hook, T. B., Liu, M.,Ma, T, *J. Electrochemicsl Soc.*, **1986,** 134, 206-209.
28. Private communication, G. Hughes, DuPont Corp., Electron Beams Division., Danbury, CT.
29. Box, G.; Hunter, W.; Hunter, J.; *Statistics for Experimenters,* John Wiley and Sons, New York, NY, 1978.
30. Taguchi, G. , *Quality Engineering in Production Systems,* McGraw-Hill, New York, NY, 1989.
31. Huang, J. P.; Pearce, E. M.; Reiser, A; In *ACS Symposium Series 412*, Reichmanis, E.; MacDonald, S. A.; Iwayanagi, T.; Eds., American Chemical Society, Washington, DC, 1989, pp. 364-384.
32. Huang, J. P.; Kwei, T. K.; Reiser, A.; *Macromolecules*, **1989**,22, 4106-4112.
33. Arcus, R. A.; *Proc. SPIE*, **1986**, 631, 124-134.
34. Garza, C. M.; Szmanda, C. R.; Fischer, R. L.; *Proc. SPIE*, **1988,** 920, 321-338.
35. Szmanda, C. R.; Trefonas, P.; *Microelectrnic Engineering*,**1991,** 13, 23-28.
36. Guo, J. Z. Y.; Chen, G.; White, V.; Anderson, P.; Cerrina, F.; *J. Vac. Sci. Technol.*, **1990,** 8(6), 1551.
37. Fukuda, H.; Okazaki, S.; *Japanese J. Appl. Phys.* **1989**, 28, 2104-2109.
38. Private communication, T. H. Fedynyshyn, Shipley Company, Marlboro, MA.
39. Fenstermaker, C. A.; McCrackin, F. L.; *Surface Sci.*, **1969**,16, 85-96.
40. Sullivan, M. ; Taylor, J. W.; Babcock, C.; *J. Vac. Sci. Technol.*,**1991**, B9, 3423-3427.
41. Garnett, J. M. *Phil. Trans.*, **1906**, 205, 237-263.

RECEIVED November 11, 1992

Chapter 16

Diazonaphthoquinone–Novolac Resist Dissolution in Composite Langmuir–Blodgett and Spin-Cast Films

Laura L. Kosbar[1,6], Curtis W. Frank[2,5], R. Fabian W. Pease[3], and John Hutchinson[4]

Departments of [1]Chemistry, [2]Chemical Engineering, and [3]Electrical Engineering, Stanford University, Stanford, CA 94305

[4]Department of Electrical Engineering, University of California at Berkeley, Berkeley, CA 94720

We have used the Langmuir-Blodgett (LB) film fabrication method to isolate the diazonaphthoquinone photoactive compound (PAC) in positive resist films in order to study the mechanism by which the PAC inhibits the dissolution of the novolac resin. The UV absorbance of the PAC when it is uniformly distributed in a novolac matrix is different from that when isolated in pseudo two-dimensional layers. These spectral differences largely disappear when the PAC layers are thermally diffused into the novolac. We found that LB layers of PAC deposited on the surface of spin-cast novolac films did little to inhibit dissolution of the resin unless the film was first baked. After thermal treatment, changes in both the development rate and induction time were measured in the uppermost region for novolac films with one to seven monolayers of PAC. These changes were sufficient to produce 1 μm thick resist images using as little as three monolayers of PAC on 1 μm spin-cast films of pure novolac.

The production of microelectronic devices is strongly dependent on accurate and reproducible control of the microlithographic process. The most commonly used microlithographic resists for the past decade have been members of the positive-acting novolac/diazonaphthoquinone class. The diazonaphthoquinone (DNQ) is a photosensitive material that inhibits the dissolution of the novolac matrix prior to photo-degradation but allows or even promotes dissolution after exposure. Even though these resists have been in common use for many years, the mechanism of novolac resin dissolution inhibition by DNQ sensitizers and the origin of a delay in the onset of dissolution, called the induction effect, are poorly understood.

It has long been assumed that the unexposed DNQ simply acts as a hydrophobic additive that retards the wetting and dissolution of the soluble novolac in the basic developer.*(1-3)* It has even been suggested *(3)* that the sensitizer forms a hydrophobic barrier at the surface upon exposure to the developer, which might help explain the existence of an induction time. Other researchers have, however, proposed that more specific interactions between the novolac and the DNQ contribute to dissolution inhibition. For example, Honda, et al.*(4)* found that hydrophobic naphthalene derivatives that did not contain the sulfonyl ester did not inhibit dissolution. They

[5]Corresponding author

[6]Current address: T.J. Watson Research Laboratory, IBM Corporation, Yorktown Heights, NY 10598

0097–6156/93/0527–0245$06.25/0

proposed that polar sites on the inhibitor are necessary to produce sufficient interactions with the resin. Huang, et al.*(5)* also suggested that the extent of dissolution inhibition of the sensitizer was significantly out of proportion compared to its volume fraction, indicating that the inhibitor interacts in a more structure-specific fashion. Finally, several researchers *(4,6,7)* have suggested that hydrogen bonding within the resin or between the resin and the inhibitor plays an important role in the inhibition mechanism.

Chemical reactions that may contribute to the changes in the dissolution kinetics caused by DNQ have also been proposed. Koshiba, et al.*(2)* observed that DNQ gradually breaks down in aqueous solutions with tetramethyl ammonium hydroxide (TMAH) and phenols. The reaction products include a DNQ-phenol compound joined through a diazo-linkage. Koshiba suggested that the interaction of the developer with the unexposed film may actually create an insoluble layer as it penetrates into the film. This inhibition mechanism was also included as part of Hanabata's *(8,9)* "stone wall" dissolution model.

From the preceding discussion it appears that the dissolution inhibition observed for DNQ/novolac systems may be due to a combination of dipolar interactions, chemical reactions, and possible formation of a protective hydrophobic layer at the resist surface. However, it is difficult with spin-cast resists to separate surface reactions from those that may occur in the bulk, or to separate the effects of chemical reactions from purely hydrophobic interactions. In this paper, we use the Langmuir-Blodgett (LB) film fabrication technique to control spatial distribution of the PAC sensitizer, thus allowing study of the effects of the PAC on dissolution inhibition. The film configuration selected for the experiments is a novel composite structure having a spin-cast underlayer and one or more LB layers of PAC deposited on the novolac surface.

Experimental

Materials. The novolac sample, which was synthesized by a polycondensation reaction of meta-cresol and formaldehyde, known as N-5 resin, was a gift from the Kodak Company. The novolac was purified by two precipitations from tetrahydrofuran into hexane. The material is quite polydisperse with M_w=13,000 and M_w/M_n=8.5, as measured by gel permeation chromatography (GPC) using polystyrene standards. Such high polydispersity is fairly typical for meta-cresols.*(1)*

The diazonaphthoquinone photoactive compound was a gift from Fairmount Chemical under their trade name of Positive Sensitizer 1010. We believe that it contains a 5-substituted sulfonyl group, with one or more hydroxyl substituted benzophenones attached to the sulfonyl ester, but the extent of esterification is unknown. This material was used as received.

Isopropyl acetate was used as the spreading solvent for the Langmuir films. It was obtained from Aldrich, and distilled prior to use. Cellosolve (2-ethoxyethanol) was used as the casting solvent for spin-cast films. It was obtained from Aldrich as a spectroscopic grade solvent, and was used as received.

Sample Preparation and Lithography. One to seven LB layers of PAC were transferred directly onto clean quartz and chrome coated silicon substrates as well as on top of spin-cast films of novolac containing 2.5% PAC, and films of pure spin-cast novolac, as described previously *(10)*. Some of the substrates received thermal treatments consisting of baking in a convection oven at 80-105°C for up to 80 minutes prior to exposure. Wafers were exposed on the Canon Model FPA(4) stepper at exposure settings from 2 to 6. Unfortunately, no absolute exposure calibration is available for this process tool; we will thus only consider qualitative trends as the exposure increases with exposure setting.

Spectroscopic Measurements. UV spectra were recorded for spin-cast films of novolac/PAC mixtures on quartz substrates and for one or more LB layers of PAC deposited on the top surface of spin-cast pure novolac. In both cases, spectra were obtained for films that had received no thermal treatment, and for films that had been baked at 90°C for 40-80 minutes prior to UV exposure on the Canon stepper. UV spectra were recorded for both the exposed and unexposed portions of the film. UV spectra were also recorded for LB films of pure PAC on quartz without a novolac underlayer. Again, one of the films received no thermal treatment, while the other film was baked at 90°C for 80 minutes. Both films received similar UV exposures on the Canon as did the films with novolac.

Development Rate and Induction Time Measurements. The substrates were developed in 0.2M KOH in a tank at room temperature, or in 1:2.5 mixtures of Shipley MP312:water in a Perkin Elmer Development Rate Monitor (DRM) at 21°C. The progress of the development was monitored in the DRM by recording the light intensity of the interference pattern from the laser beam reflecting from both the front surface of the resist film and the front surface of the wafer. As the thickness of the film changes, interference effects between the light reflected from the two surfaces cause modulation of the measured light intensity, as shown in Figure 1. We were most interested in the dissolution at the top surface of the resist film, so light intensity measurements were recorded at intervals of 0.1 sec. for the first 20 sec., up to 2 sec. intervals for development times over 300 sec. The induction times were measured directly from the interferogram, and the development rates were obtained by analyzing the data using the Parmex program developed at the University of California, Berkeley.

Results

Lithography. In order to motivate the subsequent discussion of the spectroscopic and dissolution measurements, we begin by describing some of the lithographic processing results for the composite LB PAC/spin-cast novolac films. A central feature of this research is the evaluation of the proposal that dissolution inhibition is due to increased surface concentration of hydrophobic molecules.*(3, 14)* To address this from the point of view of a potential resist, we first prepared composite films consisting of a spin-cast base layer containing 2.5% PAC in a novolac matrix that was coated with three to seven LB layers of PAC; the films were not baked. The films were exposed on the Canon at setting of five using the fine line test pattern, and developed in 0.2 M KOH.

Optical and SEM micrographs of the developed patterns are reproduced in Figure 2. In previously reported work *(10)* with alternating LB layers of novolac and PAC we saw small isolated dots of underdeveloped material when two consecutive monolayers of PAC were used. In the present experiments with top surface deposition of LB PAC layers we see much more extensive portions of resist affected by the PAC. The development of these films was very non-uniform. The LB top layer films did protect the unexposed regions in some parts, but certainly not in any continuous fashion. In fact, the protected areas appear to have a crystalline surface deposit. There were many regions in the unexposed film that were not protected at all, and severe thinning was observed in these areas. Thus, simply having a high concentration of hydrophobic molecules at the surface of the film is not sufficient to inhibit dissolution uniformly.

We have previously shown *(10)* that 1 μm thick resist images could be produced using three to seven monolayers of PAC over pure novolac, and that even one monolayer of PAC was sufficient to produce some differentiation between the exposed and unexposed regions as long as a thermal treatment prior to exposure was employed. In the present study, full height resist images could be maintained in the films with three monolayers of PAC for bake times of 5-60 minutes (Figure 3). Considerable thinning is seen for the shortest bake time of 5 minutes (Figure 3A) suggesting insufficient surface protection. Significant undercuts are observed in Figure

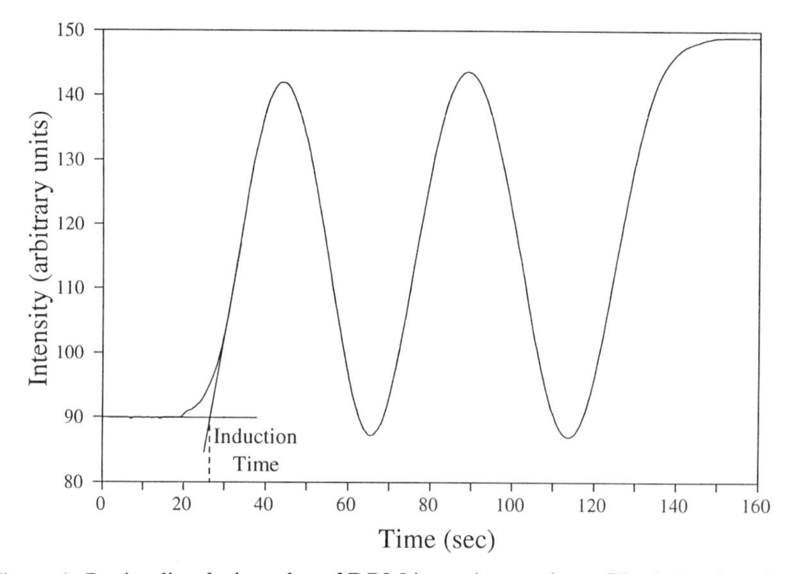

Figure 1. Resist dissolution plot of DRM intensity vs. time. The induction time was defined as the intersection of lines extrapolated from the flat initial portion of the curve and the slope of the curve once dissolution begins.

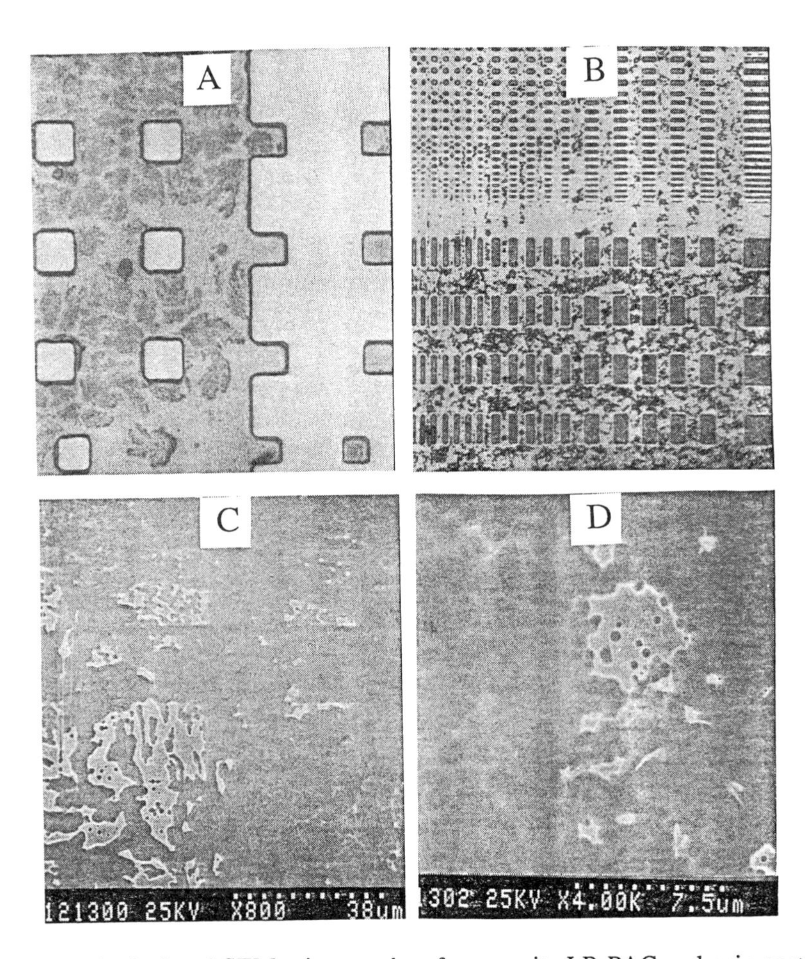

Figure 2. Optical and SEM micrographs of composite LB PAC and spin-cast resist films containing 2.5 wt% PAC after exposure and development. A) three monolayers of PAC on resist; B) seven monolayers of PAC on resist; C) and D) SEM closeups of structure in resist films with seven monolayers of PAC on the surface, after development. Figure 2C is reproduced with permission from reference 10. Copyright 1990 American Institute of Physics.

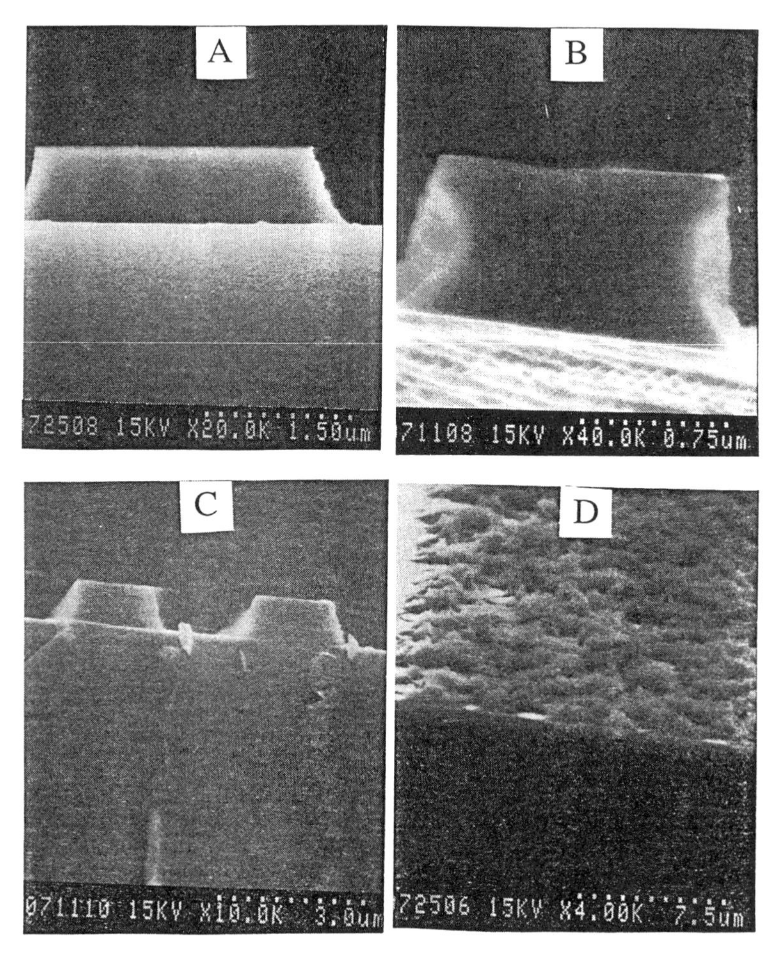

Figure 3. SEM micrographs for composite resist films with three monolayers of PAC on 1 mm of spin-cast novolac, after baking and development. The films were baked at 90°C for A) 5 minutes; B) 20 minutes; C) 60 minutes and D) 5 minutes. Sample D was overdeveloped to observe the uniformity of dissolution. Figure 3B is reproduced with permission from reference 22. Copyright 1990 Society of Photo-Optical Instrumentation Engineers.

3B for the 20 minute bake indicating surface protection over a rather well-defined top zone, with little protection of the 2.5% PAC/novolac lower spin-cast layer. Good protection with no undercut is seen for the 60 minute bake in Figure 3C, suggesting that sufficient PAC diffusion and structural reorganization has taken place to provide dissolution inhibition. We also tried overdeveloping the sample baked for the shortest time (5 minutes) and found that the dissolution was very non-uniform (Figure 3D), unlike that observed for bulk novolac/PAC resists. It is possible that this non-uniform development reflects the lack of optimization of PAC-novolac interactions that might be expected to develop as the PAC diffuses into the novolac.

Non-uniform development was also observed with samples that had large numbers of PAC layers on the surface. For films with seven monolayers of PAC, even a bake of 20 minutes resulted in a non-uniform surface after development (Figure 4A), somewhat similar to what we had observed in the samples where we had deposited seven monolayers onto resist films without thermal treatment (Figure 2B). In Figure 4A, however, all the exposed areas did develop and all the unexposed areas were protected to some degree, indicating that the thermal treatment is of some benefit. We also baked some of these samples for 60 minutes prior to exposure and development. In this case, the tops of the resist images appear to be very smooth and well protected (Figure 4B). Again, it appears that the PAC is only effective at inhibiting dissolution of the novolac when it is in intimate contact with it, such as might be achieved at a sufficiently advanced stage of the thermally-activated diffusion process.

Effect of Annealing and Exposure on UV Spectra. To explore the possibility that there may be specific PAC-PAC interactions in the top surface LB PAC layer, as proposed earlier *(10)*, we measured UV spectra for films of pure PAC, films on a novolac base layer and mixtures of PAC in a novolac matrix as a function of thermal treatment. (*See* Figure 9, page 255.)

We will first examine the influence of the mode of film formation, i.e. spin-cast vs. LB deposition. All films will be compared prior to thermal treatment. The spectra for the spin-cast novolac/PAC films prior to thermal treatment (Figure 5A) are very similar to those found in the literature *(11)*. Prior to exposure, significant peaks are observed with maxima at 340 and 405 nm due to the π-π^* and n-π^* transitions of the DNQ, respectively *(11)*. The strong absorptions observed at 290 nm and below 250 nm are primarily due to the novolac, although absorbances from the DNQ are expected around 260 and 230 nm. Upon exposure, the peaks due to the DNQ bleach, presumably due to the loss of conjugation as the DNQ is converted into the indene carboxylic acid photoproduct. At the highest exposure the DNQ has essentially all decomposed and the spectrum closely resembles that of the pure novolac.

The spectra of the LB films of PAC on pure novolac prior to thermal treatment in Figure 6A are quite different from those of the corresponding spin-cast films of Figure 5A. First, although the peaks around 400 nm are still visible, the peak at 340 nm is difficult to discern, perhaps due to an extensive broadening. Since the 340 band is nominally due to a π-π^* transition, this may indicate π interactions between the PAC molecules. Support for this suggestion comes from the observation that the minimum at 254 nm is also much higher than it was in the case of the spin-cast film. Moreover, there appears to be a significant slope to the baseline with decreasing wavelength.

When the LB film of PAC on novolac was exposed (Figure 6A), there was less bleaching observed of the 400 nm peaks and essentially no reduction of the 340 band and for the minimum at 254 nm compared to the spin-cast films of Figure 5A. Moreover, there appears to be a new band growing in at around 480 nm. Since a diazo-linkage absorbs in the 430-500 nm range *(2,11,12)*, it is possible that the new absorption band is due to formation of a diazo-linked dimer structure. Such a side reaction between two DNQ molecules has been suggested previously *(13)*, and is

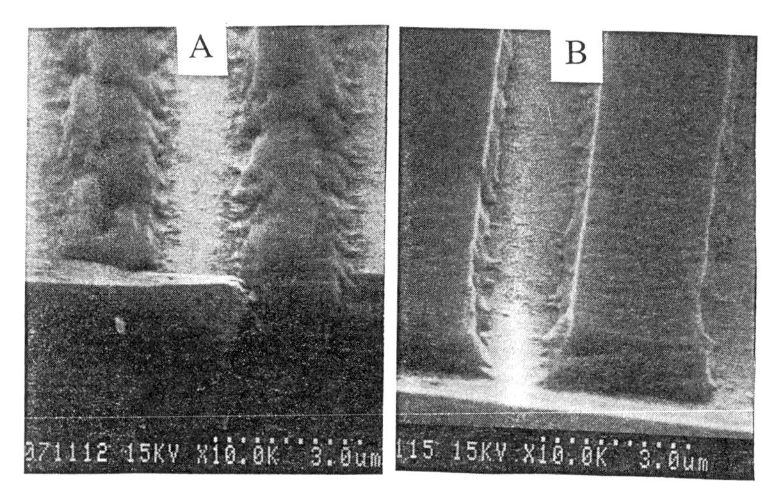

Figure 4. SEM micrographs for composite resist films with seven monolayers of PAC on 1 mm of spin-cast novolac, after baking and development. The films were baked at 90°C for A) 20 minutes and B) 60 minutes.

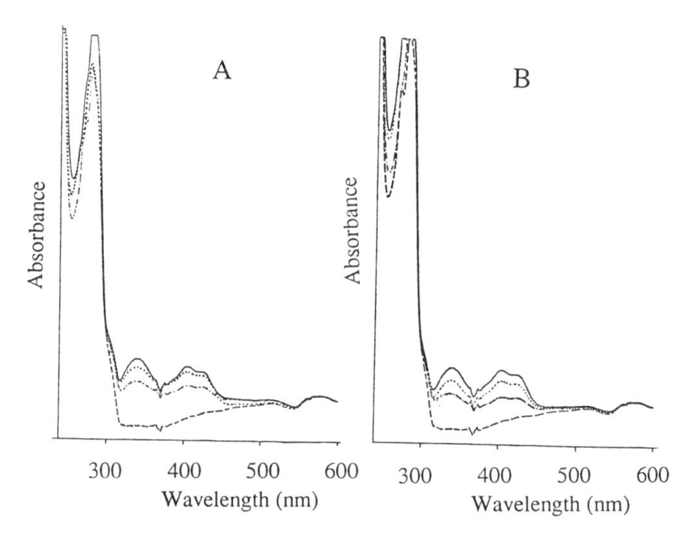

Figure 5. UV spectra for spin-cast films of PAC and novolac. The absorbance spectra are plotted for unexposed films as well as those exposed at settings of 2, 4, and 6 on the Canon stepper for films that A) did not receive any thermal treatment and B) were baked at 90°C for 40 minutes.

shown in Figure 7. Dimerization would certainly be expected to modify the diffusion properties of the PAC. Moreover, it may stabilize the second sensitizer molecule against further degradation, resulting in the observed limited bleaching. In Figure 5A the highest exposure dose leads to the greatest decrease in absorption at 400 nm; yet it also exhibits the largest increase at 480 nm. The first three curves appear to cross over at an isosbestic point.

Next, we consider the effect of thermal treatment on the spin-cast and LB films. First, we note that there is little difference in the spectra of the baked and unbaked spin-cast films in Figures 5A and B, suggesting that there are minimal thermally-induced reactions of the PAC. By contrast, the spectra of the baked LB PAC films on novolac in Figure 6B that were baked are significantly different from the unbaked ones in Figure 6A and are actually quite close to those observed for the annealed spin-cast films in Figure 5B. The thermal treatment appears to have modified the film, possibly due to diffusion of the PAC into the novolac, thus breaking up some of the interactions between PAC molecules.

A second possibility is that the thermal treatment may have degraded the PAC in some way. To distinguish between these two possible effects of temperature, we can examine the spectra for LB films of pure PAC on quartz shown in Figure 8. Here, diffusion of the PAC into a novolac sublayer is no longer a possibility, although disordering of PAC clusters or thermally-induced reactions between PAC molecules can occur. There is a slow bleaching of the 340 and 400 nm peaks, and a new peak very obviously grows in around 480 nm; there are two isosbestic points at about 300 nm and 440 nm. Below 220 nm, the absorption actually increases with exposure dose. This may explain why we observe little or no reduction of the minimum at 254 nm in the spectrum with the LB PAC on top of a novolac underlayer (Figure 6).

The UV absorption spectra show that the LB PAC films have very different absorption characteristics compared to PAC in the spin-cast films. When the PAC is allowed to diffuse into the novolac, however, the LB/spin-cast films eventually exhibit more "normal" absorption spectra. We interpret the differences in the absorption spectra as indicating that adjacent layers of PAC may have specific ground state electronic as well as photochemical interactions. It is certainly possible that these might affect the dissolution characteristics of both the exposed and unexposed regions.

Effect of Annealing on Dissolution Behavior.

Induction Effects. A characteristic feature of the development behavior of novolac/DNQ resists is the delay in the onset of resist dissolution in the unexposed regions. Not only do the unexposed films dissolve more slowly than the exposed films, but they can also exhibit a significant initial suppression of dissolution for short periods of time. There are many factors that have been proposed to have an effect on this phenomenon, including oxidation of the surface of the resin during prebaking *(19)*, differences in molecular weight distribution *(12,20)*, or the concentration of the DNQ.*(21)*

The induction effect was examined primarily with samples having one to two monolayers of PAC deposited on spin-cast novolac films that were subsequently baked for 10-80 minutes at 80-105°C. Samples of pure novolac films were also included to determine whether the bake conditions contributed to the induction effect in the pure resin. The induction time was defined as the break in the trace of light intensity vs. time, as measured on the DRM and illustrated in Figure 1. Representative effects of increasing bake time and temperature are shown in Figures 10A and B. We note that the observed induction period is not due to oxidation or other changes in the resin, as there is essentially no induction effect observed for the pure novolac, even for the longest bake times and highest temperatures.

The data for the induction times for films with one and two monolayers of PAC and various bake conditions are plotted together in Figure 11. The induction times

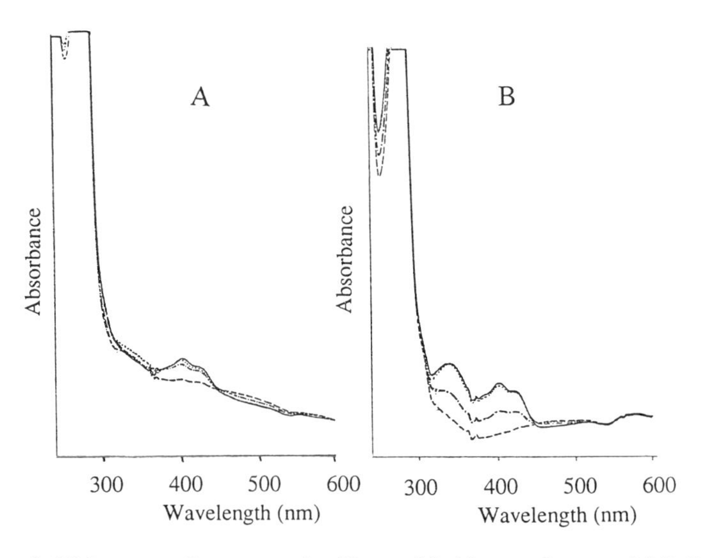

Figure 6. UV spectra for composite films with 19 monolayers of LB PAC on spin-cast novolac for films that A) did not receive any thermal treatment and B) were baked at 90°C for 80 minutes. The wafers were exposed as described in Figure 5.

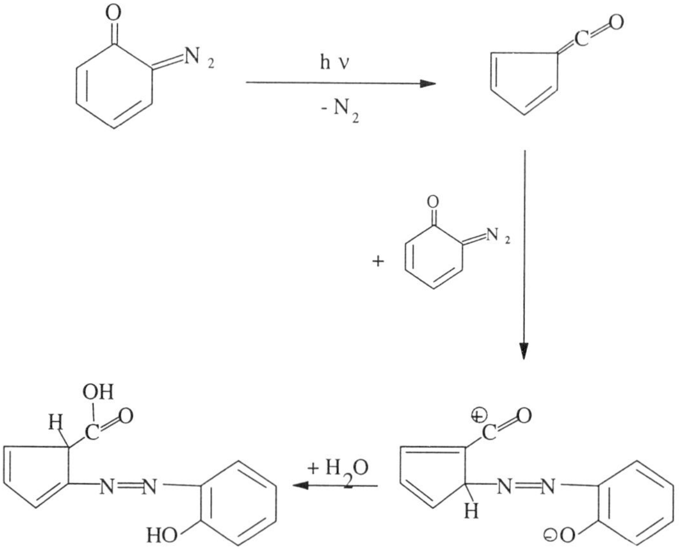

Figure 7. Proposed reaction mechanisml for photochemical reaction of two diazoquinone molecules that leads to formation of a diazo-linkage.

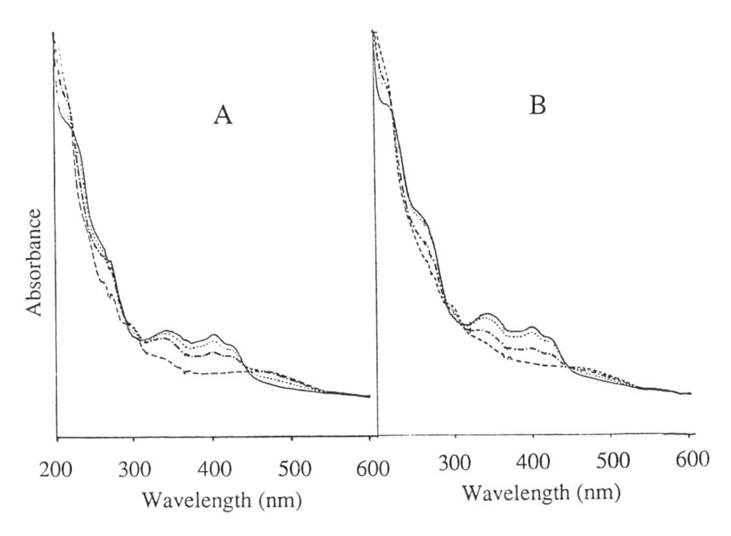

Figure 8. UV spectra for 10 monolayers of LB PAC on quartz for films that A) did not receive any thermal treatment and B) were baked at 90°C for 80 minutes. The wafers were exposed as described in Figure 5.

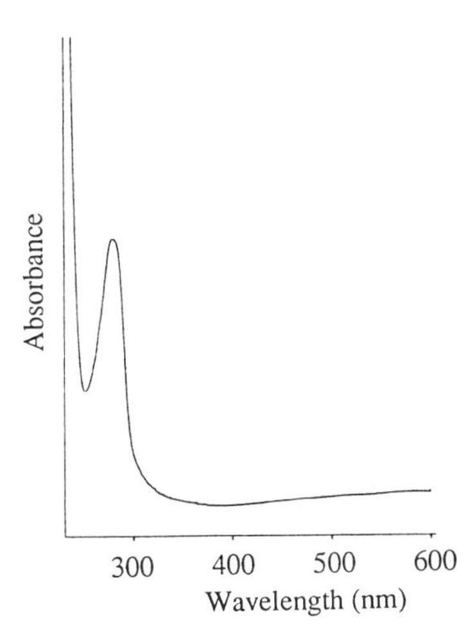

Figure 9. UV spectra of spin-cast novolac film.

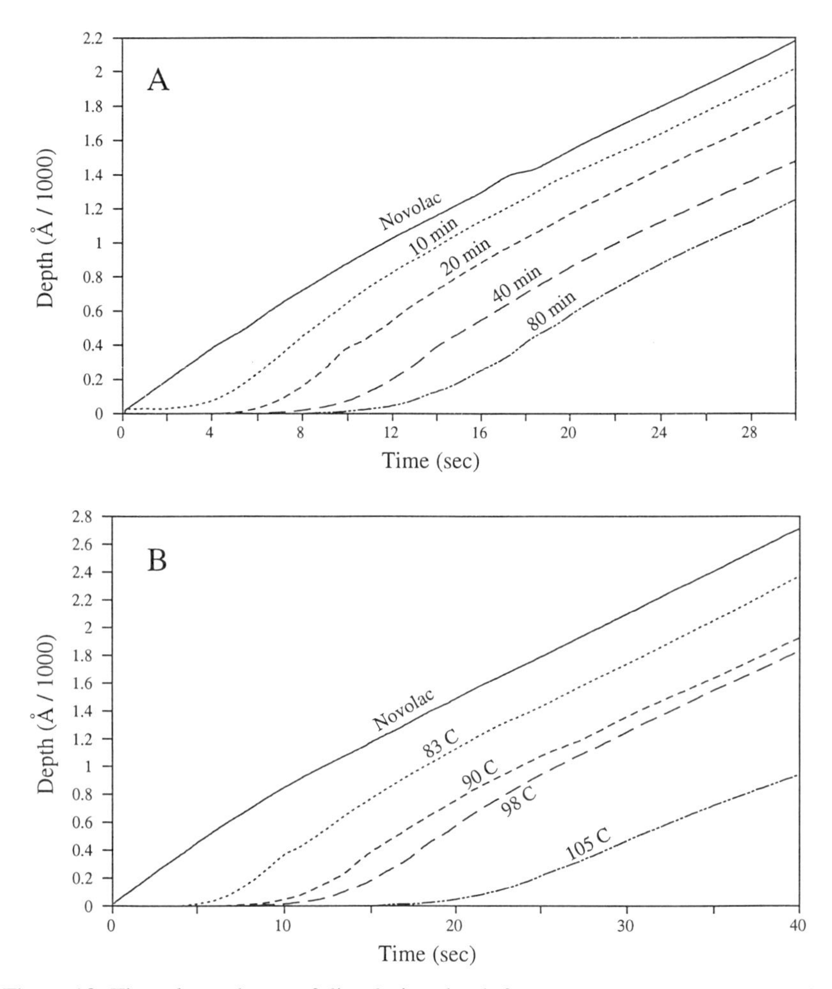

Figure 10. Time dependence of dissolution depth for composite films of LB PAC and spin-cast novolac exposed to the developer. The induction time is the delay before the onset of development. The induction time is affected by A) the time for which the films were baked at 98°C and B) the temperature at which the films were baked for a constant time of 80 minutes. Figure 10 is reproduced with permission from reference 22. Copyright 1990 Society of Photo-Optical Instrumentation Engineers.

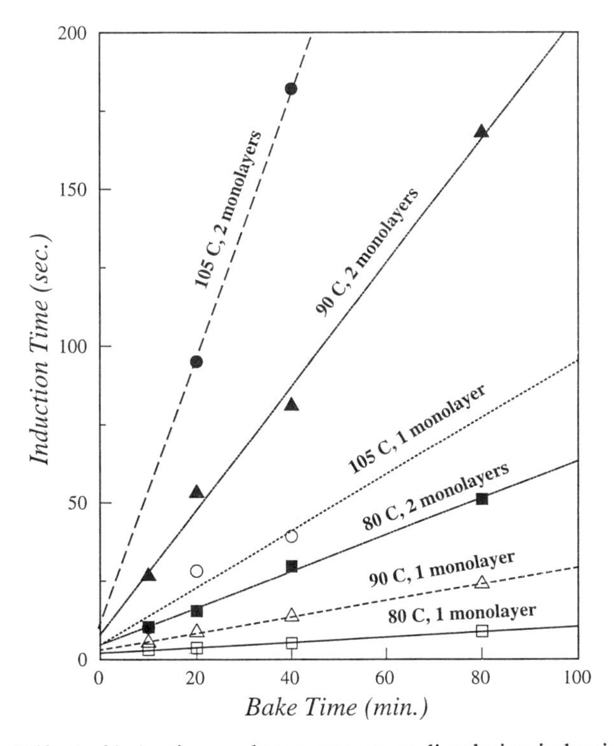

Figure 11. Effect of bake time and temperature on dissolution induction time for films that were baked at various temperatures, and with one or two monolayers of PAC initially deposited on the surface of the film.

appear to depend linearly on bake time at a given bake temperature with the slopes of the lines approximately tripling per 10°C increment in bake temperature. The induction times for samples with two monolayers of PAC are much greater than for those with only one layer. The slopes of the lines for the samples with two monolayers are about five to eight times greater than those for the samples with one monolayer, even though the concentration is only twice as great. The induction time is obviously not a simple function of the surface concentration of the sensitizer.

It is possible, however, that interactions of the PAC with the resin are making it more effective at delaying the onset of dissolution. If we assume that the induction time is related to the formation of a PAC/resin complex, that the concentration of the complex is proportional to the induction time, and that the reaction follows first order kinetics, we can find the rate constants for the reactions from the slopes of the lines in Figure 12. An Arrhenius plot of the rate constants against the inverse reaction temperature is given in Figure 13. The effective activation energies calculated from these lines are about 25 kcal/mol for the films with one PAC monolayer, and 13.5 kcal/mol for the samples with two PAC monolayers.

Dissolution Rate Effects. Most of the mathematical models for resist image formation are based on the assumption that the rate of resist dissolution may be related to the local concentration of unexposed PAC in the film *(15,16*. In general, a relationship can be established between the dissolution rate and the PAC concentration, as demonstrated in the plot of bulk development rate with time in Figure 14 for spin-cast films containing 0.1-2.0 wt% PAC. We have extrapolated these data to dissolution rates less than 20 Angstroms/sec and have attempted to use this relationship to approximate the concentration of PAC at various depths in the composite LB PAC/spin-cast novolac films after thermal treatment. Initially, the PAC is concentrated at the surface of the film, but it should diffuse into the film during baking to create a concentration gradient. The rate of development over this diffusion zone should be affected by the local concentration of PAC such that the rate will increase with decreasing PAC concentration until it reaches the dissolution rate of pure novolac. If the calibration curve established for the spin-cast films is applicable to our annealed composite films, we can then estimate the "effective" PAC concentration with depth into the film.

Various problems occurred in attempting these experiments, however. For films with more than two monolayers of PAC, the DRM trace of light intensity vs. time during development was very noisy and unusable. It appears that the surface of these films did not develop smoothly, but rather pitted and roughened significantly prior to achieving a uniform development rate. SEM micrographs of this effect are shown in Figures 3B and 4A. This may be related to the occurrence of local interactions of the PAC molecules that we have suggested previously. Development traces from these films could not be analyzed by Parmex. Even for the films with only one or two layers of PAC, the traces were occasionally poor, especially for the higher bake temperatures. As a result, it was difficult to get reproducible results.

More problematic, however, are the trends observed in the experiments for which the data were reproducible. A set of curves generated by Parmex for dissolution rate as a function of depth for films with one monolayer of PAC is shown in Figure 15. These films were baked at 98°C for 10 to 80 minutes. As expected, the dissolution rate is lower at a given depth for the films baked for longer times. This supports the idea that the PAC is diffusing into the film and retarding the dissolution rate due to increased local PAC concentrations. Using Figure 14 as a calibration curve for dissolution rate with respect to PAC concentration, we can examine the curves in Figure 15 to find the "effective" PAC concentration with depth. Such curves may be found in Figure 16.

Although all of the curves of Figure 16 suggest that the PAC concentration decreases with increasing depth, in agreement with our general intuition, there are several puzzling features. For example, the shapes of the curves do not follow the

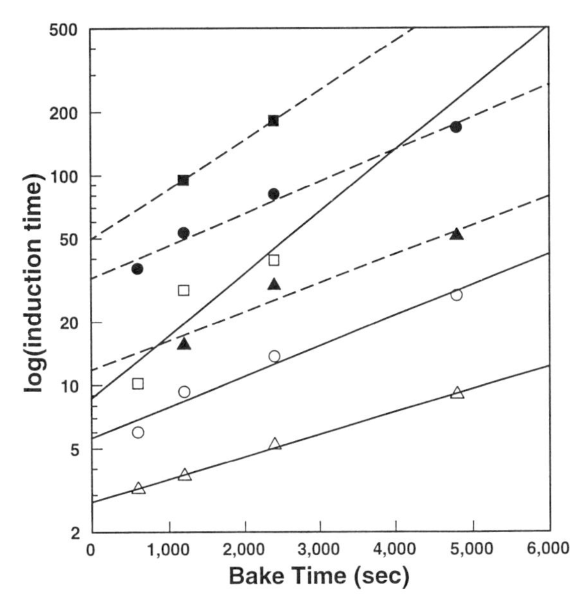

Figure 12. Plot of log (induction time) vs. bake time. The data are from films with one monolayer (solid lines, open symbols) and two monolayers (dashed lines, solid symbols). The data are for films baked at 80°C (▲); 90°C (●) and 105°C. (■)

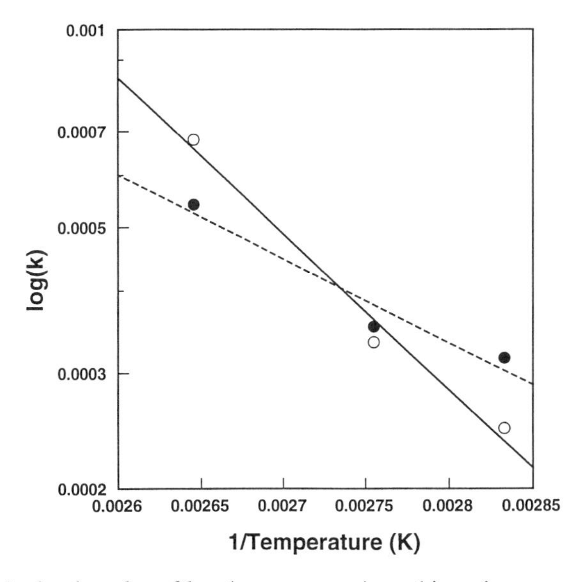

Figure 13. Arrhenius plot of log (rate constant) vs. 1/reaction temperature. The rate constants were calculated from the induction times for various bake times. The data are for films with one monolayer (solid line, open circles), and two monolayers (dashed line, solid circles) of PAC on novolac.

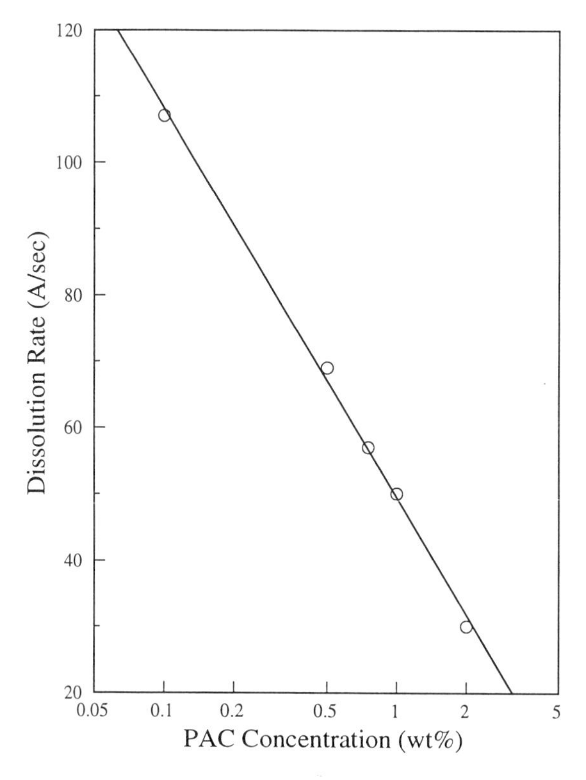

Figure 14 . Effect of PAC concentration on bulk development rate of spin-cast films.

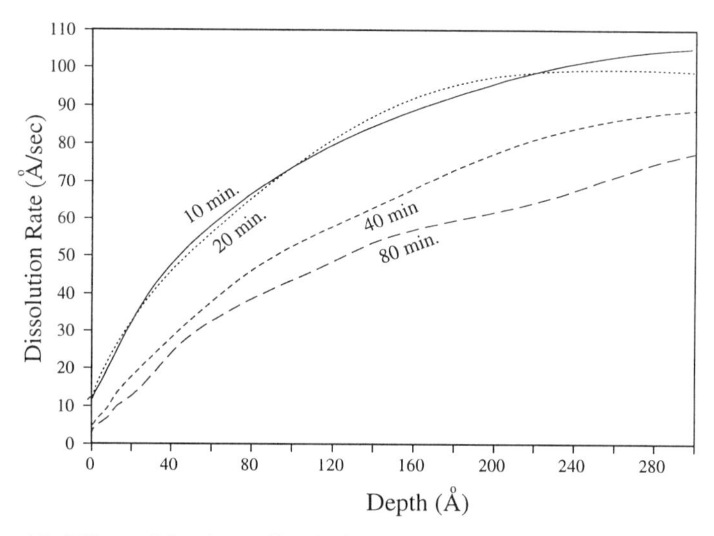

Figure 15. Effect of depth on dissolution rate for films with one monolayer of PAC-1 on spin-cast novolac. The films were baked at 98°C for 10-80 minutes (as specified on curves) prior to development in 1:2.5 Microposit 312:water.

Gaussian form expected for diffusion from an instantaneous plane source *(18)*. Not only does the effective PAC concentration drop much too abruptly for shallow depths, but the apparent zero depth concentration actually increases with the bake time. A related observation is that the area under the curves of Figure 16 increases with bake time, suggesting that there is an increase in the amount of PAC. Of course, this is unreasonable. A possible explanation for these results is that the calibration curve developed for the spin-cast films is not applicable for characterizing the early stages of the diffusion process in the composite films. We will consider this further in the Discussion.

Due to the potential uncertainties in estimating the local PAC concentration for these films, we have compared the various bake conditions simply by comparing the depth at which the film achieves a given dissolution rate. We assume that the dissolution rate is still dependent on the local concentration of some species of PAC; hence, a comparison of a given rate under different conditions is still valid. The values recorded in Table 1 are the depths into the film at which a development rate of 50 Å/sec was achieved. The films reach this development rate at about 50-400 Å into the film for samples with one monolayer of PAC, and roughly 200-500 Å for the films with two monolayers, although we could not get measurements for the films baked at the highest temperatures. The depth is approximately proportional to the square root of time for several sets of data, as would be expected for a diffusion-controlled process. However, there are several inconsistencies in the results, making any attempt at a quantitative analysis of limited utility.

Discussion

The simplest models of resist dissolution, such as those developed by Dill *(15,16)* assume that the dissolution rate of a resist film is correlated with the concentration of unexposed inhibitor. While we have found this to be true for bulk spin-cast films, the situation is unclear for our composite films. The dissolution rates of the top 1000 Å of films with one to two monolayers of PAC did vary with prebake conditions, as would be expected if the PAC diffused into the film leading to a concentration gradient with depth. However, these qualitative trends did not lend themselves to any simple quantitative analysis. For example, the dissolution rate decreased at the surface for longer bake times, which is not expected as a result of diffusion. Either our calibration procedure is flawed, or simple diffusion of the PAC is not sufficient to explain the dissolution inhibition phenomena we have observed.

The question of the validity of the calibration procedure has two parts. The first, more mechanical, aspect is that our assumption that the data of Figure 14 may be extrapolated to lower dissolution rates may not be valid. It is, of course, in this low dissolution rate region that the information about the early stages of the process is obtained. We intend to explore this point further with our new quartz crystal microbalance dissolution rate instrument.

The second aspect of the calibration procedure is of more fundamental significance. This is the issue of the importance of hydrogen bonding, which is expected to be extensive in the spin-cast novolac matrix. In the normal resist deposition process, the novolac and PAC are dissolved in a solvent that will support hydrogen bonding. We would expect that a complex combination of all of the possible binary interactions would exist in the resist solution, including solvent-novolac, solvent-PAC, PAC-novolac, PAC-PAC, novolac-novolac and solvent-solvent. These must be altered as the solvent evaporates during the initial spin-casting and the subsequent baking. Although we are unaware of literature that attempts to separate all of these possible interactions, it seems likely that the final PAC/novolac film represents some form of optimization of possible intermolecular interactions. It seems quite probable that the same state cannot be achieved for our composite LB PAC/spin-cast novolac films at early stages of the diffusion process. This problem will certainly be exacerbated by the

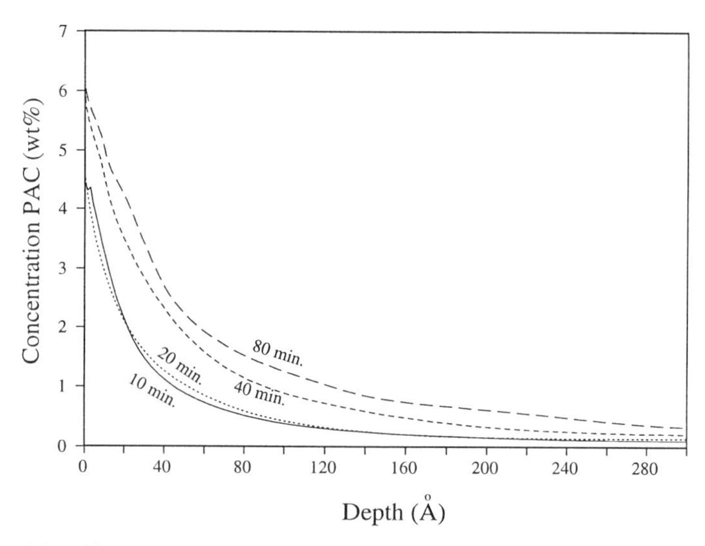

Figure 16. Effective PAC concentration as a function of depth for the films described in Figure 15.

Table 1. Depth (Å) into film at which a dissolution rate of 50 Å/sec was attained for various bake conditions

		Bake Temperature (°C)			
	Bake Time (min.)	80	90	98	105
One monolayer films	10	69	60	76	-
	20	82	63	52	-
	40	90	125	142	307
	80	-	101	188	380
Two monolayer films	10	221	349	284	-
	20	260	285	-	-
	40	492	-	340	-
	80	-	-	416	-

fact that the PAC diffusion is being performed at temperatures well below the glass transition temperature of the novolac matrix.

A second possible explanation would be if some of the PAC were being immobilized at the surface due to chemical reactions between PAC molecules or to strong interaction with the resin, which would inhibit its further diffusion. This would lead to a higher concentration of PAC at the surface than would be predicted by diffusion and might produce apparent changes in the diffusion rate with depth.

Similar problems occur in verifying the theories that dissolution inhibition is due to the formation of a "protective layer" of hydrophobic PAC molecules on the surface of the resist as the more soluble novolac gets leached out. Composite films were produced that had a pure layer of PAC deposited on the surface which, according to Murata *(3)* should result in the highest level of dissolution inhibition. This was not found to be the case in our experiments. The composite films that did not receive any thermal treatments to mix and/or react the PAC with the resin exhibited minimal inhibition to dissolution or induction effects, and the inhibition that did occur was patchy and inconsistent. Thermal treatments were required to achieve more uniform and effective dissolution inhibition.

Another result that indicates that inhibition is not solely dependent on PAC concentration is that the induction time increases with bake temperature and time, while the surface concentration of PAC should be decreasing due to diffusion. Even if some of the PAC is being immobilized due to interactions with the resin, the free PAC should still diffuse away from the surface, and the surface concentration must be lower than it was initially. Many of the theories about dissolution inhibition revolve around the accessibility of the phenolic hydrogen for removal by a basic anion *(2,4-7)*. Both the hydrophobic *(2,3,6,7,14)* and the hydrophilic *(4,5)* character of the sensitizer have been proposed as the important features that allow it to interact with the novolac and to inhibit contact of the phenolic group with the developer.

At this time, our experiments most strongly support the hypothesis of Honda, et al. *(4)*, which proposes the existence of "macromolecular complexes" between the PAC and the novolac based on hydrogen bonding between the two, or the formation of actual chemical bonds between the PAC and the resin. The activation energies calculated from Figure 16 are larger than would be expected from a single hydrogen bond. They could be due to actual covalent bonding of the PAC onto the novolac through a diazo-linkage, or formation of multiple hydrogen bonds, as proposed by Honda. Hydrophobic interactions seem an unlikely explanation for the activation energies we have observed, however, due to the small energies of the interactions.

If the interactions are hydrophilic in nature and the PAC molecules preferentially interact with the phenolic OH, they will block or directly compete with the developer for interaction with these sites. Honda's proposal that a single PAC molecule may interact with as many as 18 phenolic groups would also help explain how the very low concentration of PAC that we are using is capable of the significant inhibition effects we have observed. The differences between the calculated activation energies for the one and two monolayer samples might arise from multiple hydrogen bonds. It is possible that for films with higher PAC concentrations, the PAC may be diffusing into sites that involve fewer hydrogen bonds per molecule. The high concentration of PAC molecules at the surface may lead to a sufficiently large percentage of the phenolic groups close to the surface being involved in interactions with PAC molecules. In the films with one monolayer, the PAC molecules may have to form more extensive hydrogen bonded complexes to sufficiently impact the induction time. Thus, the increase in inhibition with increasing bake time and temperature may be explained by a large number of PAC molecules diffusing to and interacting with effective inhibition sites on the polymer.

Several researchers have proposed that inhibition by PAC molecules is at least in part due to the reaction of the PAC molecules with the resin *(2,8,9,14)*. Reaction of the PAC with the novolac would probably exhibit similar phenomena to that which would be caused by simple physical interaction between the two components. The extent of these reactions would also be expected to increase with increasing bake time and temperature. It may be difficult to distinguish between these two phenomena. The UV spectra that we have obtained suggest that there are interactions between unexposed PAC molecules in LB layers. However, these interactions are essentially broken up by diffusing the molecules into the novolac, suggesting that no chemical reactions have occurred. The spectra also indicate the possible formation of new photoproducts upon exposure of the LB films of PAC. We cannot, at this point, differentiate between the contributions of chemical interactions or chemical reactions in dissolution inhibition mechanisms.

The observation of significant induction times for the composite LB/novolac films indicates that these films may be useful lithographically, if sufficient contrast can be achieved between the dissolution of the exposed and unexposed films. These films would have several potential lithographic advantages. From our data on development rate, it appears that the sensitizer is primarily concentrated in the top 100-1000 Å of the film. This would allow for true top surface imaging, which is important because it reduces the required depth of focus of exposure systems. The depth of focus decreases with increasing numerical aperture of the lens *(1)*, and high numerical aperture lenses are required to achieve submicron resolution. LB composite films would require very little depth of focus to expose all of the sensitizer. In some applications, such as those requiring lift-off structures, or accurate reproduction of the top surface image during very directional dry etch transfer steps, it is an advantage to produce an undercut image profile. This could be accomplished by controlling the concentration of the sensitizer in various regions of the film, and matching the dissolution behavior of the resin to the PAC concentration in the surface and bulk regions of the film. By using a 0.5-1 μm spin-cast film as the underlayer of the resist, the concern with pinholes in very thin spin-cast films is also lessened.

A thin imaging layer is also of value in reducing the problems of self-absorption in the film by the novolac matrix. Even for exposures in the deep-UV, exposure of the sensitizer should not be a problem if it is all contained in the surface layer of the resist. This permits the use of standard thickness resist images (0.5-1 μm) that are important for dry etch resistance, while still maintaining the option of using highly absorbing resins such as novolac. In fact, the thin imaging layer and absorbing underlayer would help reduce the problems of standing waves and reflection off underlying structures. Furthermore, it might be possible to dye the resin to absorb the reflected light in near or mid-UV exposures. Another advantage of localization of the sensitizer is the potential for tailoring of the resist sidewalls.

Summary

The mechanism by which PAC inhibits dissolution of novolac resins is very poorly understood. We have explored this process through the use of techniques that allow the spatial control of PAC and novolac within resist films. The UV absorption spectra indicate that altered ground states occur in the unbaked LB films of PAC. Such PAC molecules deposited on the surface of the film appear to do little to inhibit novolac dissolution. However, thermal treatment, which causes the PAC to diffuse into the underlying novolac layer, results in spectra for the exposed and unexposed PAC that are similar to those found in spin-cast films. Moreover, annealed composite films exhibit increased resistance to dissolution

The induction time increased with increasing prebake time and temperature, even though these should decrease the surface concentration of PAC. It is possible that the thermal treatment allows the PAC molecules to diffuse to sites where they can most efficiently interact with the novolac resin. The increase in induction time with bake temperature is on the order of that expected for the rates of chemical reactions.

Activation energies of about 25 and 13.5 Kcal/mol were estimated for films with one and two monolayers of PAC, respectively. Whether the interaction between the sensitizer and the novolac involves polar interactions such as hydrogen bonding, or actual covalent bonding such as in the formation of diazo-linkages will require further study.

Acknowledgments

We would like to thank W. Oldham and A. Neureuther for allowing us access to the Microlab at Berkeley, and accounts on their computers. This work was supported by SRC through the Sematech Berkeley-Stanford Center of Excellence on Lithography and Pattern Transfer.

Literature Cited

1. Moreau, W.M. *Semiconductor Lithography: Principles, Practices, and Materials*, Plenum Press: New York, NY, 1988.
2. Koshiba, M.; Murata, M.; Harita, Y.; Yamaoka, T. *Polym. Eng. Sci.*, **1989**, *29*, 916.
3. Murata, M.; Koshiba, M.; Harita, Y. *Proc. SPIE*, **1989**, *1086*, 48.
4. Honda, K.; Beauchemin, Jr., B.; Hurditch, R.; Blakeney, A.; Kawabe, Y.; Kokubo, T. *Proc. SPIE*, **1990**, *1262*, 493.
5. Huang, J-P ; Kwei, T.K.; Reiser, A. *Proc. SPIE*, **1989**, *1086*, 74.
6. Templeton, M.K.; Szmanda, C.R.; Zampini, A. *Proc. SPIE*, **1987**, *771*, 136.
7. Garza, C.M.; Szmanda, C.R.; Fischer, Jr., R.L. *Proc. SPIE*, **1988**, *920*, 321.
8. Hanabata, M.; Uetani, Y.; Furuta, A. *Proc. SPIE*, **1988**, *920*, 349.
9. Hanabata, M.; Furuta, A.; Uemura, Y. *Proc. SPIE*, **1987**, *771*, 85.
10. Kosbar, L.L.; Frank, C.W.; Pease, R.F.W. *J. Vac. Sci. Technol. B*, **1990**; *8*, 1441.
11. Ershov, V.; Nikiforov, G.; de Jonge, C. *Quinone Diazides*, Elsevier, Amsterdam, 1981.
12. *Absorption Spectra in the Ultraviolet and Visible Region*, Lang, L., Ed.; Krieger: Huntington, NY, 1977; 21, 87; 1978; 22; 119; 1979; 23; 101.
13. U.S. Patent 3,639,184 IBM, 1972.
14. Koshiba, M.; Murata, M.; Matsui, M.; Harita, Y. *Proc. SPIE*, **1988**, *920*, 364.
15. Dill, F.H.; Hornberger, W.P.; Hauge, P.S.; Shaw, J.M. *IEEE Trans. on Electron Devices*, **1975**, *ED-22*, 445.
16. F.H. Dill, *IEEE Trans. on Electron Devices*, **1975**, *ED-22*, 478.
17. Kosbar, L.L. , *P.h.D. Thesis*, Stanford University: Stanford, CA, 1990.
18. Crank, J., *The Mathematics of Diffusion*; Oxford University Press: London, 1956, p.11.
19. Dill, F.H.; Shaw, J.M., *IBM J. Res. Dev.*, **1977**, *21*, 210.
20. Halverson, R.; MacIntyre, W.; Motsiff, W. *IBM J. Res. Dev.*, **1981**; *26*, 590.
21. Kaplan, M.; Meyerhofer, D. *RCA Rev.*, **1979**, *40*; 166; *Polym. Eng. Sci.*, **1980**, *20*, 1973.
22. Kosbar, L.L.; Frank, C.W.; Pease, R.F.W., *Proc. SPIE*, **1990**, *1262*, 585.

RECEIVED January 7, 1993

Chapter 17

New Photoresist Processes at UV Wavelengths Less Than 200 Nanometers

D. J. Ehrlich, R. R. Kunz, M. A. Hartney, M. W. Horn, and J. Melngailis

Lincoln Laboratory, Massachusetts Institute of Technology, Lexington, MA 02173–9108

Over the past two years rapid progress has been made in the design of high-numerical-aperture imaging systems for lithography using radiation at the edge of the vacuum ultraviolet. These systems soon will be prototyped for VLSI manufacturing applications. Because of the combination of excess optical absorption in polymers and the small depth-of-focus of the new optical systems, many traditional photoresist principles are no longer useful at these short UV wavelengths. Here, we summarize the technology issues and several recent approaches to developing new photoresist materials and new chemical mechanisms for imaging at these wavelengths. The four areas of research outlined are use of silylation processes in surface imaging, bilayer resist schemes as alternatives to single-layer resists, development of an adlayer dosing technique based on molecular layer reactions, and use of dry planarization layer and imaging layer deposition to enable an all-dry 193-nm photolithography sequence. More details on much of the work described can be found in the references cited.

Silylation Processes in Surface Imaging

High resist absorption at 193 nm makes conventional processing unrealistic for aromatic-based resists, and alternative approaches must be pursued. In surface imaging lithography, first described by Taylor et al. (*1*) nearly ten years ago, a pattern is defined in surface or near-surface regions of a resist rather than throughout the resist thickness. At longer wavelengths, surface imaging can eliminate problems such as reflectivity variations due to different substrates or topographies and can result in improved process latitude (*2*). For 193-nm lithography, however, surface imaging is an enabling technology rather than a technique for process or yield enhancement.

Of several surface imaging schemes, silylation processes have been the most extensively developed and are now being used in advanced processing lines (*3*). A schematic of a 193-nm positive-tone silylation process is shown in Figure 1. First, exposed areas of the resist are crosslinked using excimer laser irradiation. The

0097–6156/93/0527–0266$06.00/0

material is then placed in a silicon-containing gas, which diffuses into and reacts selectively with the uncrosslinked portions of the film. The selectivity for silylation is determined by the relative diffusion rates in the exposed and unexposed areas. Since the silylating agent diffuses through the top surface of the resist, only this portion of the film needs to be exposed and crosslinked to prevent silylation. Finally, the latent silylated image is developed in an oxygen reactive ion etching (RIE) plasma, which selectively removes unsilylated portions of the film. Areas in which the silylating agent has reacted with the film are converted to SiO_2 and are not etched in the plasma.

One class of acid-catalyzed resists has been designed to crosslink and would behave in the manner depicted in Figure 1. These materials were developed initially by Rohm and Haas Co. (*4*) and are commercially available from Shipley (SAL 601 and SNR 248). The mechanistic chemistry of the resists is well documented (*5,6*).

More surprisingly, conventional novolac/diazoquinone photoresists or even novolac or polyvinylphenol without photoactive compounds also crosslink efficiently when exposed to 193-nm radiation (*7*). The mechanism for this crosslinking is believed to be a free-radical reaction initiated by the homolytic cleavage of the methylene bridge between phenolic groups (*8*). When conventional resists are exposed to radiation under vacuum, as in electron beam lithography, some crosslinking can occur, which is attributed to ester formation between the ketene formed from the diazoquinone decomposition and the hydroxyl groups of the novolac polymer (*9*). However, with the novolac resins without photoactive compounds the mechanism clearly is not ester formation since the crosslinking still occurs with 193-nm irradiation. Additionally, the amount of crosslinking is comparable whether the exposure is carried out in air, under vacuum, or in inert atmospheres such as argon.

We have found (*10*) that the addition of a photoactive compound hinders the crosslinking of the novolac polymer. Exposure response curves for a series of meta-/para-cresol novolacs with varying photoactive compound loading are shown in Figure 2. Clearly, the addition of photoactive compound reduces the sensitivity of the resist. There are two possible mechanisms underlying this effect, both of which assume that the major pathway to silylation selectivity is crosslinking of the resist film upon exposure. In one of the mechanisms, the diazoquinone absorbs some of the incident radiation and is converted to carboxylic acid, which does not result in formation of a crosslink. In the other mechanism, the diazoquinone acts as a scavenger for the free radicals generated during the crosslinking or undergoes a side reaction with the resin, reducing the number of crosslinks formed. The excess dose required (relative to the dose for pure resin) as a function of photoactive compound loading is shown in Figure 3, where a linear relationship is observed. Since the conversion of a photoactive compound to carboxylic acid is linear, this conversion is likely the primary explanation for the decreasing sensitivity of the resist; Fourier-transform infrared spectroscopy of the exposed films shows evidence for the formation of carboxylic acid. However, the linear decrease in sensitivity does not preclude a first-order side reaction between the photoactive compound and resin.

In addition to decreasing the sensitivity of the resist, the presence of photoactive compound reduces the diffusion rate of the silylating agent, in this case dimethylsilyldimethylamine (DMSDMA). Rutherford backscattering data for a series of resists silylated at 100 °C for 2 min at 10 Torr are shown in Figure 4. Clearly, even the addition of a small amount of photoactive compound dramatically decreases the diffusion rate. The DMSDMA diffuses to a depth of 800 nm in pure resin, while addition of only 10 wt % of photoactive compound confines the diffusion to 230 nm. Additional increases in photoactive compound loading further reduce

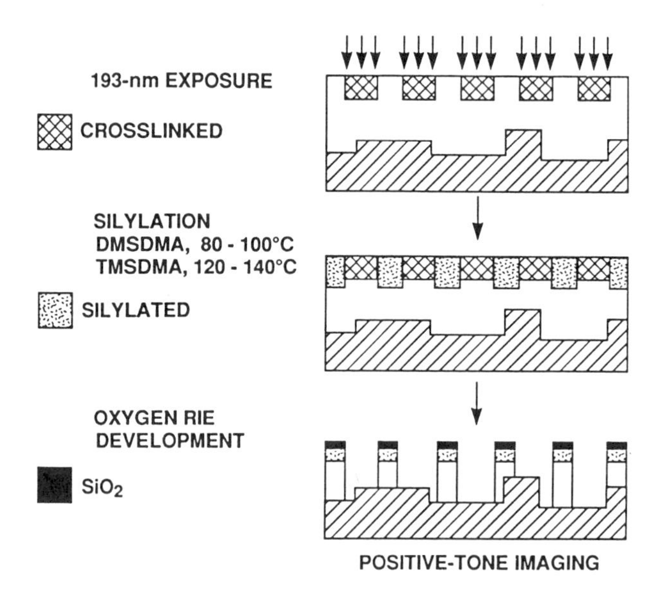

Figure 1. Schematic process flow for 193-nm silylation.

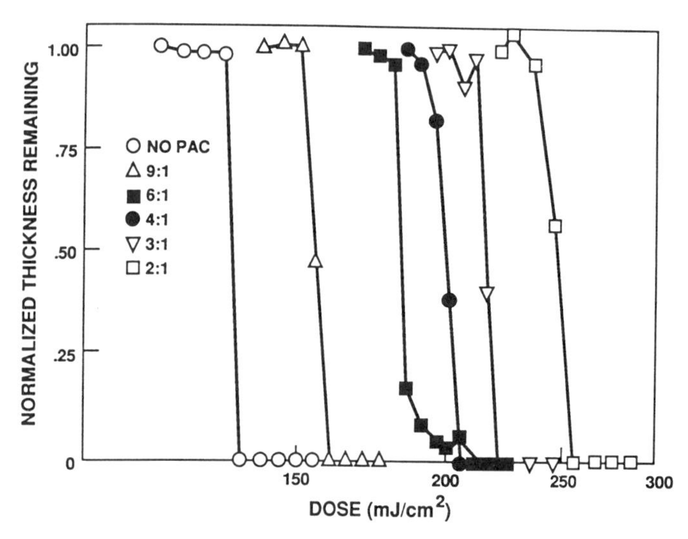

Figure 2. Lithographic response curves for resists prepared from pure resin and from 9:1, 6:1, 4:1, 3:1, and 2:1 blends of meta-/para-cresol novolacs and 2,1,4,-diazonapthoquinone photoactive compound.

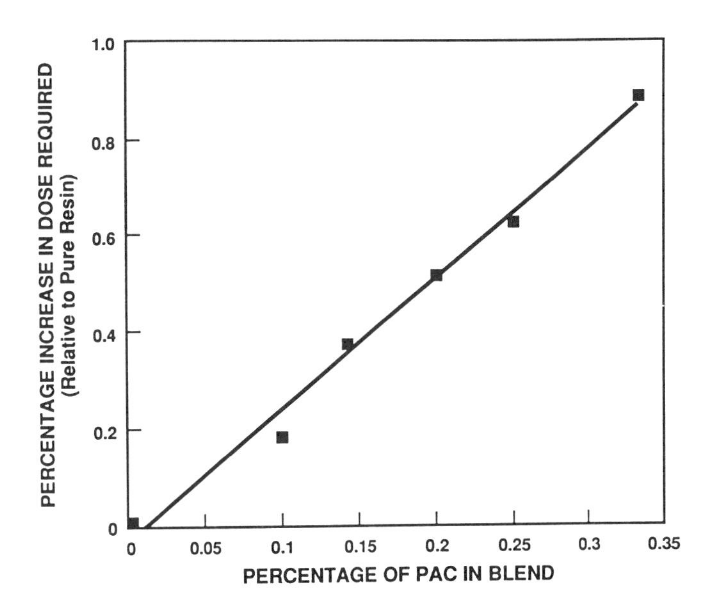

Figure 3. Increase in exposure dose required (relative to the dose for pure resin) as a function of weight % of photoactive compound. The solid line represents a linear best fit to the data.

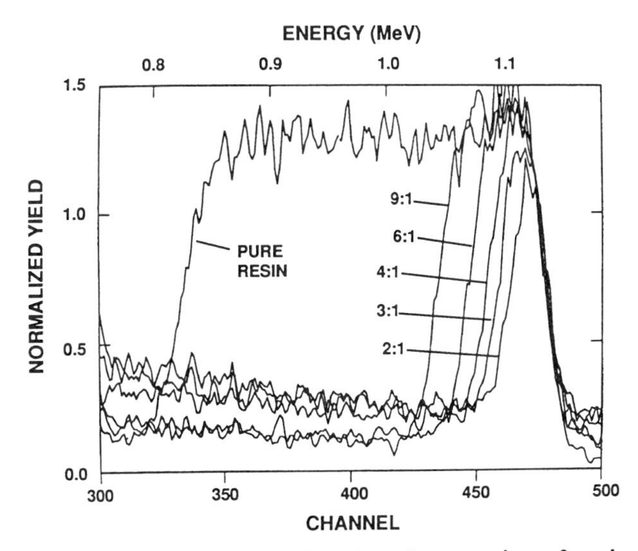

Figure 4. Rutherford backscattering data for a series of resists with photoactive compound. The films were silylated at 100 °C for 2 min with 10-Torr DMSDMA. One channel width on the x-axis corresponds to ~ 2.5 nm of silylation depth.

diffusion rate, with only 80 nm of silylation at 33 wt % loading. Presumably, this decrease in diffusion rate is due to hydrogen bonding between the novolac and the diazoquinone groups.

A scanning electron micograph (SEM) of 0.2-μm lines and spaces patterned in polyvinylphenol and projection printed with a 193-nm excimer laser is shown in Figure 5. The film was silylated at 140 °C for 1 min with 10-Torr trimethylsilyldimethylamine (TMSDMA). The pattern was then developed in an oxygen RIE plasma at 20 mTorr and –200-V bias.

Dry-Developed Bilayer Resists

Bilayer resists have been developed as alternatives to single-layer resists at 248-nm wavelength to preclude the need for antireflective coatings and to improve plasma-process etch resistance. To fulfill the etch-resistance requirements, many of these resists contain silicon. Various formulations have proposed the use of polysilanes for 248 nm (*12*). These materials have high silicon content (> 30 wt %) for etch resistance and are very photolabile in the deep UV. The three-dimensional polysilanes, or polysilynes, undergo photodegradation and rapid oxidation when irradiated at 193 nm (*13*). Since the photooxidized latent image is limited to within ~30 nm of the surface because of polymer opacity, an anisotropic development step is desirable. Although a 193-nm wet development process using these materials has been demonstrated (*14*), a surface imaging dry development process does not limit the resist thickness and can more easily be adapted to a multichamber process environment.

Auger electron spectra (AES) of the silicon LMM transition for poly(cyclohexylsilyne-co-*n*-butylsilyne) exposed to various laser doses at 193 nm are illustrated in Figure 6. The peak locations for elemental silicon and SiO_2 are designated on the figure; note the shift toward lower kinetic energies with increasing exposure dose, indicative of the oxidation of the polymer. This oxidation also occurs at wavelengths > 200 nm but is most efficient and complete at the shorter wavelengths. The atomic composition of poly(*n*-butylsilyne) exposed to 100 mJ/cm^2 was measured by AES to be $Si_{1.0}O_{1.34}C_{1.99}$. Note the surprisingly low C/Si ratio (the C/Si ratio is > 4 in the original polymer), resulting from alkyl-group photodesorption during exposure (*15*). Infrared spectroscopic studies have also been performed (*15*) and indicate that the structure of the photolyzed product is a three-dimensional siloxane, intermediate in properties between a true polysiloxane and SiO_2. This oxidized product results from oxygen insertion into each Si-Si bond, so that each silicon is bound to three oxygen atoms.

The dry-developed resist process relies on the selective etching of unexposed polysilyne, and various plasma chemistries have been studied for selective dry etching of Si over SiO_2 (*16–18*). Chlorine is a good choice for the Si/SiO_2 system (*15*), as atomic chlorine inserts into Si-Si bonds but is too weak an oxidant to attack the stronger Si-O bonds. However, using chlorine, we have observed poor selectivity for polysilynes, presumably because of the presence of the weaker Si-alkyl bonds in the exposed as well as the unexposed areas. Therefore, a weaker oxidant such as bromine is better suited to optimize etch selectivity between photooxidized and virgin polysilyne films.

The actual chemistry leading to etch selectivity in polysilynes is more complex than the case of Si/SiO_2, however. Nakamura et al. (*18*) have observed that poly-Si/SiO_2 etch selectivity can vary by a factor of 300, depending on the level of carbon in the plasma reactor. The poorest selectivity was between 10:1 and 20:1 and occurred when the carbon level in the plasma was highest (i.e., with a hydrocarbon resist masking layer). For polysilyne etching, we have found that both the amount

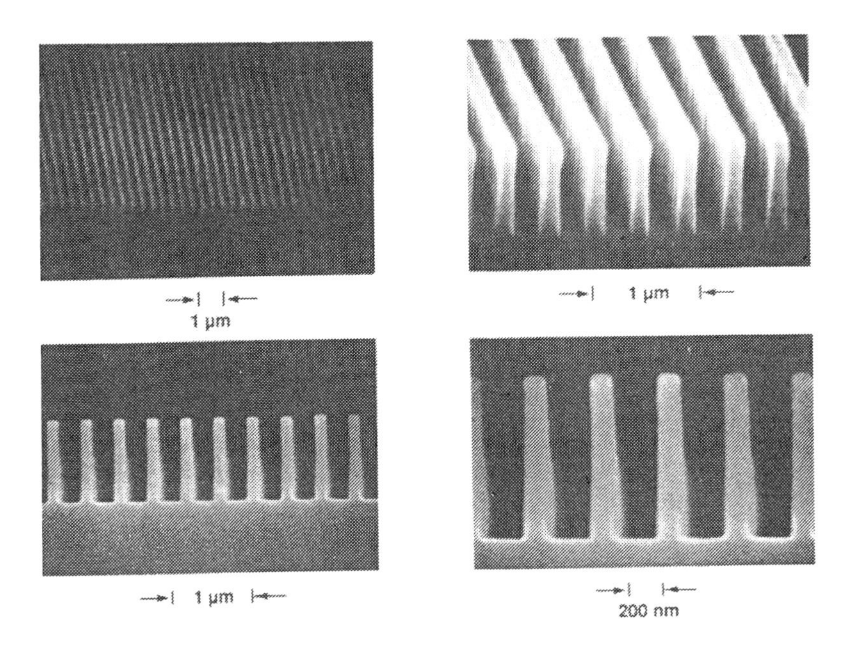

Figure 5. SEM of a 0.2-μm line-and-space pattern in high-molecular-weight polyvinylphenol exposed to 60 mJ/cm^2 and silylated at 140 °C for 1 min with 10 Torr TMSDMA.

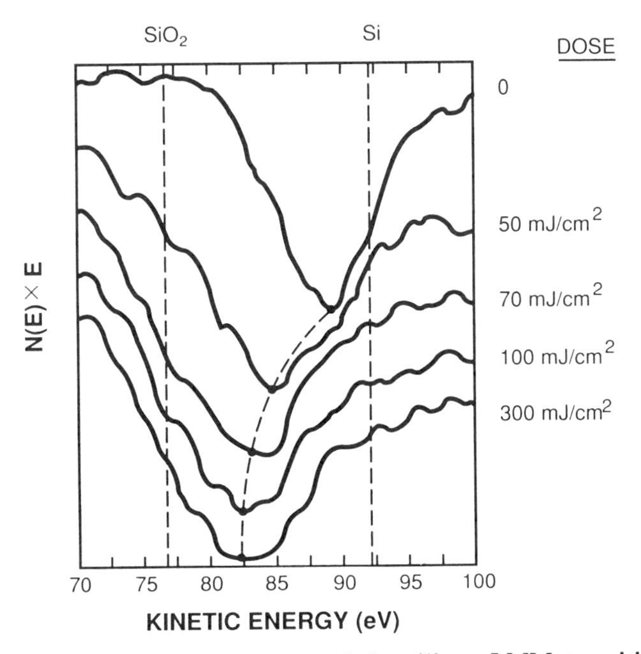

Figure 6. Auger electron spectra of the silicon LMM transition for poly(cyclohexylsilyne-co-*n*-butylsilyne) exposed to various doses at 193 nm (Reproduced with permission from reference 14. Copyright 1991 Society of Photo-Optical Instrumentation Engineers.)

and type of carbon in the film are as important as the photooxidation efficiency in optimizing resist sensitivity and contrast. For example, poly(phenylsilyne) (PPSy) exhibits very poor etch selectivity owing to the presence of the phenyl group. The aromatic carbon ring composing the phenyl group is very stable and only slowly attacked by bromine. As a result, the decomposition and etching of the phenyl moiety are rate limiting for both the oxidized and virgin PPSy. The net effect is a reduction in the HBr etch selectivity between virgin and oxidized PPSy.

Etch rate data for poly(cyclohexylsilyne-co-*n*-butylsilyne) at –60-V bias in HBr are shown in Figure 7. The HBr etch rate contrast is calculated to be 0.3. However, the O_2 RIE following HBr RIE raises overall process the contrast to 5 because of the > 50:1 etch selectivity of the carbon-based planarizer vs. the polysilyne. Figure 8 shows an SEM of poly(cyclohexylsilyne-co-*n*-butylsilyne) patterned at 90 mJ/cm^2 and developed by HBr RIE at –60-V bias followed by O_2 RIE of the planarizing layer. The bilayer consisted of 0.75-μm-thick hard-baked AZ photoresist as the planarizer and a 100-nm-thick polysilyne as the imaging layer. High-aspect-ratio, straight-walled profiles with nominal 0.2-μm lines and spaces were obtained.

Molecular Layer Dosing

A new adlayer dosing technique (*19* has been developed to provide ultrathin (10–50 nm thick) inorganic layers as etch stops and diffusion barriers for deep-UV lithography. This method is based on conditioning a substrate to a critical prescribed coverage of adsorbed water and then reacting the layer with $SiCl_4$ or $TiCl_4$. Insofar as it takes advantage of the high molecular surface mobility and the self-assembly characteristics of adsorbates, the approach is similar to that of atomic layer epitaxy. To achieve low process temperature, however, hydrolysis reactions rather than pyrolysis reactions are used. These reactions are run continuously by controlling the adsorbate species in steady-state equilibrium with a mixed vapor or by sequential cycling of ambients.

This new deposition technique is carried out in a turbomolecular-pumped vacuum system (base pressure, 10^{-7} Torr). The walls of the system and the gas manifold are heated to ~ 55 °C to prevent deposition, while the sample stage is held at 10–15 °C. Gas adsorption and film growth are monitored with a quartz crystal microbalance. The sequence of steps for deposition of SiO_2 is as follows: (1) Normal hydrophilic substrates are loaded into the system. In practice, nearly all silicon process wafers (bare, thin film, or resist coated) support the process. However, in the experiments outlined below, we standardized to oxygen-plasma-ashed wafers. (2) The substrates are dosed with water vapor at a pressure between 0.2 and ~ 3 Torr. Near room temperature, this pressure is sufficient to consistently form a multilayer of adsorbed H_2O in a few seconds, as monitored with the microbalance. (3) The excess water can then be rapidly evacuated to retard homogeneous vapor-phase reactions. (4) An ambient pressure (~ 5 Torr) of $SiCl_4$ vapor is introduced and allowed to react with the adsorbed water layer. (5) The $SiCl_4$ vapor is purged, and the cycle is repeated for layer-by-layer growth.

In our system this sequence is automated using computer control, with electrically controlled valves and closed-loop feedback from the pressure sensors. In practice, we have found that near room temperature, using $SiCl_4$, the surface reactions are dominant, and at the expense of some precision in thickness control the third step can be eliminated. In this case the reaction proceeds continuously on the substrate surface until the depletion of reactants or buildup of reaction products terminates it. The procedure for growth of TiO_2, with $TiCl_4$ vapor substituted for $SiCl_4$, is very similar to that for SiO_2 except that the third step must be retained to achieve high-quality stoichiometric films.

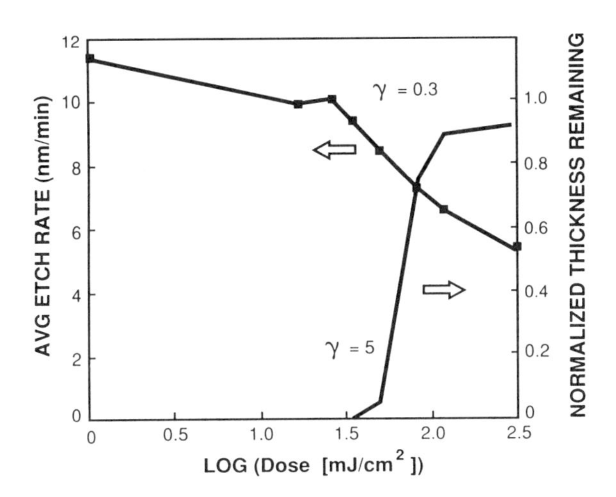

Figure 7. Average etch rate of poly(cyclohexylsilyne-co-*n*-butylsilyne) by −60-V bias HBr RIE (left axis) and the normalized thickness remaining in a poly(cyclohexylsilyne-co-*n*-butylsilyne)/novolac bilayer after HBr RIE development at −60-V bias followed by O_2 RIE (right axis). Contrast values are indicated for each curve. (Reproduced with permission from reference 14. Copyright 1991 Society of Photo-Optical Instrumentation Engineers.)

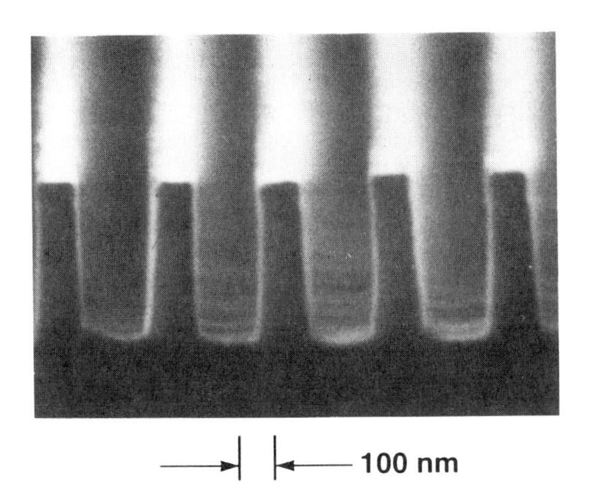

Figure 8. SEM of 0.2-μm line-and-space features patterned in 100-nm-thick poly(cyclohexylsilyne-co-*n*-butylsilyne). Development was by HBr RIE at −60-V bias, and pattern transfer was by O_2 RIE. (Reproduced with permission from reference 14. Copyright 1991 Society of Photo-Optical Instrumentation Engineers.)

Preliminary characterizations of the SiO_2 films have been made by various methods. The index of refraction under most growth conditions is 1.45. At least to the sensitivity of AES and x-ray photoemission spectroscopy (~ 1 at. %), films are stoichiometric. Secondary ion mass spectrometry shows chlorine levels to be only two times higher than for device-quality thermal SiO_2. Qualitative measurements of hermeticity performed using pinhole-decoration etching on films 30 to 50 nm thick show extremely low process-related pinhole densities. These could not be quantified in our experiments as they were below the nominal ambient dust level in the laboratory. Wafers overgrown with SiO_2 can be immersed in acetone for 1 h with no visible attack of the resist by the solvent.

In addition to resist technology, an important potential application of the adlayer dosing technique is trench filling for VLSI. This application stems from the unusual ability of the dosing reaction to conformally coat internal surfaces. An SEM of typical overgrowths obtained by filling silicon trenches with deposited oxides are shown in Figure 9.

All-Dry Photoresist Scheme

Four distinct processes enable an all-dry bilayer resist scheme at 193 nm. Two of these, dry development and dry pattern transfer, are discussed in previous sections. The other two are dry planarization layer deposition and dry imaging layer deposition, both of which have been demonstrated using a conventional rf-powered, parallel-plate plasma-enhanced chemical vapor deposition (PECVD) system.

As an alternative to conventional spin-on organic planarization layers, deposition of amorphous carbon (*a*-C:H) planarization films from hydrocarbon precursors by PECVD has recently been demonstrated (*19,20*). The films are dry deposited at low temperature (< 50 °C), low bias (< 25 V), and a high deposition rate (300–800 nm/min). Depending on the source gas and subsequent processing steps, a post-deposition hardening step may be required (*21*). These PECVD layers yield excellent local planarization and are compatible with surface-imaging resist systems using acetate-based solvents and with toluene-solvated polysilynes.

Besides yielding excellent local planarization, PECVD carbon films show extended planarization as good as or better than that obtained with the best available spin-on planarization coatings. Figure 10 illustrates the exceptional planarization that can be achieved from a PECVD *a*-C:H film made from isoprene. The features shown are 1.5-μm-deep trenches in silicon covered by a plasma-deposited oxide and then planarized by the PECVD *a*-C:H layer.

An alternative to the adlayer dosing technique described earlier is the use of PECVD to form organosilicon imaging layers (*22*). These layers are deposited from liquid precursors such as (i) hexamethyldisilazane (HMDS), $(CH_3)_3Si\text{-}NH\text{-}Si(CH_3)_3$, (ii) hexamethyldisilane (HM_2S), $(CH_3)_3Si\text{-}Si(CH_3)_3$, (iii) tetramethysilane (TMS), $(CH_3)_4Si$, and (iv) TMSDMA, $(CH_3)_3Si\text{-}N\text{-}(CH_3)_2$. The films are deposited at relatively high pressure (300–950 mTorr) and low power density (0.06–0.55 W/cm^2) at a deposition rate of 1–25 nm/min. Similar to the polysilynes, these organosilicon films undergo photooxidation when exposed at 193 nm. In particular, films deposited from TMS, when dry developed by the same HBr RIE process described previously for the polysilynes, show a sensitivity of ~ 30 mJ/cm^2. A typical all-dry resist process module is shown in Figure 11. The planarization layer was PECVD *a*-C:H, the imaging layer was PECVD TMS, the development was by HBr RIE, and the pattern transfer was by O_2 RIE.

Conclusion

Recent progress in photoresist processes at wavelengths < 200 nm has been

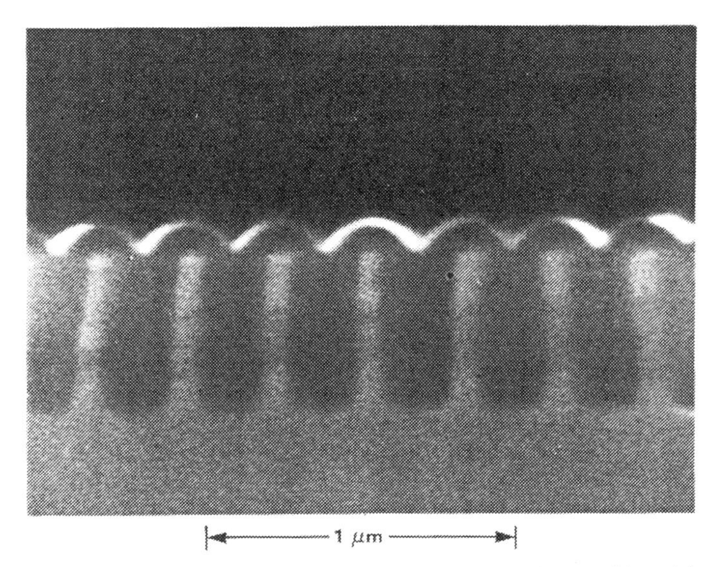

Figure 9. SEM of silicon trenches filled by deposited oxides (cleaved profile). Typical growth conditions are 2-Torr H_2O, 5-Torr $SiCl_4$, and ~15-min total growth time. Note the excellent filling and planarization of the substrate topography. (Reproduced with permission from reference 19. Copyright 1991 Americam Institute of Physics.)

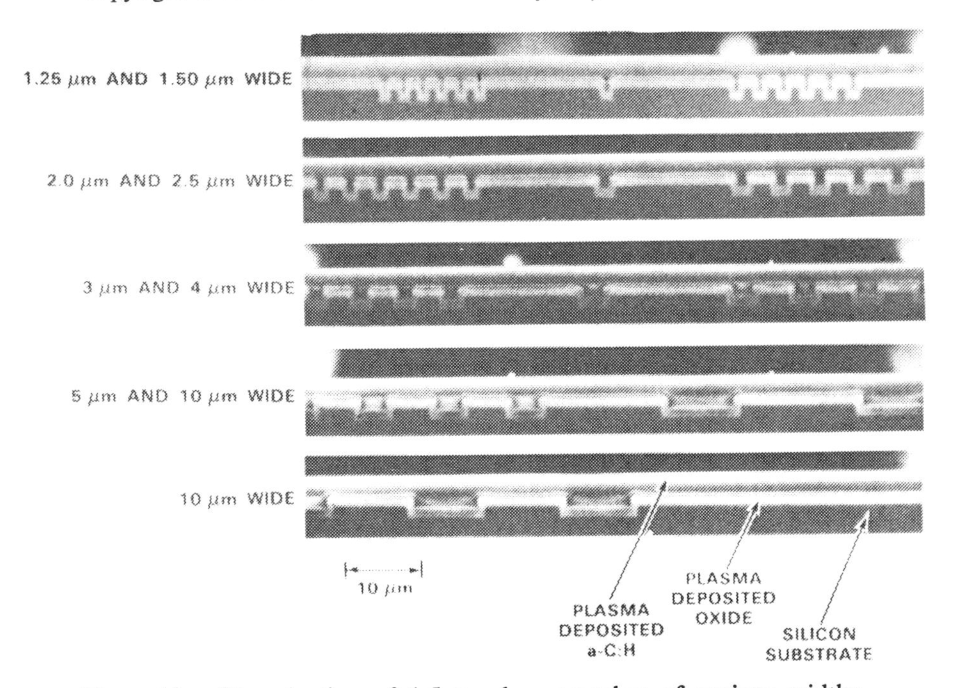

Figure 10. Planarization of 1.5-μm-deep trenches of various widths. Trenches etched in silicon at various widths were covered by 1.5 μm of plasma-deposited oxide and then planarized by a PECVD *a*-C:H film made from isoprene.

DRY PLANARIZATION LAYER:	PECVD a-C:H (1.4 µm)
DRY IMAGING LAYER:	PECVD TMS (0.13 µm)
DRY DEVELOPMENT:	HBr RIE (–60 V, 35 mTorr)
DRY PATTERN TRANSFER:	O_2 RIE (–150 V, 40 mTorr)

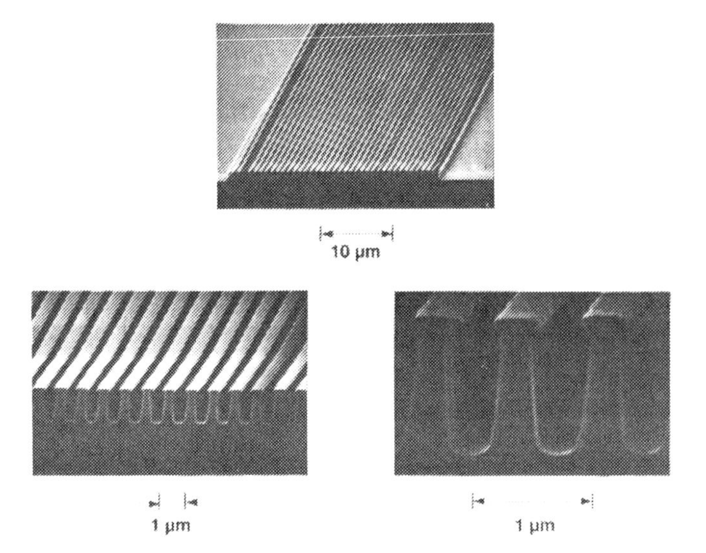

Figure 11. All-dry resist process module showing 0.4-μm features patterned using a dry-developed PECVD resist on a dry-deposited planarization layer. Pattern transfer was by O_2 RIE.

reviewed. Although much work remains, these approaches have already been shown capable of well-controlled lithography with low dose requirements, good edge definition, and adequate sidewall slope control and reproducibility for use at linewidths < 0.25 μm.

Acknowledgments

We would like to thank K. Stoy-Pavelle, C. A. Marchi, C. L. Dennis, D. M. Craig, R. W. Otten, S. G. Cann, R. S. Uttaro, B. E. Maxwell, C. A. Bukowski, R. B. Goodman, R. R. Paladugu, and W. F. DiNatale for expert technical assistance. This work was sponsored by the Defense Advanced Research Projects Agency and the Air Force Office of Scientific Research.

Literature Cited

1. Taylor, G. N.; Stillwagon, L. E.; Venkatesan, T. *J. Electrochem. Soc.* **1984,** *131,* 1658.
2. Lin, B. J. In *Introduction to Microlithography*; Thompson, L. F.; Willson, C. G.; Bowden, M. J., Eds.; ACS Symposium Series 219; American Chemical Society, Washington, DC, 1983.
3. Garza, C. M.; Misium, G. R.; Doering, R. R.; Roland, B.; Lombaerts, R. *Proc. SPIE* **1989,** *1086,* 229.
4. Liu, H. Y.; deGrandpre, M. P.; Feely, W. E. *J. Vac. Sci. Technol. B* **1988,** *6,* 379.
5. Thackeray, J. W.; Orsula, G. W.; Pavelchek, E. K.; Canistro, D. L.; Bogan Jr., L. E.; Berry, A. K.; Graziano, K. A. *Proc. SPIE* **1989,** *1086,* 34.
6. deGrandpre, M. P.; Graziano, K. A.; Thompson, S. D.; Liu, H. Y.; Blum, L. *Proc. SPIE* **1988,** *923,* 158.
7. Hartney, M. A.; Kunz, R. R.; Ehrlich, D. J.; Shaver, D. C. *Proc. SPIE* **1990,** *1262,* 119.
8. Johnson, D. W. *Proc. SPIE* **1984,** *469,* 72.
9. Pacansky, J.; Waltham, R. J. *J. Phys. Chem.* **1988,** *92,* 4558.
10. Hartney, M. A.; Johnson, D. W.; Spencer, A. C. *Proc. SPIE* **1991,** *1466,* 238.
11. Schellekens, J. P. W.; Visser, R. J. *Proc. SPIE* **1989,** *1086,* 220.
12. Wallraff, G. M.; Miller, R. D.; Clecak, N.; Baier, M. (Ref. 10) p. 211.
13. Kunz, R. R.; Horn, M. W.; Goodman, R. B.; Bianconi, P. A.; Smith, D. A.; Freed, C. A. *J. Vac. Sci. Technol. B* **1990**, 8, 1820.
14. Kunz, R. R.; Bianconi, P. A.; Horn, M. W.; Paladugu, R. R.; Shaver, D. C.; Smith, D. A.; Freed, C. A. (Ref. 10) p. 218.
15. Kunz, R. R.; Horn, M. W.; Bianconi, P. A.; Smith, D. A.; Freed, C. A. *J. Vac. Sci. Technol. A* **1991,** *9,* 1447.
16. Schwartz, G. C.; Schiable, P. M. *J. Vac. Sci. Technol.* **1979,** *16,* 410.
17. Bestwick, T. D.; Oehrlein, G. S. *J. Vac. Sci. Technol. A* **1990,** *8,* 1696.
18. Nakamura, M.; Iizuka, K.; Yano, H. *Jpn. J. Appl. Phys.* **1989,** *28,* 2142.
19. Ehrlich, D. J.; Melngailis, J. *Appl. Phys. Lett.* **1991,** *58,* 2675.
20. Pang, S. W.; Horn, M. W. *J. Vac. Sci. Technol. B* **1990,** *8,* 1980.
21. Pang, S. W; Horn, M. W. *IEEE Electron Device Lett.* **1990,** *11,* 391.
22. Pang, S. W.; Goodman, R. B.; Horn, M. W. In *Proceedings of the 7th International IEEE VLSI Multilevel Interconnection Conference*; IEEE: New York, 1990; p. 435.
23. Horn, M. W.; Pang, S. W.; Rothschild, M. *J. Vac. Sci. Technol. B,* **1990,** *8,* 1463.

RECEIVED October 23, 1992

Chapter 18

Space Environmental Effects on Selected Long Duration Exposure Facility Polymeric Materials

Philip R. Young and Wayne S. Slemp

Polymeric Materials Branch, Mail Stop 226, NASA Langley Research Center, Hampton, VA 23665–5225

The National Aeronautics and Space Administration's Long Duration Exposure Facility (LDEF) provided a unique environmental exposure of a wide variety of materials. The effects of 5 years and 9 months of low-Earth orbit (LEO) exposure of these materials to atomic oxygen (AO), ultraviolet and particulate radiation, meteorid and debris, vacuum, contamination, and thermal cycling is providing a data base unparalleled in the history of space environment research. Working through the Environmental Effects on Materials Special Investigation Group (MSIG), a number of polymeric materials in various processed forms have been assembled from LDEF investigators for analysis at the NASA Langley Research Center. Specimens include silvered perfluorinated ethylene propylene (FEP) Teflon thermal blanket material, polysulfone and epoxy matrix resin/graphite fiber reinforced composites, and several high performance polymer films. These samples came from numerous spacecraft locations and, thus, received different environmental exposures. This paper reports the chemical characterization of these materials.

The results of infrared, thermal, x-ray photoelectron, and various solution property analyses have shown no significant change at the molecular level in the polymer that survived exposure. However, scanning electron and scanning tunneling microscopies show resin loss and a texturing of the surface of some specimens which resulted in a change in optical properties. The potential effect of a silicon-containing molecular contamination on these materials is addressed. The possibility of continued post-exposure degradation of some polymeric films is also proposed.

The NASA Long Duration Exposure Facility (LDEF) was placed into low-Earth orbit (LEO) in April 1984 by the Space Shuttle Orbiter Challenger. It was retrieved on January 10, 1990, by the Orbiter Columbia. Shortly after that event, the most detailed analysis began of materials that had been in space in the history of the U.S. Space Program. The vehicle provided a unique environmental exposure of a wide variety of spacecraft materials *(1,2)* during its approximately 34,000 orbits or three-quarters of a billion mile journey. The effects of 5 years and 9 months of exposure to atomic oxygen (AO), ultraviolet and particulate radiation, meteoroid and debris, vacuum, contamination, and thermal cycling on the LDEF and its contents is providing a data base which will set the standard for space environmental effects on materials well into the 21st Century.

The objective of the current work is to assess the response of selected polymeric materials to the extended LEO environment provided by LDEF. The approach has been to characterize molecular level effects in addition to more obvious visual, physical and mechanical effects. This approach should provide fundamental information for use in developing new and improved materials for long-term LEO missions.

Specimens selected for this study came from Langley experiments and from materials made available to the Space Environmental Effects on Materials Special Investigation Group (MSIG) during LDEF deintegration activities at the Kennedy Space Center in the January-May 1990 time period. They include silvered perfluorinated ethylene propylene (FEP) Teflon thermal blanket material, Kapton film, and graphite fiber reinforced polysulfone and epoxy polymer matrix composites. The chemical characterization of these materials using infrared spectroscopy, dynamic and thermomechanical analyses, solution property measurements, and x-ray photoelectron spectroscopy is reported. The potential effect of a silicon-containing contaminant on the performance of these materials is discussed. In addition, the possibility of continued post-exposure degradation of some polymeric materials is proposed. This study is intended to add to the body of knowledge on space environmental effects on materials being derived from the LDEF mission.

Experimental

Infrared spectra were recorded on a Nicolet 60SX Fourier Transform Infrared Spectrometer System using a diffuse reflectance technique (DR-FTIR) *(3)*. Ultraviolet-Visible (UV-VIS) spectra were run on a Perkin-Elmer Lamda 4-A spectrophotometer. Dynamic mechanical spectra were obtained on a DuPont Model 982 Dynamic Mechanical Analyzer/Model 1090 Thermal Analyzer (DMA). Glass transition temperature (Tg) determinations were made on a Perkin-Elmer Model 943 Thermomechanical Analyzer (TMA). Additional thermal analyses was performed using a DuPont Model 1090/Model 910 Differential Scanning Calorimeter (DSC). The equipment and techniques used to make solution property measurements have been previously reported *(4)*. Gel permeation chromatography (GPC) was performed on a Waters Associates system in chloroform using a $10^6/10^5/10^4/10^3$ Å Ultrastyragel column bank.

X-ray Photoelectron Spectroscopy (XPS) measurements were conducted at the Virginia Tech Surface Analysis Laboratory, Department of Chemistry, VPI&SU, Blacksburg, VA *(5)*. Measurements were made on a Perkin-Elmer PHI 5300 spectrometer equipped with a Mg X-ray Kα source (1253.6 eV), operating at 15 kV/120mA. Typical operating pressures were $<10^{-7}$ torr. Analyses were made at take-off angles of 45° or 90°.

Scanning electron microscopy (SEM) and energy dispersive X-ray (EDS) analyses were conducted both at Virginia Tech and at Langley. An ISI SX-40 SEM (ISI, Milpitas, CA) equipped with a Tracor Northern Z-MAX 30 EDS analyzer (Tracor, Madison, WI) was used at Virginia Tech. A Cambridge StereoScan 150 SEM (Cambridge Instruments, Deerfield, IL) equipped with an EDAX S150 detecting unit (EDAX International Inc., Prairie View, IL) was used at the Langley Research Center. Scanning tunneling microscopy (STM) was performed on a NanoScope II instrument (Digital Instruments, Inc., Santa Barbara, CA). The visual appearance of selected specimens was documented using various photographic techniques.

Fiber/resin weight percents were determined by acid digestion according to ASTM D371. Mechanical property measurements on $[\pm 45]_s$ tensile test specimens were performed according to ASTM D3039-76. Test specimens were pre-cut to standard dimensions and tapered end-tabs machined from epoxy/glass cloth were adhesively applied prior to integration onto the experimental tray and subsequent exposure.

DISCUSSION

The LDEF structure and orbital orientation are depicted in Figure 1. As described in Reference 1, the spacecraft was 30 feet long, 14 feet in diameter, and had 12 sides or rows with 6 experimental trays per row. One end of the gravity gradient stabilized vehicle faced space, and one end faced earth. Additional experimental trays were mounted on the space and earth ends. Fully integrated with experiments, it weighed about 11 tons.

Rows are numbered 1 through 12 in Figure 1 and trays are lettered A through F. Thus, the location of specimens discussed in this report should follow from this tray and row notation scheme. For example, B9 denotes the location of specimens on Tray B at Row 9. The orbital orientation of the satellite was such that Row 9 nominally faced the RAM direction and Row 3 faced the WAKE direction throughout the 5.8-year flight. Recent LDEF supporting data analysis have determined that the actual RAM direction was 8° of yaw from the perpendicular to Row 9, in the direction of Row 10.

LDEF provided a very stable platform for LEO exposure of materials. The environment a specimen experienced depended upon its location on the vehicle. Two significant environmental effects of concern for polymeric materials are atomic oxygen (AO) and ultraviolet (UV) radiation. The total AO fluence and equivalent UV sun hours for each LDEF row and tray are given in Figures 2 and 3, respectively. The calculation of these two parameters can be found in Reference *6*. AO fluence ranged from about 8.8×10^{21} atoms/cm^2 for Row 9 to about 1.1×10^{3} atoms/cm^2 near Row 3. Equivalent UV sun hours for both rows was about 11,100 hours. Additional integrated environmental effects pertinent to this report are included with information in Table I. NASA Conference Publication 3134 documents the LEO environment for the LDEF and its contents in detail *(7)*. A condensed version of the present paper may also be found in that publication *(8)*.

During the months preceding the LDEF retrieval, a significant effort went into planning for the chemical characterization of polymeric materials onboard the vehicle. Much of this effort was summarized at the LDEF Materials Data Analysis Workshop which ran concurrent with activities surrounding the successful return of the spacecraft to the Kennedy Space Center in early 1990 *(2)*. Chemical, physical and mechanical changes were anticipated as the result of environmental exposure. Among the anticipated chemical effects were modification to molecular structure, molecular weight (MW) and molecular weight distribution (MWD) changes, Tg and crystallinity effects, and changes in surface chemistry. Resin loss, changes in transparency and α_s/ε, and contamination were also expected to be observed.

An approach for characterizing these effects was adopted. Where appropriate, solution property measurements including GPC, low angle laser light scattering photometry (LALLS), differential viscometry (DV), GPC-LALLS, and GPC-DV were considered to be invaluable in determining how the MW and MWD responded to exposure. FTIR analyses were expected to play a key role in following molecular level changes in the polymer backbone. Various thermal analyses including DSC, DMA and TMA would also be employed to examine Tg, Tm, and crystallinity effects. Finally, extensive use was planned for XPS, EDS, SEM, and STM for examining environmental effects on surface chemistry. Thus, the examination of selected

LDEF IN ORBIT

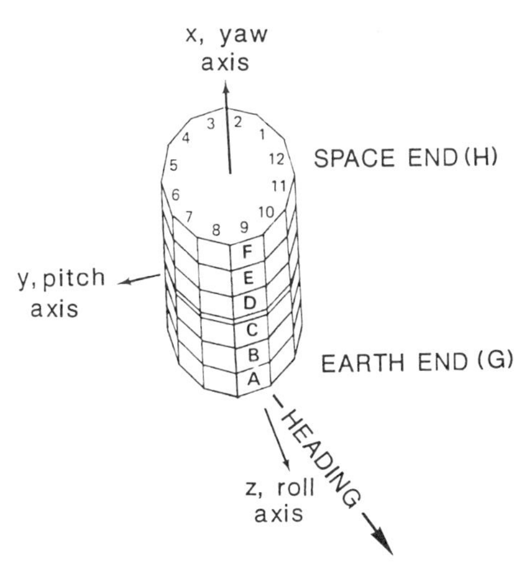

Figure 1. LDEF photograph, sketch, and orbital orientation.

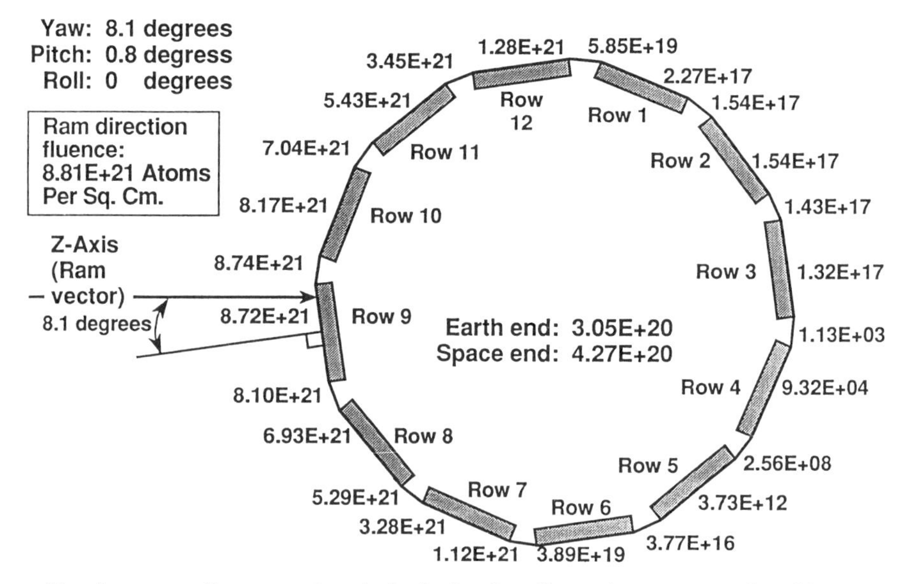

Figure 2. Atomic oxygen fluence for each LDEF tray location.

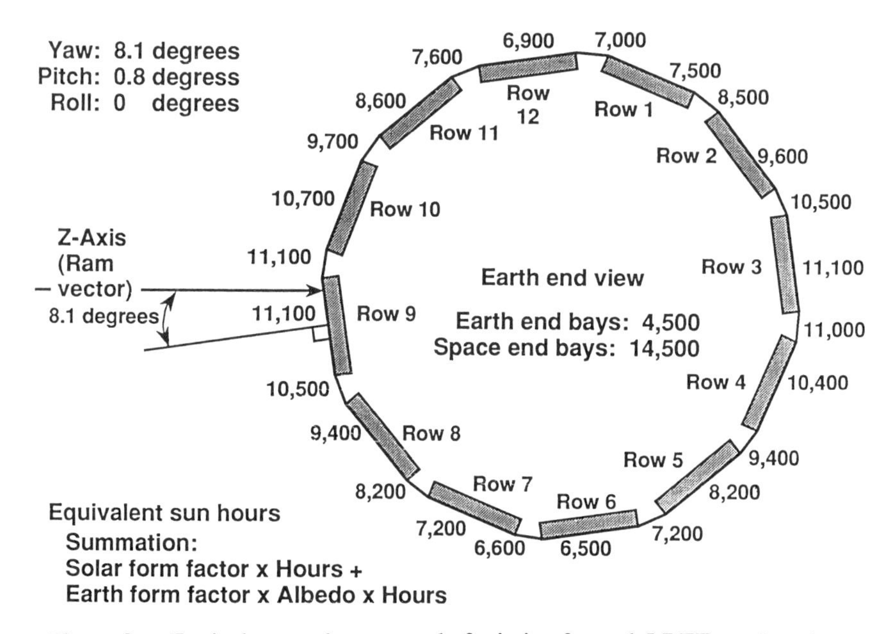

Figure 3. Equivalent sun hours at end of mission for each LDEF tray location.

exposed materials proceeded in a deliberate step-by-step manner as the various specimens became available for analysis. As evidenced by the foregoing discussion, the present effort was, indeed, focused on characterizing the molecular level effects of environmental exposure.

Preliminary Observations. The environmental effects of 69 months of LEO exposure were studied during extensive inspection and photo-documentation activities at the Kennedy Space Center in the January-May 1990 time period. This inspection confirmed many of the anticipated effects of exposure and gave confidence that planned chemical characterizations would be appropriate.

The visual effects of AO, UV, meteoroid and debris, vacuum, contamination, thermal cycling, and environmental synergisms were dramatic. For example, degradation of polymeric films, coatings, and composites were observed on many leading edge (Row 9) experiments. Similar materials on the trailing edge (Row 3) and other locations tended to exhibit UV-induced yellowing. The appearance of several Ag/FEP teflon thermal blankets will be discussed in a later section.

A series of white paint specimens located around the LDEF structure had discolored on trailing and space end surfaces. A higher meteoroid and debris impact density was observed on leading edge surfaces. Thin polymeric films were usually penetrated by these impacts. The outgassing of various specimens was probably greater than had been anticipated. The combined effects of thermal cycling, AO and UV on these outgassed products apparently produced a thin molecular contamination layer on selected surfaces. AO erosion was also observed to negate UV and contamination effects on many specimens located near Row 9. Finally, particulate radiation exposure was deemed insufficient to produce significant effects on both non-metallic and metallic materials. As many environmental effects and materials as is practical will be characterized. An initial report on part of this effort is included herein.

Specimens selected for this study are described in Table I. They were included as part of larger experiments intended to evaluate the LEO environment on a variety of materials for advanced spacecraft application *(1)*. The location of specimens on LDEF is included in Table I along with a summary of pertinent exposure conditions.

Ag/FEP Film. The thermal blanket material, also known as flexible second-surface mirror (SSM) thermal control coating, is used in a variety of space applications where thermal control is a consideration *(9)*. Figure 4 gives a schematic of the SSM describing its composition and function. The outer layer consists of a nominal 5 mil Type A FEP copolymer Teflon film. Solar radiation passes through this transparent film and is reflected away by the silver backing. This yields a low absorption of solar radiation, α_S. The thermal emittance, ε, is a function of the polymer bulk infrared absorption and, therefore, is largely dependent on material thickness. These properties provide for a variable α_S/ε ratio, desirable for certain spacecraft thermal considerations. The material has applications as both a thermal blanket cover and as an adhesively bonded thermal coating.

Visual inspection of the fully integrated LDEF at the Kennedy Space Center revealed that the appearance of all the SSM material was not the same. Thermal blankets located near the Row 3 trailing edge exhibited a highly specular appearance while those located near the Row 9 leading edge exhibited a diffuse or "frosted" appearance. The change from specular to diffuse was essentially graduated as the location changed from trailing to leading edge. Since this observation correlated with the anticipated atomic oxygen fluence at these locations, the two phenomena were assumed to be related.

DR-FTIR analysis of specular and opaque blanket specimens revealed surprisingly similar spectra *(10)*. Extensive thermal analysis of selected materials

TABLE I SPECIMENS, LDEF LOCATION AND PRELIMINARY ENVIRONMENTAL EXPOSURE CONDITIONS

SPECIMEN	LOCATION	Exposure[a] AO (atoms/cm^2)	UV (sun hours)
Silvered FEP Teflon	A[b] 10[c]	8.17×10^{21}	10,700
Thermal Blanket	C 8	6.93×10^{21}	9,400
	C 5	3.73×10^{12}	8,200
	F 2	1.54×10^{17}	9,600
KAPTON Film	H 7	4.27×10^{20}	14,500
Adhesively Bonded	B 9 (10 month)[d]	2.60×10^{20}	- -
Silvered FEP Teflon	B 9	8.72×10^{21}	11,200
934/T300 Epoxy Composite	B 9		
P1700/C6000 Polysulfone Composite	B 9		
KRS-5 Window	B 9		

[a] Boeing Aerospace supplied MSIG calculations; NAS1-18224, TASK 12.
[b] LDEF tray location (see figure 1).
[c] LDEF row location (see figure 1).
[d] Environmental Exposure Control Cannister.

GENERAL INTEGRATED LDEF ENVIRONMENT

- Particulate Radiation
  e^- and p^+: 2.5×10^5 rad
- Vacuum
  10^{-6} - 10^{-7} torr
- Altitude
  255-180 nautical miles
- Micrometeoroid and Debris
  34,336 impacts (0.5mm to 5.25mm)
- Thermal Cycles
  ~34,000 (-40 to 190°F extremes)
- Orbital Inclination
  28.5°

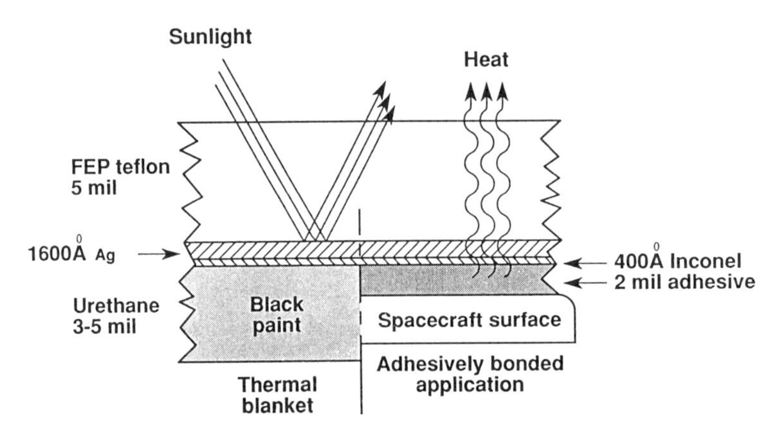

Figure 4. Flexible second-surface mirror thermal control coating in thermal blanket and adhesively bonded applications.

also failed to detect significant differences. The FEP film layer was delaminated from the urethane paint layer backing and analyzed. The vapor deposited silver remained on the urethane layer. Figure 5 shows the DSC thermogram from -120°C to 340°C for a specular specimen which flew on Row 5. Inflections in the trace around -10°C are likely associated with the Tg of the FEP teflon. The melt endotherm (Tm) around 250°C is also apparent. All analyzed specimens showed essentially the same DSC thermogram. No significant differences were noted in Tg, Tm, or the heat of fusion for any specimen.

XPS analysis also failed to reveal major differences in the surface chemistry of opaque flight specimens as compared to control materials. Table II summarizes XPS results for selected opaque specimens. No appreciable amount of oxygen was incorporated into these specimens nor had the surface concentration of carbon and fluorine changed.

Thus, based on FTIR, DSC, and XPS analyses, we concluded that the opaque appearance of the thermal blankets was not a result of a change in chemical properties. However, subsequent SEM and STM analysis showed that the appearance was likely due to "texturing" or "carpeting" of the FEP surface by atomic oxygen erosion *(11)*. Figure 6 shows SEM photomicrographs of a Row 5 (specular) and a Row 8 (diffuse) blanket specimen. Figure 7 gives STM line plots for control (specular) and exposed (diffuse) specimens. The texturing noted for the diffuse sample is essentially perpendicular to the plane of the film and in the direction of the atomic oxygen flow. This phenomenon has been observed on previous shuttle flights *(11)*.

As will be discussed in a later section, the surface of several blanket specimens which remained specular was found by XPS to be contaminated with a silicon-containing species. For example, data for the tray F2 material given in Table II shows about 1.5% atomic concentration of silicon on the film surface. This contamination tends to complicate the interpretation of selected experimental results in this study. For example, Figure 8 shows DR-FTIR spectra for specimen F2 and a control. The subtraction of the two spectra revealed the presence of a carbonyl group at $1724 cm^{-1}$ in the exposed specimen. Determining whether this new carbonyl group is associated with contamination or, perhaps, AO-induced oxidation of the FEP surface is difficult to establish.

The multiple XPS carbon peaks and elevated oxygen concentration associated with the F2 specimen was initially assumed to be associated with contamination. Although no silicon was found on the specular C5 specimen, the multiple carbon peaks for that example, shown in Figure 9, were also assumed to be associated with contamination. However, emerging research is suggesting that the FEP surface is crosslinked by deep UV exposure *(12)*. The possibility of UV crosslinking of fluoriniated polymers is currently being investigated *(13)*. This phenomenon most likely was not observed with opaque blanket specimens because atomic oxygen had eroded away the UV-crosslinked layer.

Kapton Film. Kapton film from the space end of LDEF was also examined. The 5 mil thick 0.5 inch wide strips were included as part of an experiment to evaluate the effect of the space environment on polymers considered for space-based radar phased-array antenna material *(14)*. The location of these films at position H7 was such that the flow of atomic oxygen was perpendicular to the 5 mil film edge and parallel to the film surface.

Figure 10 shows SEM micrographs of one of the films. The sketch at the top of the figure identifies a portion of the film with the appropriate SEM micrograph. For example, 10a. shows the leading film edge directly exposed to AO while 10b. shows the trailing edge which received only sweeping exposure. 10c. and 10d. show exposed and protected film surfaces. Texturing is noted on surfaces exposed to

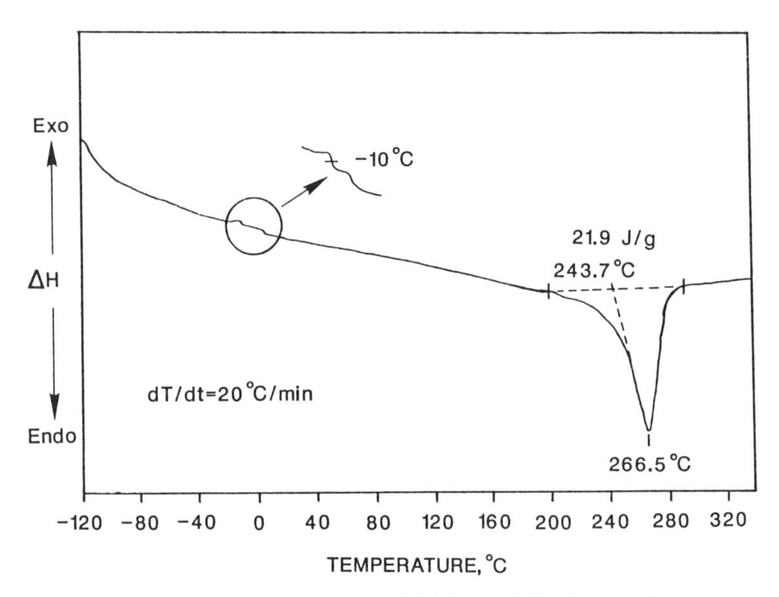

Figure 5. DSC thermogram of C5 thermal blanket specimen.

TABLE II XPS RESULTS FOR SELECTED SILVERED FEP TEFLON MATERIALS

	REFERENCE SPECIMENS			OPAQUE SPECIMENS			SPECULAR SPECIMENS		
PHOTO PEAK	STANDARD[a]	CONTROL[b]	STANDARD[c]	A10	C8	B9/5.8 yr[d]	B9/10 mo[e]	F2	C5
C 1s B.E. (eV)[f]	290.9	290.9	290.9	290.9	290.9	290.9	290.9	290.9/289.0/284.2	290.9/289.0/284.5
A.C. (%)[g]	31.6	29.6	31.3	28.5	30.4	31.4	30.8	39.8	37.5
F 1s B.E.	689.0	688.9	688.6	688.5	688.8	688.6	688.4	686.8	686.9
A.C.	65.6	66.4	68.7	69.8	66.2	67.5	68.7	47.8	56.5
O 1s B.E.	--	--	--	530.9	--	532.2	532.5	531.0	531.0
A.C.	NSP[h]	<1.0	NSP	1.3	<0.5	1.1	0.5	7.4	2.8
Si 2p B.E.	--	--	--	--	--	--	--	102.7	--
A.C.	NSP	NSP	NSP	NSP	NSP	NSP	NSP	1.5	NSP

[a] Commercially obtained blanket.
[b] Flight control blanket.
[c] Commercially obtained, adhesively bonded.
[d] Adhesively bonded, full exposure.
[e] Adhesively bonded, enviromental control canister exposed.
[f] Binding energy, electron volts.
[g] Atomic concentration, percent.
[h] No significant peak.

C5 (SPECULAR)

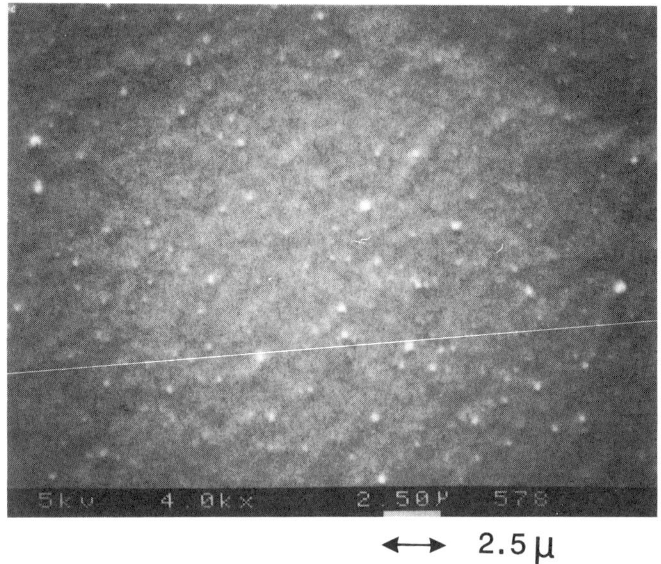

C8 (DIFFUSE)

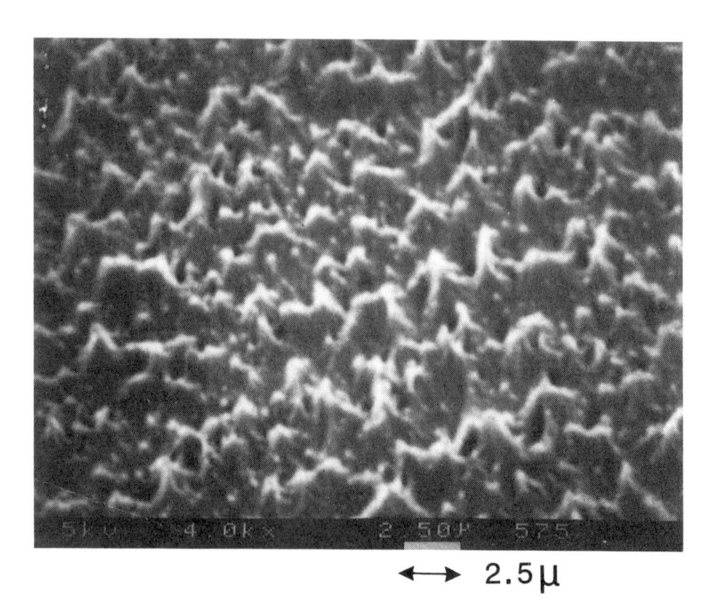

Figure 6. SEM of C5 and C8 silvered FEP teflon thermal blanket specimens (X3200).

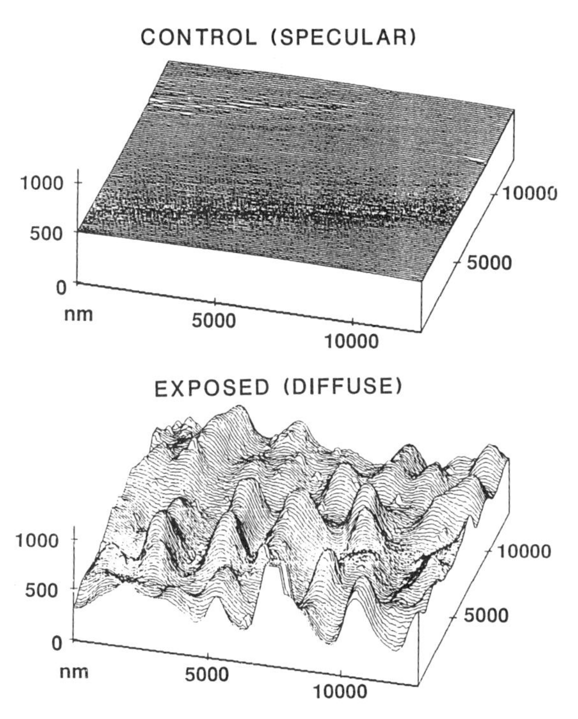

Figure 7. STM analysis of control and exposed FEP teflon thermal blanket specimens.

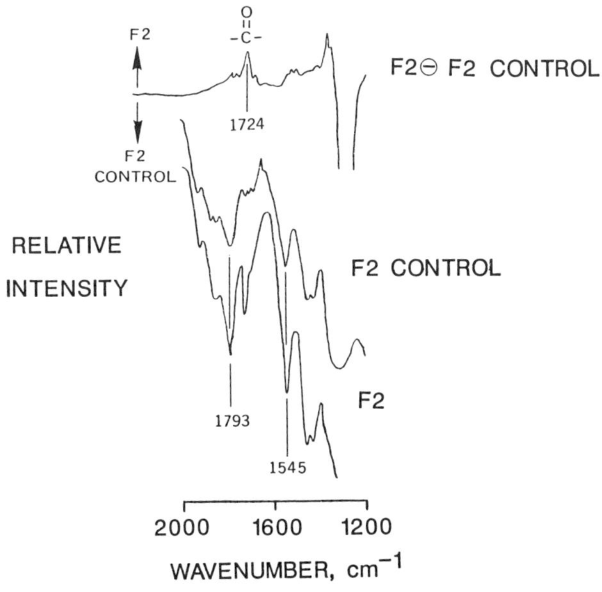

Figure 8. DR-FTIR spectra of F2 flight and control thermal blanket specimens.

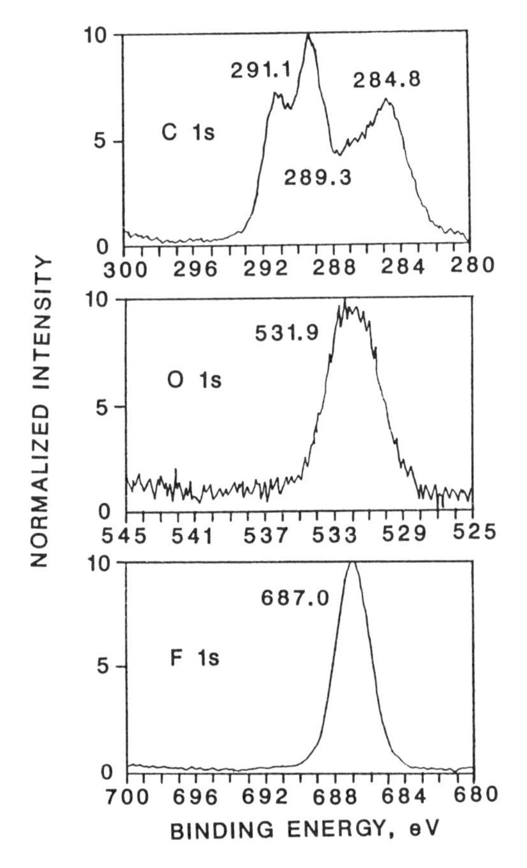

Figure 9. Narrow scan XPS spectra for C5 thermal blanket specimen.

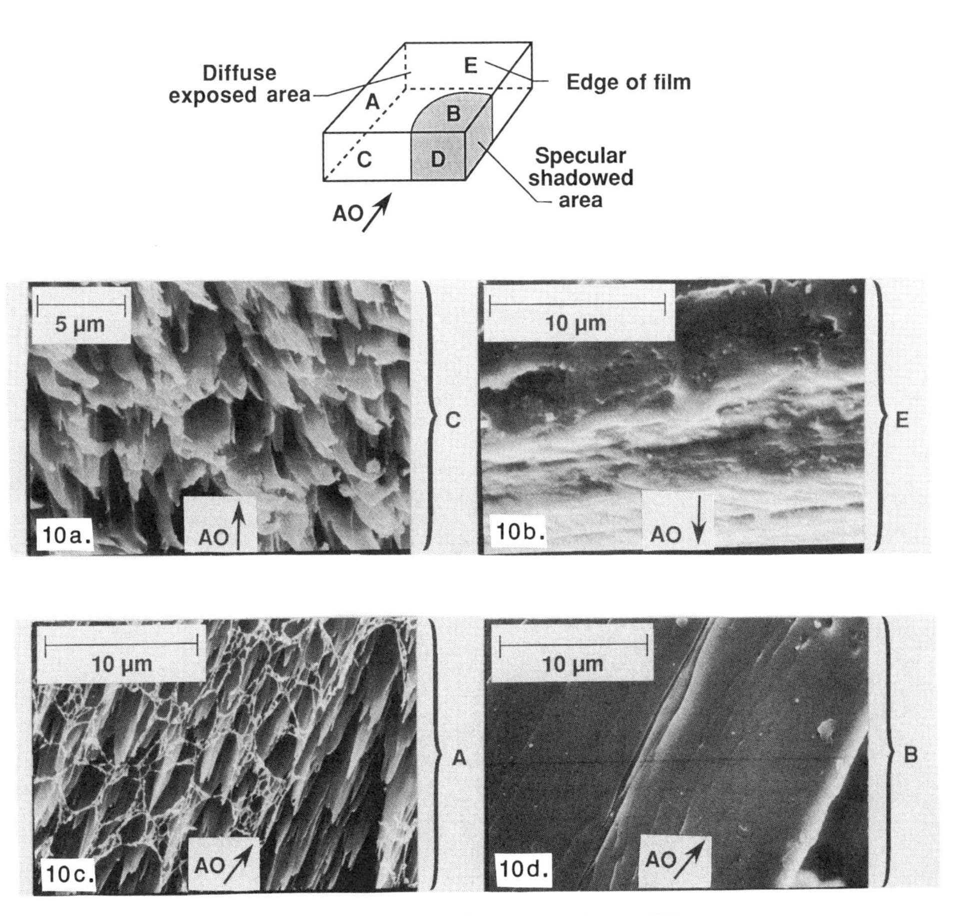

Figure 10. SEM of LDEF-exposed Kapton Film.

atomic oxygen. However, in contrast to the perpendicular projections shown in Figures 6 and 7, the texturing in Figure 10c. is in the plane of film (and AO flow). The "spider web" feature observed in Figure 10c. to be connecting the tips of "christmas trees" has been observed with other film specimens *(15)*. Chemical analysis suggests that this web effect is due to residual Kapton.

Additional FTIR, UV-VIS, DMA, TGA, and XPS analyses were conducted on exposed and control Kapton films. No interpretable differences were noted between specimens by these techniques except for a substantial decrease in UV-VIS transmission for exposed, diffuse-appearing films *(15)*.

Composites: Chemical Characterization. A series of epoxy and polysulfone composites flew on a Row 9 experiment and received 5.8 years of LEO exposure *(16)*. Figure 11 shows a photograph of a P1700/C6000 polysulfone/graphite flight specimen. The dark circular eroded region received direct exposure for the duration of the flight. The remaining outer surface was protected by an aluminum template which held the specimen in place. This appearance was typical of that observed for other composite specimens.

Figure 12 gives DR-FTIR spectra of protected and exposed regions of the polysulfone composite. A portion of "protected" surface was cut from the specimen and placed at the focal point of the diffuse reflectance optics to obtain that particular spectrum. This technique failed to produce a satisfactory "exposed" spectrum due to the insufficient matrix resin on the specimen surface. However, the spectrum shown was obtained by filing into the exposed surface and mixing the resultant powder with KBr *(3)*. The two spectra in Figure 12 are essentially identical, suggesting that the molecular structure of the polymer which survived exposure is comparable to the original molecular structure.

DMA also suggested a loss of resin due to exposure. Figure 13 shows DMA curves for polysulfone composite samples cut from larger specimens. The magnitude of the damping peak, a function of the amount of matrix resin, decreased after exposure. The fact that DMA damping peaks for protected and exposed composites remained at the same temperature suggests that the Tg for this material had not changed significantly due to exposure. Addition study of this specimen tends to confirm that the Tg remained unchanged after exposure *(17)*.

The possibility that the Tg of the exposed side might be different from the protected side was examined by thermomechanical analysis. The TMA probe was carefully placed in contact with exposed and protected surfaces of several polysulfone specimens. Any movement in the probe as a function of temperature was recorded. Table III summarizes the results of this evaluation. The Tg of a non-flight control specimen was essentially independent of the side examined. The slight variation in reported values can be explained by the judgment the analyst used in determining the inflection point in the TMA curve indicative of the glass transition. The Tg of flight samples given direct and protected flight exposure is also given in Table III. A careful examination of all data suggests there is no significant difference in Tg for any of these samples.

Solution property data was obtained on selected polysulfone composites. The matrix resin was extracted using chloroform and quantified by a previously reported procedure *(4)*. Figure 14 shows GPC molecular weight distributions for polysulfone resin like that used to prepare prepreg, a control composite that remained at Langley, one which flew protected from direct exposure, and a composite which received direct exposure. The distributions are virtually superimposable. There is also no discernable difference in various molecular weight averages for these four materials. Table IV summarizes molecular weight data obtained by GPC relative to polystyrene, and absolute values obtained by GPC-DV and GPC-LALLS. As might be expected, slight differences are noted in values obtained by different techniques. However,

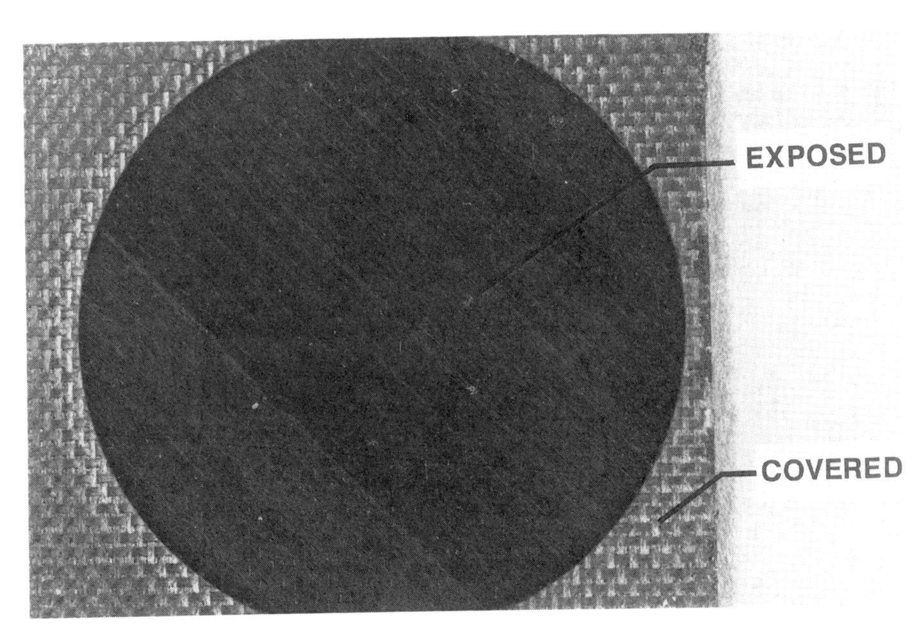

Figure 11. LDEF exposed P1700/C6000 polysulfone composite [± 45°]s.

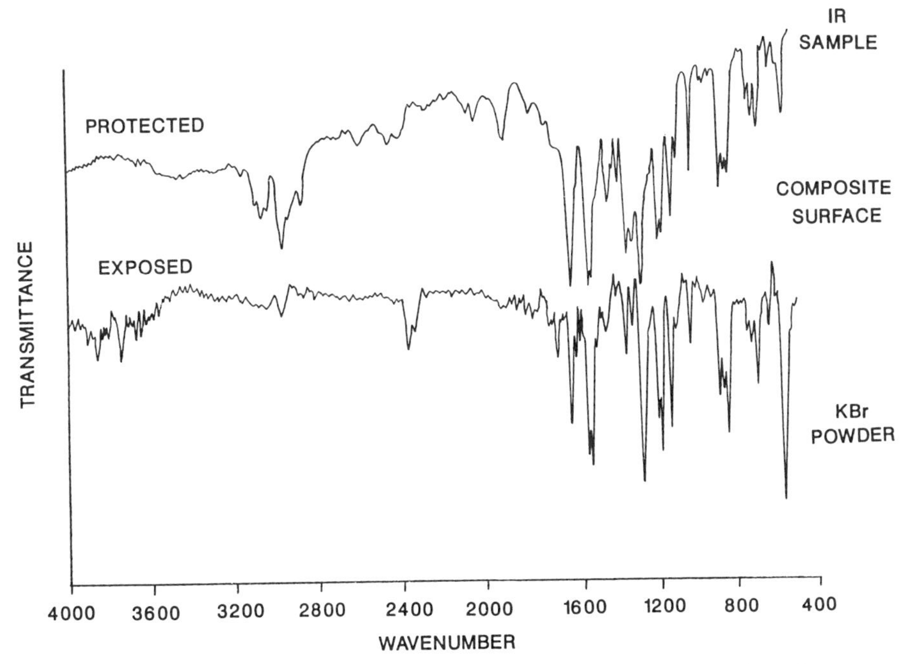

Figure 12. DR-FTIR spectra of LDEF-exposed P1700/C6000 polysulfone composite.

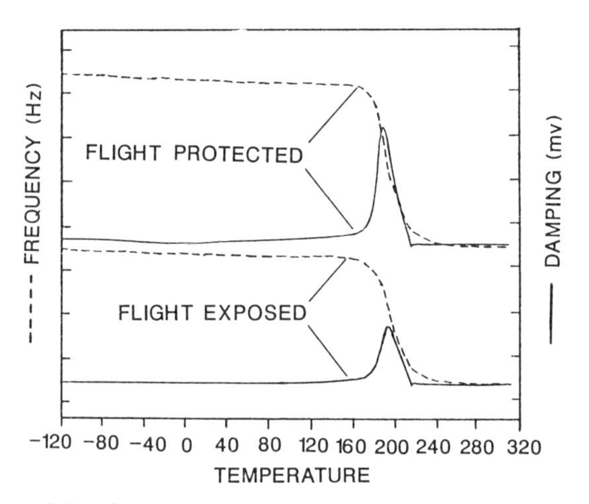

Figure 13. DMA of LDEF-exposed P1700/C6000 polysulfone composites.

TABLE III THERMOMECHANICAL ANALYSIS OF P1700/C6000 POLYSULFONE COMPOSITES

Sample	Tg (°C)	Contacted side
Langley Control	167°	Random
	167°	
	170°	
	166°	
Flight Control	164°	Side A
	166°	Side B
Flight Exposed	170°	Exposed side
	171°	
	169°	Nonexposed side
	171°	

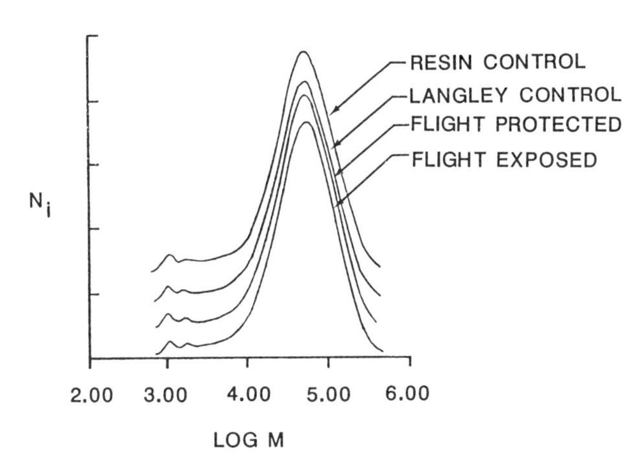

Figure 14. GPC molecular weight distribution of P1700 polysulfone composite matrix resins.

TABLE IV MOLECULAR WEIGHT OF POLYSULFONE MATRIX RESIN

Sample, Technique	M_N[a]	M_W	M_Z	M_W/M_N	Intrinsic[b] Viscosity
Resin control					
GPC	19,000	70,000	117,000	3.8	
GPC-DV	18,000	50,000	83,000	2.8	0.45
GPC-LALLS	21,000	46,000	74,000	2.2	
Static LALLS		45,900			
Langley control					
GPC	18,000	68,000	113,000	3.9	
GPC-DV	16,000	50,000	81,000	3.1	0.43
GPC-LALLS	18,000	40,000	65,000	2.1	
Flight protected					
GPC	19,000	68,000	114,000	3.7	
GPC-DV	17,000	53,000	87,000	2.9	0.40
GPC-LALLS	21,000	40,000	66,000	1.8	
Flight exposed					
GPC	18,000	68,000	114,000	3.7	
GPC-DV	17,000	48,000	80,000	2.9	0.44
GPC-LALLS	21,000	46,000	71,000	2.2	

[a] Molecular weight in g/mole
[b] Intrinsic viscosity in dL/g

these data support the general conclusion that there is no significant difference at the molecular level between polymer which survived exposure and the original polymer.

Composites: Physical and Mechanical Characterization. Selected physical property measurements show a difference between LDEF-exposed and control composites. Table V gives fiber and resin weight percents for four different composite materials. Data are given for composite specimens which remained at Langley as controls, specimens which flew protected from direct exposure, and specimens which received full RAM exposure for 5.8 years. The location of the flight protected specimens was such that they experienced only vacuum, thermal cycling, and contamination. In addition to these effects, the flight exposed specimens experienced AO, UV, meteoroid and debris, and particulate radiation. Data in Table V for the Langley control and flight protected composites probably agree to within the repeatability of the determination. However, subject to qualifications discussed below, composites which received direct exposure appear to have lost about 10% resin.

Figure 15 shows SEM photographs of an exposed 4-ply $[\pm 45]_s$ 5208/T300 epoxy/graphite fiber composite specimen. The photograph on the left of Figure 15 shows the edge of a sample which had been cut in preparation for analysis. Visible in this photograph are the +45 bottom ply, the two -45 symmetrical center plies, and what is left of the +45 top ply. The photograph on the right gives a higher magnification of the interface between the first and second plies. Both resin and fiber loss are apparent. Thus, actual resin loss was probably greater than the data in Table V indicates. Fiber weight percent was determined gravimetrically after resin had been digested in a hot H_2SO_4/H_2O_2 bath. The weight of the recovered fiber was divided by the original specimen weight. Resin content was then obtained by subtracting the fiber content from unity. Any fiber loss due to environmental exposure, as documented in Figure 15, would not be reflected in this determination. Thus, the uncertainty in how much fiber was lost places the 10% resin loss in question. A loss of one complete ply of the 4-ply composite would result in an actual 25% resin/fiber loss.

Figure 16 gives tensile strength and modulus data for five composite materials. Baseline values were obtained when the composites were fabricated in 1983. The Langley control data is a 1990 repeat of the earlier data. Measurements on flight protected and flight exposed specimens are also given. The baseline, Langley control, and flight protected data for respective composites are probably all the same. However, specimens which flew exposed to the LDEF environment experienced a deterioration in tensile and modulus properties. No doubt, more than resin and fiber loss contributed to this phenomena. The synergistic effects of thermal cycling, vacuum, and exposure to UV, in addition to AO erosion, have not been determined.

The surface of these composites was left unprotected in order to maximize the effects of the LDEF environment. Thus, resin loss was severe. Figure 17 illustrates that a thin coating can effectively protect a composite. A thin SiO_2/Ni vapor deposited coating appears to have prevented resin loss for this epoxy composite. The chemical and physical properties of this material have not been evaluated.

Contamination. The possibility that molecular contamination may bias some of the observations made in this study must also be addressed. Visual inspection of the LDEF at KSC revealed that a brown "nicotine-like" stain was deposited on selected surfaces. This contaminant may have resulted from outgassing from various experiments as well as from the LDEF structure itself. XPS and EDS analyses confirmed the presence of silicon in this residue.

Table VI summarizes XPS data on selected specimens discussed in this paper. As much as 20.5% atomic concentration of silicon was detected on specimen

TABLE V FIBER/RESIN CONTENT OF COMPOSITE MATERIALS

	FIBER/RESIN WEIGHT PERCENT						
COMPOSITE	LANGLEY CONTROL		FLIGHT PROTECTED		FLIGHT EXPOSED		APPARENT RESIN LOSS (%)
P1700/C3000	63.6	36.4	64.0	36.0	67.8	32.2	11.5
934/T300[1]	72.2	27.8	73.1	26.4	74.0	26.0	6.5
934/T300[2]	69.6	30.4	70.8	29.2	72.0	28.0	7.9
5208/T300	71.7	28.3	72.3	27.7	74.3	25.7	9.2

[1] 145 g/m^2 fiber areal weight.
[2] 95 g/m^2 fiber areal weight.

Figure 15. SEM of LDEF-exposed 5208/T300 epoxy composite [± 45°]$_S$.

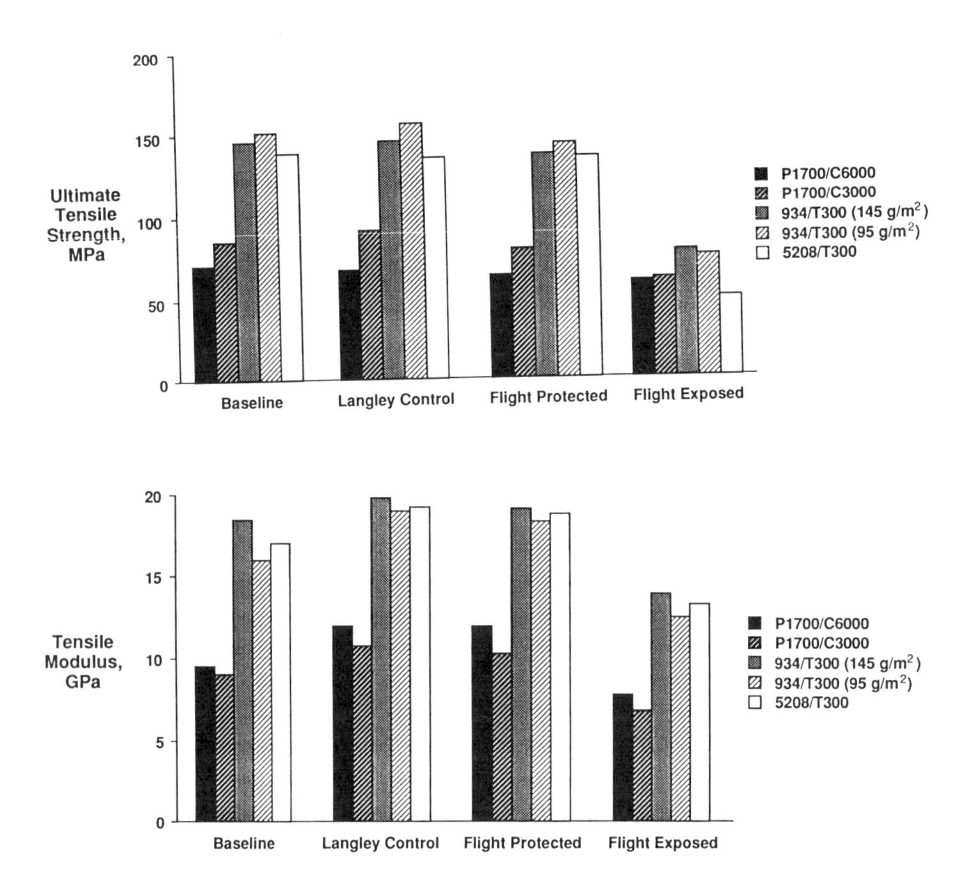

Figure 16. Mechanical property measurements on LDEF composite materials.

Figure 17. Photographs of coated and uncoated LDEF-exposed 934/T300 epoxy composites [± 45°]$_S$.

surfaces. Additional XPS and FTIR analyses suggested that the silicon is in the form of an organic silicone on surfaces not exposed to AO, and an inorganic silicate on surfaces receiving AO exposure.

KRS-5 infrared optical windows were flown as contamination monitors on a Row 9 experiment. Figure 18 shows how the windows were mounted and summarizes XPS results on a specimen which received 5.8 years of flight time. Note 20.0% silicon on the exposed side, 3.3% on the vented side, and no silicon on the control specimen. Further, the 102.9 eV binding energy of the AO-exposed silicon is the binding energy of a silicate. The 101.9 eV B.E. for the protected side is that of a silicone. The conversion of silicone to the silicate upon exposure to atomic oxygen is an accepted phenomenon. Further, silica/silicate coatings have been shown on LDEF and elsewhere to be effective barriers to atomic oxygen erosion *(18, 19)*. Thus, surfaces covered with this contamination layer may have responded differently to the LDEF environment than had they not been contaminated.

Post-Exposure Effects. Some polymeric materials which received exposure on LDEF may exhibit post-exposure effects. Environmentally exposed films and coatings have been qualitatively observed in this laboratory to continue to change in appearance with time. An appreciation of this phenonmenon may be necessary in order to analyze LDEF specimens in an efficient manner.

Prior to the LDEF flight, a series of thin films in a Langley experiment received 40 hours of LEO exposure onboard STS-8. Those films were photographed and characterized in 1983 upon their return to Langley. In February 1991, the flight specimens were removed from a desiccator where they had been stored in tin containers. Two of four films had dramatically changed in appearance.

Figure 19 shows photographs of controls and flight specimens taken in 1983 and repeat photographs taken in 1991. PEN-2,6 shown in 19a, is a state-of-the-art polyester designed to exhibit improved radiation stability *(20)*. The film had cracked and turned opaque during storage. PMDA-DAF, shown in 19b, is an experimental polyimide expected to exhibit unusual environmental stability *(21)*. That film turned opaque and lost much of its integrity.

Additional post-exposure effects were reported at the First LDEF Post-Retrieval Conference held June 2-8, 1991, in Kissimmee, Florida *(7)*. Among these were comments by the presenters that some paints were bleaching to a lighter shade, that certain degraded optical fiber bundles were returning to their original color, and that the pastel colors associated with micrometeoroid and debris impacts on AO-exposed silvered teflon thermal blankets had changed to a dull brown. These distressing observations keynote the urgency in analyzing non-metallic materials onboard LDEF in an expedient manner.

A Perspective. The evidence developed in this study suggests that the molecular structure of selected polymers which survived 5.8 years of LEO exposure on LDEF is similar to the structure of the original polymer. Most of the materials in this study received significant exposure to atomic oxygen. Current research at the Langley Research Center is focused on a series of specimens which flew inside an environment control canister and received only limited exposure. In an effort to avoid possible contamination associated with the shuttle, this canister opened in space 1 month after LDEF was deployed and closed 10 months later *(16)*. Thus, these 10-month specimens received exposure early in the LDEF mission.

The chemical characterization of these 10-month samples, specifically FTIR and solution property measurements, shows interpretable differences at the molecular level as a result of exposure *(15, 17, 22)*. This potentially contradictory observation between the performance of polymers after 5.8 years and after 10 months in space is best understood by considering the orbit of the spacecraft during its flight. LDEF was deployed in an essentially circular orbit of 257 nautical miles on April 7, 1984

TABLE VI XPS RESULTS FOR SELECTED LDEF POLYMER SPECIMENS

PHOTO PEAK	A10 (BLANKET) STAINED[a]	A10 (BLANKET) UNSTAINED[a]	F2 (BLANKET)	H7 (KAPTON) EXPOSED	H7 (KAPTON) PROTECTED[b]	B9 (EPOXY COMPOSITE) EXPOSED	B9 (EPOXY COMPOSITE) PROTECTED[b,c]
C 1s B.E. (eV)[d]	284.6	284.6	290.9/289.0/284.2	284.6/288.0	284.6/288.0	284.6	284.6/288.0
A.C. (%)[e]	31.7	30.8	39.8	31.4	62.9	54.3	62.8
F 1s B.E.	689.3	689.6	686.8	- -	- -	- -	- -
A.C.	1.2	2.3	47.8	- -	- -	- -	- -
O 1s B.E.	533.0	532.8	531.0	532.5	531.8	532.8/535.0	532.5
A.C.	47.9	46.5	7.4	46.7	25.5	33.0	24.8
N 1s B.E.	- -	- -	- -	400.1	399.9	399.9	399.7
A.C.	- -	- -	- -	3.8	8.7	5.2	3.4
Si 2p B.E.	103.4	103.2	102.7	103.2	102.5	102.9/105.7	102.7
A.C.	19.2	20.5	1.5	17.1	2.2	7.5	3.4

[a] Visual observation.
[b] Protected from direct exposure.
[c] 2.0% F, 2.0% S, 1.7% Na also detected.
[d] Binding energy, electron volts.
[e] Atomic concentration, percent.

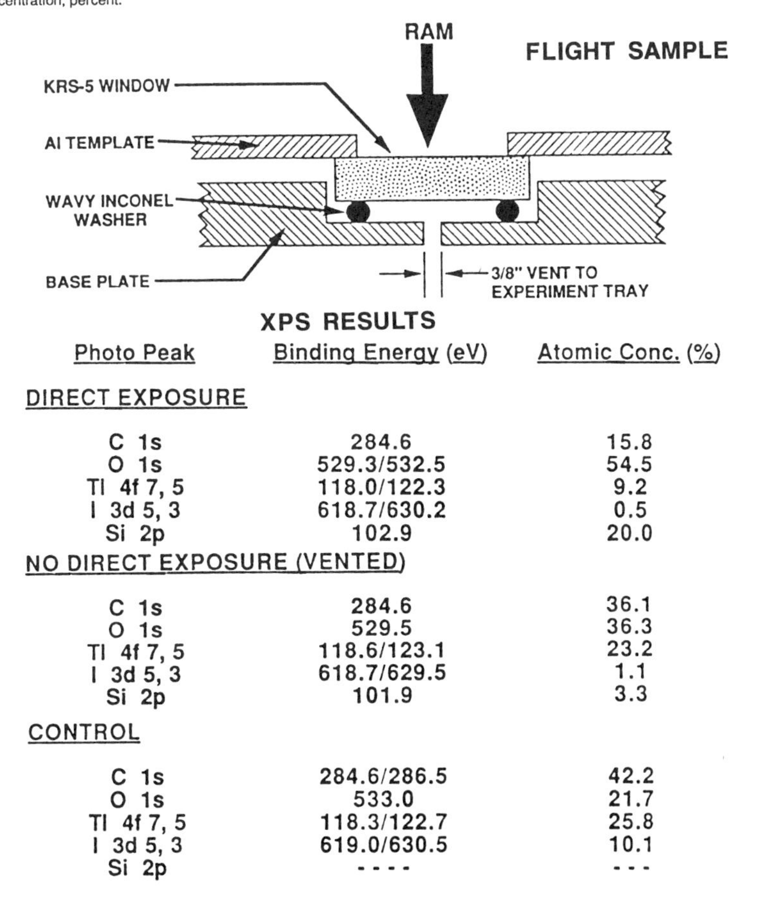

XPS RESULTS

Photo Peak	Binding Energy (eV)	Atomic Conc. (%)
DIRECT EXPOSURE		
C 1s	284.6	15.8
O 1s	529.3/532.5	54.5
Tl 4f 7, 5	118.0/122.3	9.2
I 3d 5, 3	618.7/630.2	0.5
Si 2p	102.9	20.0
NO DIRECT EXPOSURE (VENTED)		
C 1s	284.6	36.1
O 1s	529.5	36.3
Tl 4f 7, 5	118.6/123.1	23.2
I 3d 5, 3	618.7/629.5	1.1
Si 2p	101.9	3.3
CONTROL		
C 1s	284.6/286.5	42.2
O 1s	533.0	21.7
Tl 4f 7, 5	118.3/122.7	25.8
I 3d 5, 3	619.0/630.5	10.1
Si 2p	- - - -	- - -

Figure 18. LDEF-exposed KRS-5 optical window.

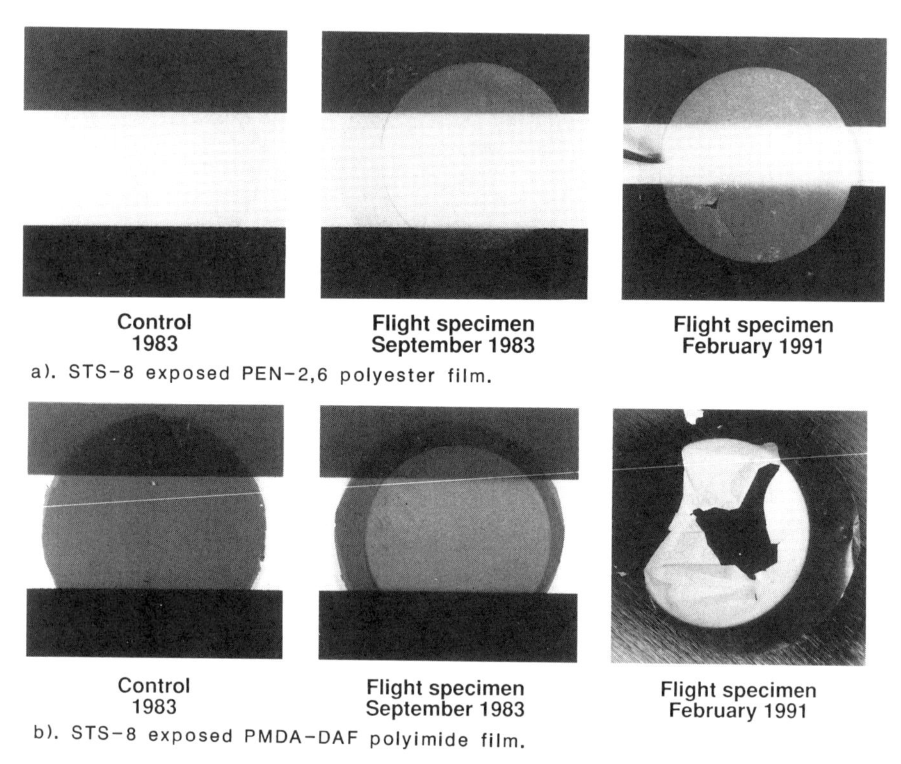

Figure 19. Post-exposure effects in two experimental polymer films.
a. STS-8 exposed PEN-2,6 polyester film.
b. STS-8 exposed PMDA-DAF polyimide film.

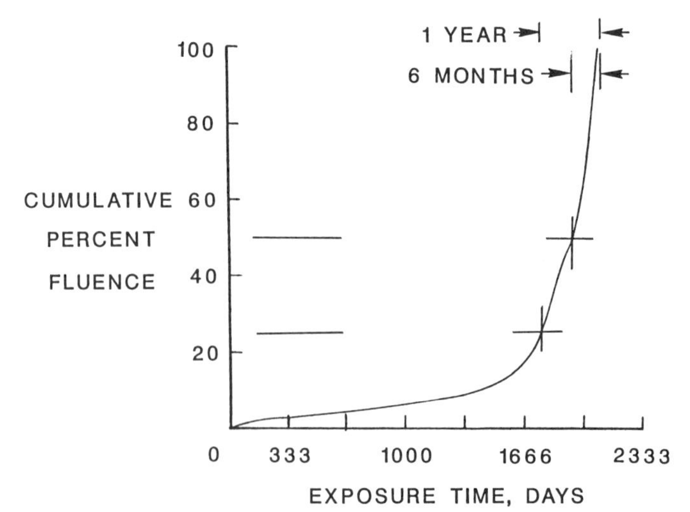

Figure 20. Approximate LDEF cumulative percent atomic oxygen fluence as a function of exposure time.

(23). It was retrieved 69 months later at an altitude of 179 nautical miles. Only about 2 months of orbit lifetime remained at retrieval. The atomic oxygen fluence differs greatly at these two altitudes.

Figure 20 is the approximate cummulative percent RAM AO fluence as a function of LDEF exposure time. Exact AO exposure for all specimens is given in reference *6*. The 10-month specimens were exposed early in the mission when AO fluence was at a minimum. The 5.8 year specimens received significant exposure near the end of the mission. As much as 50% of total AO exposure was received during the last 6 months in orbit. The molecular level effects observed after 10 months, primarily related to changes in surface chemistry, had most likely been eroded away by the time the satellite was retrieved. An earlier retrieval from a higher orbit may have provided different results.

CONCLUDING REMARKS

The LDEF is providing a wealth of information on the effects of extended exposure of spacecraft materials to the space environment. The present study examined how selected uncoated polymeric materials responded to almost 6 years of LEO exposure. Dramatic visual effects were observed. However, chemical characterization of these samples suggests that there is no significant change in the molecular structure of the surviving polymer. Any molecular level effects attributable to exposure were probably lost in the layer of material eroded away by atomic oxygen. The quantitative analysis of similar materials which received different LDEF exposures is anxiously awaited. The ultimate benefit will be a more fundamental understanding of space environmental effects and increased confidence in future spacecraft design and in the performance of spacecraft materials.

ACKNOWLEDGMENTS

This report was not possible without the capable assistance of a number of individuals. The authors thank Carol R. Kalil, A. C. Chang, Judith R. J. Davis, Karen S. Whitley, Ann C. Van Orden, and Reta R. Barrett. The contribution of Kapton films by James B. Whiteside is gratefully acknowledged.

LITERATURE CITED

1. Clark, L. G.; Kinard, W. H.; Carter, D. J. and Jones, J. L., eds.: The Long Duration Exposure Facility (LDEF). NASA SP-473, **1984**.
2. Stein, B. A. and Young, P. R. compilers: LDEF Materials Data Analysis Workshop. NASA CP-10046, **1990**.
3. Young, P. R.; Stein, B. A. and Chang, A. C.: *SAMPE Intl. Symp.*, **1983**, *28*, 824.
4. Young, P. R.; Davis, J. R. J. and Chang, A. C.: *SAMPE Intl. Symp.*, **1989**, *34(2)*, 1450.
5. NASA Grant NAG-1-1186, Virginia Polytechnic Institute and State University, Blacksburg, VA.
6. Bourassa, R. J. and Gillis, J. R.: NASA CR 189554 and CR 189627, performed under NASA Contracts NAS1-18224 and NAS1-19247 entitled "LDEF Materials Analysis" Boeing Defense and Space Group, Seattle, WA.
7. A. S. Levine, Ed., LDEF-69 Months in Space, First Post Retrieval Symposium, Kissimmee, FL, June 2-8, NASA CP-3134, Parts 1, 2, and 3 **1991**.

8. Young, P. R. and Slemp, W. S.: NASA CP-3134, **1991**, Part 2, 687.
9. Slemp, W. S.: NASA Conference Publication 3035, **1989**, Part 2, 425-446.
10. Young, P. R. and Slemp, W. S.: NASA TM 104096, December **1991**.
11. Koontz, S. H.: NASA Conference Publication 3035, **1989**, Part 1, 241-253.
12. Stiegman, A. E.; Brinza, D. E.; Anderson, M. S.; Minton, T. K.; Laue, E. G. and R. H. Liang: JPL Publication 91-10, May **1991**.
13. NASA Grant NAGW-2495, University of Queensland, Brisbane, Qld, Australia.
14. DeIasi, R. J.; Rossi, M. L.; Whiteside, J. B.; Kesselman, M.; Hever, R. L. and Kuehne, F. J.: NASA SP-473, **1984**, 21-23.
15. Young, P. R. and Slemp, W. S.: unpublished LDEF results.
16. Slemp, W. S.: NASA SP-473, **1984**, 24-26.
17. Young, P. R.; Slemp, W. S.; Siochi, E. J. and Davis, J. R. J.: *SAMPE Tech. Conf.*, **1992**, *24*, xxx.
18. Young, P. R.; Slemp, W. S.; Witte, W. G. and J. Y. Shen: *SAMPE Intl. Symp.*, **1991**, *36(1)*, 403.
19. Slemp, W. S.; Santos-Mason, B. Sykes, Jr., G. F. and Witte, W. G.: NASA TM 100459, **1988**, *1*, Section 5-1 to 5-15
20. Bell, V. L. and Pezdirtz, G. F.: *J. Polym. Sci. Polym. Chem. Ed.*, **1983** *21*, 3083.
21. Bell, V. L.: *J. Polym. Sci. Polym. Chem. Ed.*, **1976**, *14*, 225.
22. Young, P. R.; Slemp, W. S. and Gautreaux. C. R.: SAMPE Intl. Symp., *37*, 159 **(1992)**.
23. O'Neal, R. L. and Lightner, E. B.: NASA CP-3134, **1991** Part 1, 3.

RECEIVED November 11, 1992

Chapter 19

Characterization of Matrix Polymers for Electron-Beam-Cured Fiber-Reinforced Composites

C. B. Saunders, A. Singh, V. J. Lopata, W. Kremers, and M. Tateishi

Whiteshell Laboratories, AECL Research, Pinawa, Manitoba R0E 1L0, Canada

Developing products and processes for electron-beam (EB) curing is currently the focus of several industrial research groups, including the Radiation Applications Research Branch (RARB) of AECL Research. Electron-beam treatment initiates polymerization and crosslinking reactions in suitable matrices, curing the polymer and enhancing specific physical and chemical properties. Resin characterization has been the focus of our recent composite research. The effects of EB dose, dose rate and moisture on the properties of carbon fabric laminates are being studied as part of the program. Several EB-curable resins and blends were evaluated as matrices for carbon fiber-reinforced composites. Their curing doses ranged from 30 to 150 kGy. The maximum glass transition temperature (T_g) was obtained in a proprietary epoxy diacrylate polymer (204°C). The mechanical properties of the composite were also affected by the EB dose and dose rate.

Electron-beam (EB) processing uses high-energy electrons from an accelerator to initiate polymerization and crosslinking reactions in suitable matrices, curing the polymers and enhancing their specific physical and chemical properties. Electron-beam curing is being evaluated as an alternative to autoclave curing for manufacturing fiber-reinforced composite structures. Many benefits have been identified for using EB treatment rather than thermal curing to manufacture carbon- and aramid-fiber-reinforced composites, which translate to improved product quality and/or lower production costs *(1)*. Some of these benefits include *(1,2)*:

1. Curing at ambient or sub-ambient temperatures. Electron-beam curing eliminates thermal expansion in both the tool and the product, thus reducing dimensional changes and internal stresses in the final product.

0097–6156/93/0527–0305$06.00/0
Published 1993 American Chemical Society

2. Reduced curing times for individual components. A 50-kW electron accelerator can cure about 1200 kg of composites per hour (2). This production speed is several times higher than for thermal curing with a typical autoclave or oven, even though the products are cured one at a time rather than in large batches.

3. Improved material handling. EB processing is a continuous operation and components can be EB-treated immediately after they are produced. Production scheduling and inventory control become easier and the number of identical molds needed to manufacture products is reduced.

4. Improved resin stability at ambient temperature. The shelf-life of EB-curable resins can be much longer than the shelf-life of formulations for thermal curing because EB-curable formulations do not normally auto-cure at room temperature, making low-temperature storage unnecessary.

5. Reducing the amount of volatiles produced. Thermal curing of composites often produces volatile degradation products that can be hazardous and require proper control procedures *(3)*. EB processing eliminates the production of thermal degradation products, though very small amounts of gases such as hydrogen and carbon dioxide may be produced *(4,5)*. The disposal of these gases would not cause problems and would require much less effort than that required to deal with the volatiles from thermal curing.

Free-radical-initiated matrix systems are well suited to EB curing. The epoxies commonly used in the advanced carbon-fiber-reinforced composites polymerize by a cationic mechanism, and this mechanism is inhibited by trace amounts of water during EB treatment *(1)*. Electron-beam-curable matrices have been developed over the past 10 years with properties comparable to these epoxies. Epoxy acrylates offer a good balance of properties as an EB-curable resin. Mono- and multifunctional monomers, used as diluents, form an essential component of most EB-curable formulations; they are used to control viscosity and to lower the required curing dose. Difunctional acrylates are the most widely used class of multifunctional diluents.

An accelerator produces a beam of electrons that are accelerated to speeds approaching that of light, focussed into a beam, and directed onto a product. The high-velocity electrons deposit energy into the product by interaction with the electrons of the target materials, causing the desired chemical reactions. Electron penetration is a function of beam energy and product density. A 10-MeV EB can penetrate about 25 mm in a typical carbon-fiber composite (density $\sim$ 1500 kg/m^3) with one-sided irradiation, and 60 mm with two-sided irradiation, with the top and bottom surface doses being equal. The penetration limit is inversely proportional to density. Products with thickness greater than the EB penetration limit can be cured by converting the EB to greater-penetrating X-rays *(1)*.

Products manufactured using traditional fabrication methods, including filament/tape winding, resin transfer molding, pultrusion and hand lay-up, can be

EB-processed. Filament/tape winding is particularly well suited because it requires limited or no external pressure, and high winding rates can make efficient use of the available energy. Research in EB-curable composites over the past 15 years, together with the recent availability of industrial, high-energy, high-power accelerators, has made possible the industrial production of filament-wound, carbon-fiber-reinforced motor cases for aerospace applications using EB curing *(6)*.

Experimental

The objective of our program was to characterize selected EB-curable polymers and blends, and to determine the effect of dose and dose rate on the properties of carbon-fiber-reinforced composites made from these resins.

Materials

Table 1 lists the polymers and blends examined for this project *(7)*. All of the materials are commercially available from the Sartomer Company, except for ED-1, a proprietary blend designed for a specific composite application. An unsized, plain-weave, AS-4 carbon fabric, from Hercules, Inc., was used for all of the laminate samples.

Table 1. Selected Electron-Beam-Curable Resins and Blends (7)

Supplier Code	*Description*	*Supplier Information*[1]			
		MW	*Visc.*	*Sp Grav*	*Inhib.*
CN104	Epoxy diacrylate oligomer	NA[2]	650[3]	1.15	100
CN114	Epoxy diacrylate oligomer	NA	500[3]	1.15	100
C2000	Acrylated diol (C_{14}-C_{15})	300	150	0.93	125
C5000	Polybutadiene diacrylate	3000	5000	0.94	400
ED-1	Proprietary mixture of epoxy acrylates and acrylates	NA	NA	~1.10	-
S399	Dipentaerythritol pentaacrylate	525	20,000	1.192	300
CN104A	Blend of 90% CN104 and 10% S399				
CN104B	Blend of 75% CN104 and 25% S399				
CN114A	Blend of 90% CN114 and 10% S399				
CN114B	Blend of 75% CN114 and 25% S399				
C5000A	Blend of 90% C5000 and 10% S399				
C5000B	Blend of 75% C5000 and 25% S399				

[1] All resins supplied by Sartomer except ED-1: MW: molecular weight (g/mole); Visc.: average viscosity at 25°C (mPa·s); Sp Grav: specific gravity at 25°C (g/cm^3); Inhib.: concentration of selected hydroquinone (ppm) used as an inhibitor.

[2] NA: Not available

[3] Viscosity of blend of 50% oligomer and 50% tripropylene glycol diacrylate.

Procedure

Each resin system was characterized as part of a laminate. The loading of the resin or resin blend on the carbon fabric was 35% by mass. The fabrics were solvent-prepregged using a standard organic solvent such as acetone. Panels were prepared by laying up seven plys of material, all in the same fiber orientation. Samples were cured to the various doses (30 to 300 kGy) at ambient temperature, using a 10-MeV, 1-kW EB accelerator (average dose rate = 5400 kGy/h) or a gamma source (9 kGy/h). Dynamic mechanical analysis (DMA) was used to characterize the samples (1 Hz). The glass transition temperature (T_g) and the flexural modulus as a function of temperature were measured by the DMA.

Results

The curing doses for the various resins/blends were determined in an inert atmosphere and the results are shown in Figure 1. The lower the curing dose, the more cost-effective the EB curing. The curing dose was defined as the dose required to maximize the T_g of the sample. The resins that exhibited both high T_g and high modulus exhibited reasonably low EB curing doses of about 100 kGy. The curing doses ranged from 30 to 150 kGy. Blending the highly reactive pentaacrylate (S-399) with the various resins (25%) reduced their required curing dose by as much as 30%.

Both the dry and wet T_g values were measured for each resin system. Figure 2 shows that resin ED-1 had the highest T_g at 204°C, followed by the epoxy diacrylate/pentaacrylate blends (CN104B and CN114B; 155 and 158°C respectively). The epoxy diacrylates had the highest T_gs of the neat resins (121°C). Samples were immersed in water at 100°C for 24 h, and their T_gs were immediately measured as an indication of the wet T_g values. This immersion decreased the T_gs in all cases. Water acts like a plasticizer and decreases the T_g by increasing the free volume. Water introduces defects in the system, which increases the mobility of the backbone of the molecule and allows transition to the rubbery state at a lower temperature. The original T_g was recovered for each sample when they were dried.

Figure 3 shows the maximum flexural modulus of each composite below its T_g. A high modulus is recommended for the material being used to manufacture carbon-fibre-reinforced composites. The figure shows that the epoxy acrylates and their corresponding S399 resin blends exhibit flexural moduli of approximately 40 GPa. The ED-1 composite exhibits the highest modulus at 50 GPa.

Both the EB dose and the dose rate affect the curing and the final properties of the selected composite material. These effects must be understood to ensure uniform curing throughout a part, particularly when both EB and X-ray treatment are employed. The final properties of a cured composite are a function of the degree of crosslinking in the matrix polymer, and are therefore affected by any large changes in dose rate, such as when going from EB to X-ray treatment to obtain the necessary beam penetration. The magnitude of these changes must be

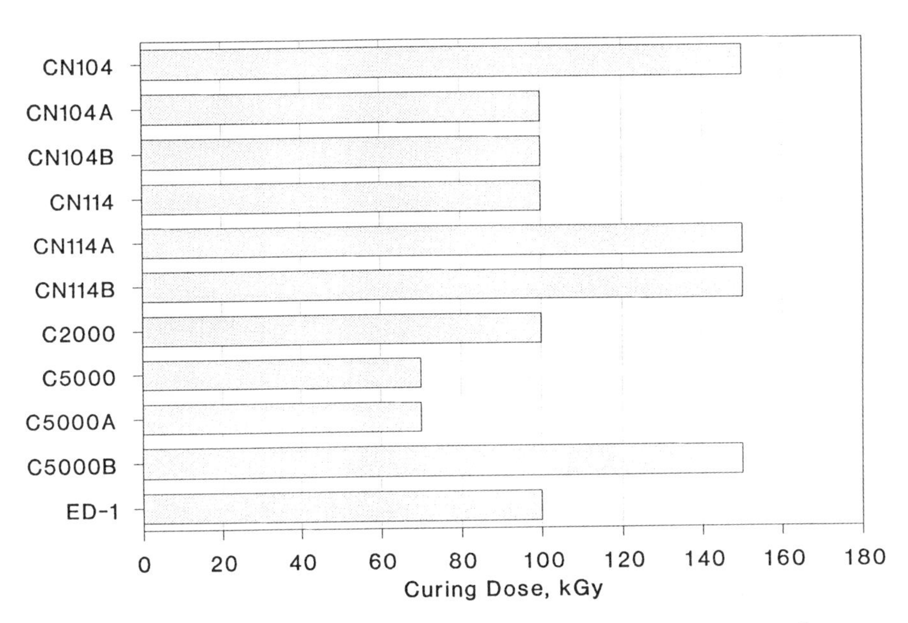

Figure 1. Required Electron Beam Dose to Cure Selected Resins and Resin Blends To Maximum Gel Fraction(35% resin; 65% carbon fabric)(A: 10% pentaacrylate blend; B: 25% blend)

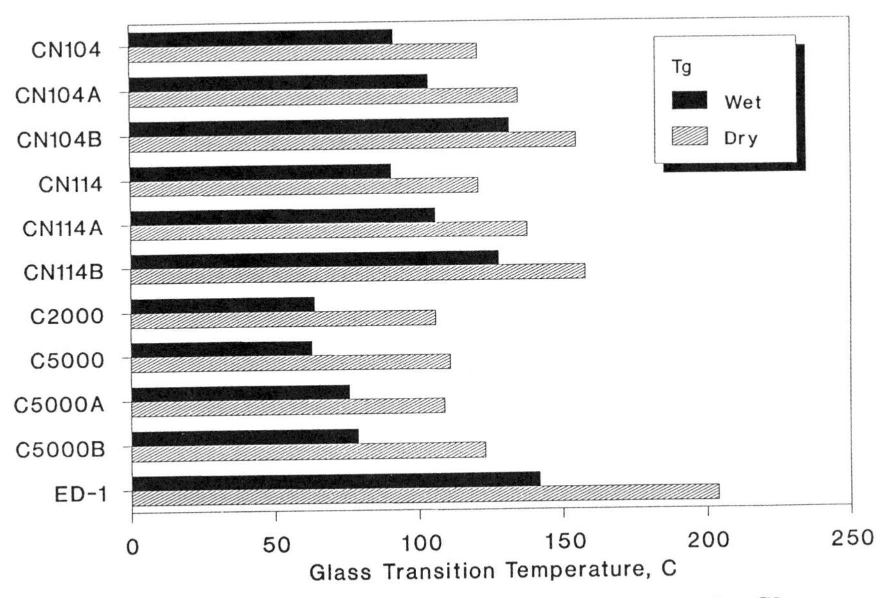

Figure 2. Effect of Moisture and Temperature on the Glass Transition Temperature of Selected Electron-Beam-Cured Composites(35% resin; 65% carbon fabric)(Immersion in water at 100°C for 24 h)(A: 10% pentaacrylate blend; B: 25% blend)

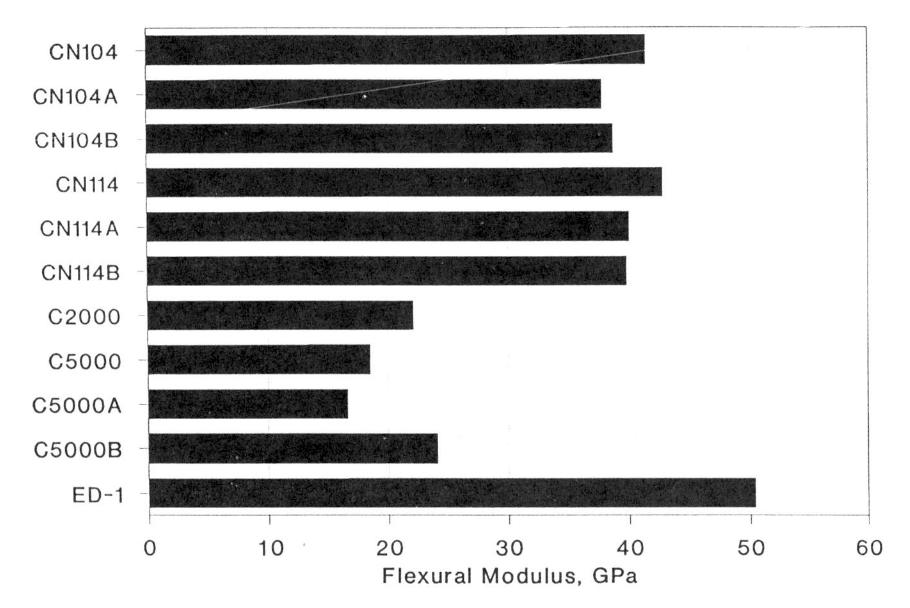

Figure 3. Maximum Flexural Modulus of Selected Carbon Fabric Laminates(35% resin; 65% carbon fabric) (A: 10% pentaacrylate blend; B: 25% blend)

known, so that the required dose can be delivered at each selected dose rate to ensure uniform properties throughout an entire product. Figure 4 shows the typical effect of dose rate on the flexural modulus of composites made using CN104 resin. At a dose rate of 5400 kGy/h (EB), a dose of 70 kGy was required to achieve the maximum modulus, and a dose of 100 kGy was needed for the 9 kGy/h (gamma) dose rate. A similar result was obtained from gel fraction analysis. The modulus was as much as 10 GPa higher for the high-dose-rate EB-cured composite, over the temperature range of 20 to 200°C, as shown in Figure 4. Figure 5 shows a similar dose rate effect on the T_g of a typical epoxy diacrylate as a function of dose. An increase in the T_g of about 10% was seen over the dose range from 30 to 300 kGy in the high-dose-rate sample.

Thermal energy is added to material during EB processing by radiation-induced exothermic reactions and by the net energy of the absorbed radiation *(1)*. Excessive heating in composites can be avoided by proper selection of the resin formulation to minimize the thermal energy released from exothermic reactions, and by reducing the required curing dose by using crosslinking promoters. Figure 6 shows the effect of blending a highly reactive pentaacrylated crosslinking promoter with a typical epoxy diacrylate resin on the T_g. The pentaacrylate increases the T_g by as much as 30°C up to a dose of 200 kGy, while reducing the required curing dose by 30%. Figure 7 confirms that mixing the pentaacrylate with the diacrylated epoxy resin has little or no effect on the flexural modulus of the composite laminate.

Summary

Electron beam (EB) processing uses high-energy electrons and X-rays to initiate polymerization and crosslinking reactions in suitable matrices, curing the polymer and enhancing specific physical and chemical properties. Many benefits have been identified for curing composites with EB treatment rather than thermal curing. These include the ambient curing temperatures, reduced curing times, improved material handling and resin stability. Electron-beam-curable matrices have been developed over the past 10 years with properties comparable to the epoxies and polyesters currently used for composite manufacturing. Resin characterization has been the focus of our recent composite research. As part of our research program to develop EB processes to cure specific composite products, the effects of many processing variables, including dose, dose rate and curing temperature, are being studied as they relate to the final composite properties. Both the EB dose and the dose rate affect the curing and the final properties of an EB-curable composite material. These effects must be understood to ensure uniform curing throughout a part, particularly when both EB and X-ray treatment are used. The composite temperature during curing is a processing variable when using EB treatment. Unlike thermal curing, where the curing temperature is set by the reaction chemistry, EB processing allows components to be cured at their end-use temperature, thus reducing the expected internal stresses.

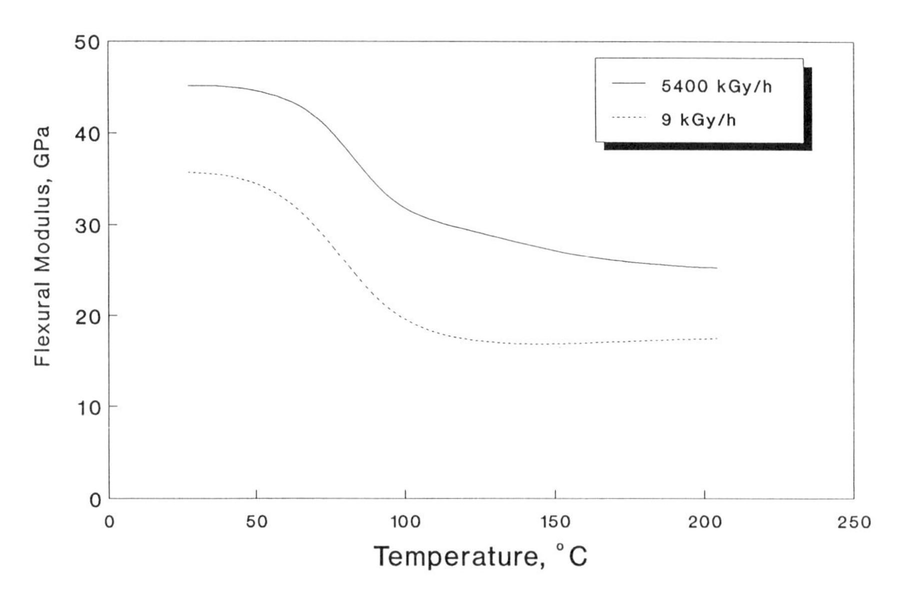

Figure 4. Effect of Dose Rate on the Flexural Modulus for CN104 Epoxy Diacrylate Composite(35% resin; 65% carbon fabric)

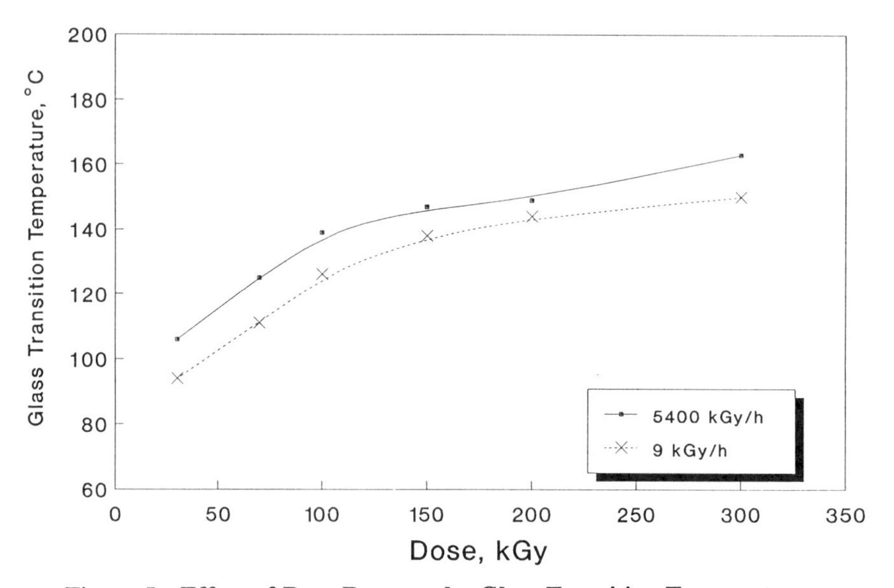

Figure 5. Effect of Dose Rate on the Glass Transition Temperature as a Function of Dose (35% CN104 polymer; 65% carbon fabric)

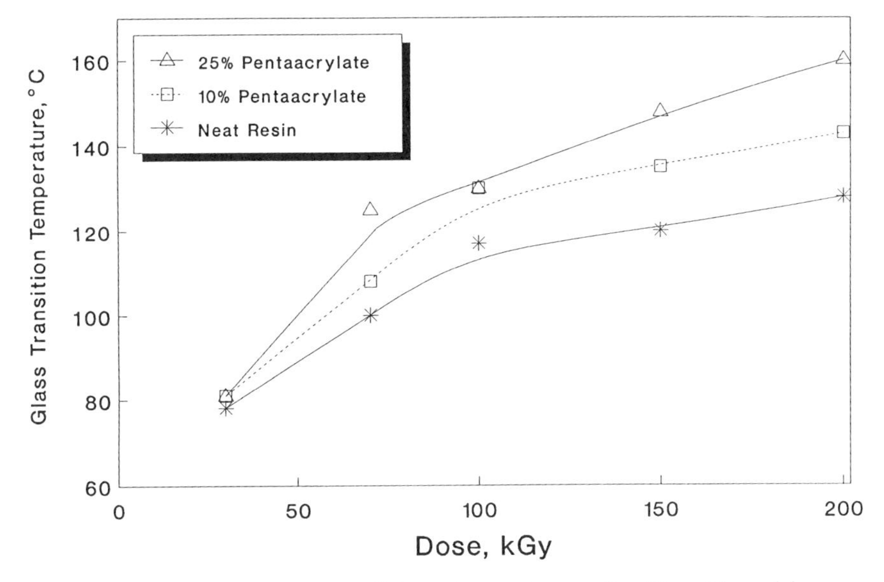

Figure 6. Effect of a Crosslinking Promoter on the Glass Transition Temperature of a Typical EB-Cured Carbon-Fabric Laminate(35% CN104B blend; 65% carbon fabric)

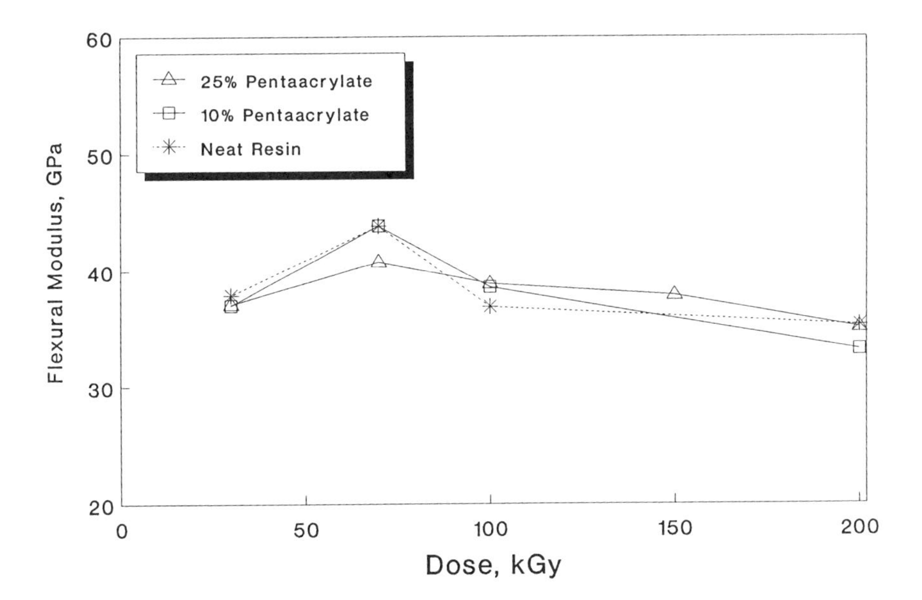

Figure 7. Effect of a Crosslinking Promoter on the Flexural Modulus of a Typical EB-Cured Carbon-Fabric Laminate (35% CN104B blend; 65% carbon fabric)

Literature Cited

1. Saunders, C.B.; Lopata, V.J.; Kremers, W.; Seier, S., *Proceedings of 36th International SAMPE Symposium, 36,* 187 **1991**.
2. Saunders, C.B.; Singh, A. *The Advantages of Electron Beam Curing of Fibre-Reinforced Composites.* Atomic Energy of Canada Limited Research Company Report, RC-264, 1989. Report available from SDDO, AECL Research, Chalk River Laboratories, Chalk River, Ontario K0J 0J0.
3. Weeton, J.W.; Peters, D.M.; Thomas, K.L. *Engineer's Guide to Composite Materials,* ASM International, Metals Park, Ohio (1987).
4. Bradley, R. *Radiation Technology Handbook,* Marcel Dekker, New York, N.Y. (1984).
5. Iverson, S.L.; Saunders, C.B.; McDougall,T.E.; Kremers, W.; Lopata, V.J.; Singh, A.; Kerluke, D. *Proceedings of International Symposium on Applications of Isotopes and Radiation in Conservation of the Environment,* International Atomic Energy Agency, Karlsruhe, Germany (1992).
6. Beziers, D.; Capdepuy, B. *Proceedings of 35TH International SAMPE Symposium,* 35, 1221 (1990).
7. Sartomer Company, Commercial Literature, Sartomer Company, West Chester, PA (1990).

RECEIVED October 7, 1992

Chapter 20

Radiation Effects on Polymers of Biomedical Significance

Shalaby W. Shalaby

Bioengineering Department, 301 Rhodes Research Center, Clemson University, Clemson, SC 29634–0905

Sterility of biomedical products based on synthetic, biosynthetic and natural (modified and unmodified) polymers is usually achieved by exposure to heat, ethylene oxide or high energy radiations. To a lesser extent, ultraviolet radiation, gas plasma and ultra filtration are also used for sterilization. However, with increased complexicity in many biomedical products, application of gamma radiations would be the most preferred sterilization method, if the quality of these products remain essentially uncompromised by undesirable radiation effects. Radiation effects on synthetic polymers, both of the absorbable (e.g. poly-2-hydroxy-acids) and non-absorbable types (e.g. polypropylene) are reviewed with emphasis on key absorbable polyesters such as polyglycolide (PG), poly-p-dioxanone (PD) and their copolymers. Examples of successful approaches to the radiostabilization of PG and PD in copolymeric forms are discussed and hypothesis on the mode of radiostabilization is provided. Effects of high energy radiations on biosynthetic polymers (e.g. poly-3-hydroxybutyrate, hyaluronic acid, growth factors) as well as natural and modified natural polymers are noted, briefly.

This report addresses the topic of radiation effects in polymers of biomedical significance, primarily, from perspective of sterilization using gamma radiation. In most of the studies cited in this review a Cobalt-60 source with an average dose rate of about 0.25 Mrad/hr (2.5 kGy/hr), was used for gamma radiation at a source temperature of about 35°C. Since its commercial application in 1960 for sterilization of medical devices a nominal sterilizing dose of 2.5 Mrad (25 kGy) is commonly used. However, as a result of recent development of scientific dose setting procedures, which accurately account for heterogeneous microbial populations, a sterilizing dose of less than 2.5 Mrad have been considered effective in a number of instances. Interest in radiation sterilization as a reliable method, with potentially broad applications, has increased significantly with the exceptional growth in the development of complex biomedical products, for which other traditional (dry heat, steam, ethylene oxide) and less traditional (gas plasma and chlorine oxide) techniques are hardly practical (1-3). Additionally, the growing number of environmentally sensitive biomedical products (e.g. absorbable devices) has increased the demand for hermetically sealed, easy-to-assemble packages, which are well-suited for gamma sterilization. This and the safety concerns associated with the use of ethylene oxide justified a careful review of the scope and limitations of gamma-sterilization (GS).

0097–6156/93/0527–0315$06.00/0

A major limiting factor to the broad use of GS is the degradation of certain polymeric substrates by gamma radiation during a typical sterilization cycle. This provided an incentive to identify synthetic, biosynthetic and natural polymers which are viewed as key components of biomedical products of emerging significance, and develop or explore means to increase their radiostability.

Key Types of Biomedically Important Polymers and Effect of Radiation

Of the many polymers employed in the production of biomedical products a few are selected for discussion. Selection of these polymers is based on (a) their successful GS and need to develop basic understanding of important structural features which may be associated with practical radiostability or (b) recognized instability to radiations and pressing need to increase their radiostability, because of their growing industrial importance. These are divided into three major groups and their response to radiation is outlined below.

Synthetic Thermoplastic Polymers. Among the widely used member of this group is polyethylene, which is available in several commercial grades. Most of these grades are GS as biomedical devices with minimum or no discernable effect on their properties. However, in critical applications such as the use of ultrahigh molecular weight polyethylene (UHMW-PE) as a component of joint prostheses, effect of radiation on long term performance is not fully understood (4). This may be related to the processing constraints associated with its exceptionally high molecular weight and the thermo-oxidative properties of the virgin polymer microparticles (5). Isotactic polypropylene is perhaps the most important member of this group, which is used as a suture and in several forms of molded articles, can not be GS (in most cases) without compromising its properties (6-11). The patented use of chain mobilizers to stabilize polypropylene does offer some promise and is yet to be widely accepted on a commercial scale (10).

Although biomedical products made of nylon 6, nylon 66, and segmented polyether-esters and polyether-urethanes have been noted for their successful GS. Careful studies of their long term performance may uncover undesirable consequences of GS, at least in few instances where the implants are characterized by high surface-to-volume ratio, and/or experience cyclic stresses (12). For many biomedical products and packaging materials, based on polyvinyl chloride or copolymers, special attention should be given to the radiation degradation of the chlorinated moieties and generation of hydrogen chloride (9). Among all the thermoplastic products, the partially aromatic high Tg, polyethylene terephthalate has been used successfully as fibrous and molded biomedical products, which are gamma-sterilizable.

Natural and Modified Natural Polymers. Cellulose and cellulose derivatives are known to undergo substantial degradation in presence of gamma radiation. This is primarily due to the presence of the aliphatic ether linkage and tertiary hydrogen in their repeat units. Hence, most cellulosic products are sterilized by ethylene oxide. Development of radiation-sterilizable cellulosics will certainly be a major breakthrough. Collagen-based products have been GS, successfully, for many years and this practice is suspected to continue.

Biosynthetic Polymers. These are produced by direct fermentation (e.g. polyhydroxy alkanoates) or as recombinant DNA products (e.g. growth factors. Sterilization of these polymers have been achieved by aseptic filtration and subsequest processing is conducted in an aseptic environment. Obviously, there is a definite need to develop GS biomedical products based on biosynthetic polymers. Future and currently investigations on the application of these polymers include implants or pharmaceutical preparation, for soft and hard tissues regeneration (13,14).

Contemporary Approaches to Radiostabilization of Emerging Polymers

Addressed in this section are new approaches to radiostabilized absorbable polymers and certain biosynthetic polymers of fast growing importance. In almost all cases this entail chemical modification of main chain or functional side groups of the polymers. Successful strategies to the radiostabilization of certain absorbable polymers and an up-date of on-going studies on other absorbable and biosynthetic polymers are outlined below.

Radiostabilization of Synthetic Absorbable Polymers. Poly(glycolide) (PG) and poly(p-dioxanone) (PD) are the base polymers of several important commercial biomedical products and many future ones (15-19), (12), (6-8), (20-22). In a successful effort to develop radiostabilized polyesters, Shalaby and coworkers incorporated certain partially aromatic moieties into the chain of the traditionally absorbable polymers PG and PD (15-17), (19), (21). This was achieved by the reaction of glycolide or p-dioxanone with polyesters made by the condensation of ethylene or trimethylene glycol with dimethyl 1,4-phenylene-bis-oxyacetate or methyl 4-carbomethoxy benzoate to produce radiostabilizer P-RSI and P-RSII, respectively (Figure 1). Analysis of typical examples of the glycolide copolymers indicated the presence of segmented copolymeric chains (23). Due to the high reactivity of absorbable polymers, melt-blending of PDS with P-RSI where m = 2 or 3 produced microscopically isotropic melts containing substantial fractions of copolymeric PD chains (18).

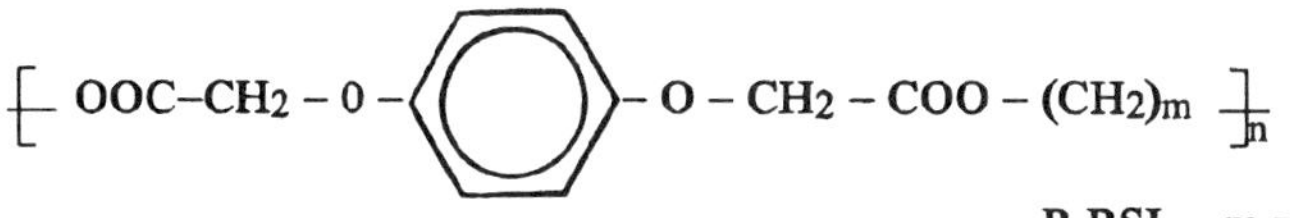

$$\left[OOC-CH_2-O-C_6H_4-O-CH_2-COO-(CH_2)_m \right]_n$$

P-RSI , m = 2,3

$$\left[OOC-C_6H_4-O-CH_2-COO-(CH_2)_m \right]_n$$

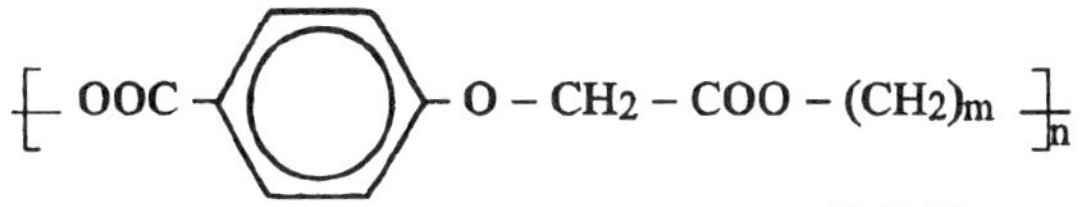

P-RSII , m = 2,3

Figure 1 Structure of Polymeric Radiostabilizers type P-RSI and P-RSII

Melt-blends of the polymeric radiostabilizers or their segmented copolymeric products with glycolide, (from a mixture of glycolide and 1,5-dioxepan-2-one or p-dioxanone) were extruded into monofilaments, using an Instron capillary rheometer. The monofilaments were drawn and annealed to strong filaments and their breaking strength was measured, periodically, after exposure to suitable degrading environments. This parameter was concluded earlier to be most reflective of the long term performance of post-irradiated absorbable polymers (19), (21), (23). The in-vitro breaking strength retention (BSR) studies were conducted in a phosphate buffer at 37° and pH of 7.26. In-vivo evaluation of the monofilaments by implanting them sub-cutaneously in rats. Sterilized monofilaments were exposed to a nominal dose of 2.5 Mrad. Typical results of sterile and non-sterile monofilaments made from a range of absorbable compositions are summarized in Table I . The results in Table I reflect the general stabilizing effect of P-RSI and P-RSII moieties in the examined systems. This may be attributed the well known effect of the aromatic ring as well as coupling of the -O-C H-C00- radicals.

Earlier Studies on alkylene oxalate polymers indicated, in certain instances, that radiation-sterilized monofilaments display fast decrease in breaking strength when implanted in rats (20,21). Presently, different types of potential radiostabilizers are being explored to determine their relative effects on poly(hexamethylene oxalate) (24). Results from these studies will be used in selecting a radiostabilizing system for further performance optimization with other high melting polymers.

Radiostabilization of Biosynthetic Polymers. Of the many biosynthetic polymers, three types are considered to be of great significance, biomedically. These are the poly(3-hydroxy alkanoates) such as poly(3-hydroxy butyrate) (PHB), hyaluronic acid (HA) and growth factors exemplified by epidermal growth factor, (EGF). Only scattered data are available in the literature which pertain primarily to the propensity of these polymers to undergo gamma-induced degradation. This is not surprising since (a) HA, is a polysaccharide with ether-bearing main chain, (b) EGF is a protein with two disulfide linkages and three primary amine groups and (c) PHB is an aliphatic polyester with a tertiary hydrogen in its repeat unit. Available data on the (a) radiostabilization of sythetic absorbable polymers (23), (b) the interaction of proteins with high energy radiations (25-26), and (c) modification of EGF without compromising its biological activity (14) may be used to develop experimental strategies for the radiostabilization of the three types of biosynthetic polymers.

Table I. In-Vivo Breaking Strength Retention Data of Gamma Sterilized (2.5 Mrad) monofilaments of homopolymers, Segmented Copolymers and Melt-Blends

Polymer(a)	Radiostabilizer(b)		% Strength Retention at			
	Type(b)	molar ratio	1wk	2wk	3wk	4wk
I-A	-	0	75	16	0	-
I-B	P-RSI-2	10	-	-	62	42
I-C	P-RS-II-2	10	91	73	55	-
II-A	-	0	-	43	30	25
II-B	P-RSI-3	12	-	62	55	40
II-C	P-RSI-3	11	-	79	72	57
III-A	P-RSI-2	0	16	-	-	-
III-B	P-RSI-2	5	33	-	-	-
IV-A	-	0	-	55	32	-
IV-B	P-RSI-2	10	-	85	51	-

(a) I-A = Polyglycolide (PG)
II-B = Copolymer, 90/10 PG/P-RSI-2
I-C = Copolymer, 90/10 PG/P-RSII-2
II-A= PD
II-B = Copolymer, 88/12, PD/P-RSI-2
II-B = Copolymer, 89/11, PD/P-RSI-3
III-A = Copolymer, 76/24 PG/Polydioxepanone
III-B = Copolymer, PG/Polydioxepanone/P-RSI-2
IV-A = PDS melted for 10 min.
IV-B = Melt Blended for 20 min., 90/10, PD/P-RSI-2 (By wt.)

(b) PRS-II-2 was made with ethylene glycol; P-RSI-2, P-RSI-3 were made with ethylene glycol and trimethylene glycol, respectively

Literature Cited

1. Gaughran, E.R.L. and Goudie, A.J., Eds., Sterilization of Medical Products by Ionizing Radiation, Multiscience Publ., Montreal, Canada, 1978.
2. Harris, L.E. and Skopek, A.J., Eds., Sterilization of Medical Products, Vol. III, Johnson & Johnson Pty. Ltd., Botany, New South Wales, Australia, 1985.
3. Kondakova, N.W., Ripa, N.V. and Sakharova, V.V., Radiobiol. (Russ.), 28 (6), 748 (1988).
4. Jahan, M.S., Wang, C., Schwartz, G. and Davidson, J.A., J. Biomed. Mater. Res., 25, 1005 (1991).
5. Deng, M. and Shalaby, S.W., Polym. Adv. Tech., 1992 (Submitted).
6. Shalaby, S.W., Encyclopedia of Medical Devices and Instrumentation, (Webster, J.E., Ed.), 4, 2324 (1988).
7. Shalaby, S.W., Encyclopedia of Pharmaceutical Technology," 1, 465 (1988).
8. Shalaby, S.W., Polym. News, 16, 238 (1991).
9. Shalaby, S.W. and Williams, B.L., Encyclopedia of Pharmaceutical Technology, Vol. 3, (1991) in Press.
10. Williams, J.L., Dunn, T.S. and Stannett, V.T., Radiat. Phys. Chem., 19, 291 (1982).
11. Williams, J.L., Polym. Prepr., 31 (2), 318 (1990).
12. Shalaby, S.W., Macromol. Revs.," 14, 406 (1979).
13. Hollinger, H.O., Schmitz, J.P., Mark, D.E., and Seyfer A.E., Surgery, 107, 50 (1990).
14. Njieha, F.K. and Shalaby, S.W., Polym. Prepr., 12 (2), 233 (1991).
15. Bezwada, R.S., Shalaby, S.W. and Jamiolkowski, D.D., U.S. Pat. (to Ethicon Inc.), 4, 510, 295 and 4, 532, 928 (1985).
16. Bezwada, R.S., Jamiolkowski, D.D. and Shalaby, S.W., Trans. Soc. Biomater, 14, 186 (1991).
17. Kafrawy, A., Jamiolkowski, D.D. and Shalaby, S.W., J Bioact. Biocomp. Polym., 2, 305 (1987).
18. Koelmel, D.F., Jamiolkowski, D.D. and Shalaby, S.W., U.S. Pat. (to Ethicon, Inc.), 4, 559,945 (1984).
19. Jamiolkowski, D.D. and Shalaby, S.W., Polym. Prepr., 32 (2), 327 and 329 (1990).
20. Shalaby, S.W. and Jamiolkowski, D.D., U.S. Pat. (to Ethicon Inc.), 4, 140, 678 and 4, 141, 087 (1979).
21. Shalaby, S.W. and Jamiolkowski, D.D., U.S. Pat. (to Ethicon Inc.), 4, 435, 590 (1984).
22. Tormala, P., Vasenius, J., Vainiopaa, S., Laiho, J., Pohjonen, T. and Rokkanen, P., J. Biomed. Mater. Res., 25, 1 (1991).
23. Jamiolkowski, D.D. and Shalaby, S.W., in Radiation Effects on Polymers, (Clough, R. and Shalaby, S.W., Eds.), Vol. 475, A.C.S. Symp. Series, Amer. Chem. Soc., Washington, D.C., 1991.
24. Johnson, R.A., Drew, M.J. and Shalaby, S.W., Polym. Prepr., 33 (2), 450 (1992).
25. Manori, L., Kushelevsky, A., Segal, S. and Weinstein, Y., J. Natl. Cancer Ins., 74, 1215 (1985).
26. Verma, S.P. and Pastugi, A, Rad. Res., 122, 130 (1990).

RECEIVED December 2, 1992

Chapter 21

Copolymerization of Glycolide with Polymeric Radiostabilizers for the Preparation of Radiation Sterilizable Braided Sutures

Dennis D. Jamiolkowski[1], Rao S. Bezwada[1], and Shalaby W. Shalaby[2]

[1]Ethicon, Inc. (Johnson & Johnson), P.O. Box 151, Route 22, Somerville, NJ 08876
[2]Bioengineering Department, 301 Rhodes Research Center, Clemson University, Clemson, SC 29634–0905

Dimethyl 1,4-phenylene-bis-oxyacetate (PEPBO) or dimethyl 4-carboxymethoxybenzoate (PECMB), appears to impart gamma stability to polyglycolide (PG) without compromising physical and biological properties. Incorporation of about 10 mole % of the polymeric radiostabilizers into the PG chain leads to high strength fibers well suited for use as braided sutures. The properties of gamma sterilized braided sutures of PEPBO/PG and PECMB/PG copolymers are equivalent to or better than the EO sterilized PG braided sutures. The stabilizers were distributed uniformly and could be processed without leaching, by being made to be part of the polymeric chain. A strategy was followed to incorporate the radiostabilizers in a block-like fashion allowing for good crystallization and excellent mechanical properties.

Implantable surgical devices must be sterilized prior to their use. Sterilization of such devices is usually accomplished by the use of heat, ethylene oxide, or gamma radiation (employing a Co60 source). In many cases, gamma irradiation is the most convenient and effective method of sterilization. However, this method of sterilization cannot be used with any of the commercially available synthetic absorbable sutures because they degrade significantly upon irradiation.

Radiation sterilization of absorbable sutures was studied by Shalaby and Jamiolkowski, *(1-4)*. This present paper deals with a specific class of polymeric radiostabilizers which can be incorporated into the polyglycolide chain to provide copolymers which can be processed into braided absorbable sutures which are efficacious, even after a sterilizing dose of gamma radiation.

The radiostabilizers used in this study are based on aromatic and glycolide moieties to provide radiostability as well as absorbability. In particular, methyl glycolate moieties were reacted with hydroquinone and 4-hydroxybenzoic acid in

0097–6156/93/0527–0320$06.00/0

separate Williamson synthesis procedures, producing dimethyl 1,4-phenylene-bis-oxyacetate (DMPBO) and dimethyl 4-carboxymethoxybenzoate (DMCMB), respectively. These two diesters, in turn, were polycondensed with excess ethylene glycol to form poly(ethylene 1,4-phenylene-bis-oxyacetate) (PEPBO) and poly(ethylene 4-carboxymethoxybenzoate) (PECMB). In a final step, the aromatic polyesters were then copolymerized with glycolide to obtain segmented copolyesters, PEPBO/PG and PECMB/PG as shown in Figure 1.

Experimental

Monomer Synthesis. A detailed preparation of DMPBO and DMCMB were described previously *(1,5)*. These monomers were purified by recrystallization from methanol. DMPBO and DMCMB have melting points of 99-101°C and 94-95°C, respectively. The structures were verified by IR and NMR spectroscopy.

Polymer Synthesis.

Preparation of poly(ethylene 1,4 phenylene-bis-oxyacetate) (PEPBO). Using the previously described procedure (1), PEPBO with an inherent viscosity (I.V.) of 0.63 dl/g was prepared. The I.V. was determined in hexafluoroisopropyl alcohol (HFIP) at 25°C and at a concentration of 0.10 g/dl.

Preparation of PEPBO/PG Copolymer. A flame dried 1 liter round bottom flask, fitted with a mechanical stirrer and nitrogen/vacuum inlet was charged with 313.4 grams (2.7 moles) of glycolide and 75.5 grams of PEPBO containing 0.0197 weight percent of dibutyltin oxide as catalyst. The reaction flask was purged and vented with nitrogen, and heated in an oil bath at 120°C to melt the glycolide and to start the dissolution of PEPBO in molten glycolide. The oil bath temperature was increased to 150°C in about 6 minutes and maintained there for about 8 minutes to continue the dissolution process. The temperature of the oil bath was raised to 195°C in about 6 minutes and maintained there for 8 hours. The copolymer was isolated and the unreacted monomer was removed by heating the ground polymer at 110°C/0.01mm Hg for 16 hours. A weight loss of 0.2% was observed. The copolymer had an I.V. of 1.69 dl/g in HFIP at 25°C and the composition of PEPBO/PG was found to be 10.5/89.5 mole% by proton NMR spectroscopy.

Preparation of poly(ethylene 4-carboxymethoxybenzoate) (PECMB). A thoroughly dried, mechanically stirred 1.5 gallon stainless steel reactor was charged with 2018 grams (9.0 moles) of DMCMB, 1676 grams (27.0 moles) of ethylene glycol, and 0.3360 grams of dibutyltin oxide. The reactor was purged and vented with nitrogen. The contents of the reactor were heated to 190°C within two hours; a reaction temperature of 190-200°C was then maintained for about 6 hours, during which time the theoretical quantity of methanol was collected. The reactor was allowed to cool to room temperature overnight and then reheated slowly under reduced pressure (0.05-1.0 mm) to 200°C. The reactor was maintained at 200-220°C/0.05-1.0 mm Hg for about 16 hours, during

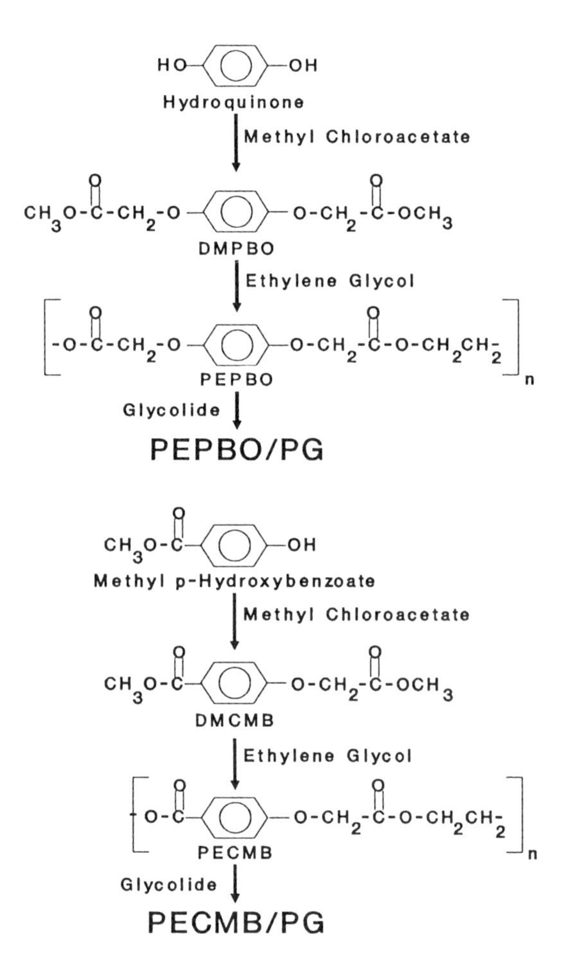

Figure 1. Synthesis scheme for the preparation of the two copolymers, PEPBO/PG and PECMB/PG.

which time ethylene glycol was removed. The prepolymer (PECMB) was isolated and characterized. The I.V. of PECMB in HFIP at 25°C was found to be 0.51 dl/g.

Preparation of PECMB/PG Copolymer. A thoroughly dried, mechanically stirred 5 gallon stainless steel reactor was charged with 329.7 grams (10 mole %) of PECMB and 1670 grams (90 mole %) of glycolide. The reactor was purged and vented with nitrogen. The mixture was heated to 100°C and maintained at 100-120°C for 30-60 minutes without stirring to ensure complete melting of the glycolide. While stirring, the reactor was then heated stepwise to 150°C, 170°C, 190°C and 210°C over a period of one hour and then maintained for 2 hours at 210°C under nitrogen. The polymer was isolated, ground and dried at 110°C (0.1 mm Hg) for 16 hours to remove unreacted glycolide. The copolymer (PECMB/PG) had an I.V. of 1.40 dl/g in HFIP at 25°C and a melting point range of 208-216°C.

Fiber Spinning, Physical Properties, and Biological Performance

The copolymers, PEPBO/PG and PECMB/PG, were melt spun and yarns assembled into braids in a manner similar to one previously described *(4, 6)*. A sample of polyglycolide was similarly extruded, braided, and post-treated for use as a control. The braided fibers of copolymers were cut into 28" lengths, placed in individual paper folders and vented heat-sealable metal foil envelopes. The materials were dried by *in-vacuo* treatment at 50°C. The braids were sterilized by gamma radiation (2.6-2.9 Mrads) by using standard industrial conditions, whereas the controls were sterilized with ethylene oxide.

The physical properties of both braids were determined; the results are given in Table I.

Table I. Comparison Fiber Properties of Uncoated 2-0 Suture Braid

Polymer Type	PG Homopolymer		PEPBO/PG Copolymer		PECMB/PG Copolymer	
Sterilization	Non-Irrad.	Co-60	Non-Irrad.	Co-60	Non-Irrad.	Co-60
Straight Tensile Strength, Kpsi	131	113	106	100	152	154
Knot Tensile Strength, Kpsi	74	60	65	61	92	89
Elongation at Break, %	19	16	24	22	18	17

The performance of various synthetic fibers as potential suture material depends on many factors. Besides initial mechanical properties, important characteristics of absorbable sutures include their biological performance with regards to absorption and tissue reaction. Of particular importance is the breaking strength retention (BSR) profile that the suture will exhibit once implanted. This latter characteristic is important because the suture must provide sufficient holding power to the traumatized tissue while it is healing.

The biological performance with regards to absorption and BSR of both braids were determined; the results are given in Table II.

Table II. Comparison Biological Performance of Uncoated 2-0 Suture Braid

Polymer Type	PG Homopolymer		PEPBO/PG Copolymer		PECMB/PG Copolymer	
Sterilization	Non-Irrad.	Co-60	Non-Irrad.	Co-60	Non-Irrad.	Co-60
In Vivo Absorption Data, Cross-Sectional Area Remaining,						
% at 5 days	100	-	-	100	-	100
91 days	12	-	-	49	-	94
119 days	0	-	-	1	-	0
182 days	0	-	-	0	-	0
In Vivo BSR,						
% at: 14 days	74	38	105	88	-	73
21 days	51	0	97	72	-	55

Results and Discussion

Incorporation of about 10 mole % of the polymeric radiostabilizers, PEPBO or PECMB, appears to impart gamma stability to PG without compromising physical and biological properties. The properties of gamma sterilized braided sutures of PEPBO/PG and PECMB/PG copolymers are equivalent to or better than the EO sterilized PG braided sutures *(3,7)*.

The exact nature of the mechanism of stabilization has not been elucidated. These radiostabilizers, however, were designed with a number of requirements in mind. Based on the structures of existing radiostable polymers, incorporation of an aromatic moiety was foremost. This however could not be done haphazardly as mechanical and biological performance could not be compromised. To insure that the stabilizers were distributed uniformly and could not be leached out during processing, they were made to be a part of the polymeric chain. To insure the compatibility and hydrolyzability of these aromatic moieties, it was felt that they should be polyester based. To further enhance the hydrolyzability of the ester groups, at least one of the carbonyls was activated by placing an oxygen in the beta position; this could be viewed as a modified glycolate moiety. Random incorporation of these radiostabilizer groups would retard the chain's ability to crystallize, thus dramatically altering the dimensional stability of fibers so produced. To address this problem, a strategy was followed to incorporate the radiostabilizers in a block-like fashion allowing for good crystallization and excellent mechanical properties.

Acknowledgment

The authors wish to thank S.C. Arnold for his review of this manuscipt.

Literature Cited

1. Shalaby, S.W.; Jamiolkowski, D.D. U.S. Pat. 4,435,590 (1984) assigned to ETHICON, INC.
2. Shalaby, S.W.; Jamiolkowski, D.D. U.S. Pat. 4,689,424 (1987) assigned to ETHICON, INC.
3. Jamiolkowski, D.D.; Shalaby, S.W. *Polym. Prepr.* **1990,** *31, (2),* 329 .
4. Jamiolkowski, D.D.; Shalaby, S.W. In *Radiation Effects on Polymers*; Clough, R.L. and Shalaby, S.W., Eds.; Symposium Series 475; American Chemical Society: Washington, DC, **1991**, 300-309.
5. Bezwada, R.S.; Shalaby, S.W.; Jamiolkowski, D.D. U.S. Pat. 4,532,928 (1985) assigned to ETHICON, INC.
6. Bezwada, R.S.; Jamiolkowski, D.D.; Shalaby, S.W. U.S. Pat. 4,510,295 (1985) assigned to ETHICON, INC.
7. Bezwada, R.S.; Jamiolkowski, D.D.; Shalaby, S.W. *Transactions of the Society of Biomaterials,* (Annual Meeting, May 1-5, **1991**) *XIV,* 186.

RECEIVED November 11, 1992

Author Index

Affiliation Index

Subject Index

G

H

I

K

L

S

Production: Margaret J. Brown
Indexing: Deborah H. Steiner
Acquisition: Anne Wilson
Cover design: Robert Sargent

Printed and bound by Maple Press, York, PA